Urban Water Cycle Modelling and Management

Urban Water Cycle Modelling and Management

Special Issue Editors

Meenakshi Arora
Hector Malano

MDPI • Basel • Beijing • Wuhan • Barcelona • Belgrade

MDPI

Special Issue Editors
Meenakshi Arora
University of Melbourne
Australia

Hector Malano
University of Melbourne
Australia

Editorial Office
MDPI
St. Alban-Anlage 66
Basel, Switzerland

This is a reprint of articles from the Special Issue published online in the open access journal *Water* (ISSN 2073-4441) from 2017 to 2018 (available at: http://www.mdpi.com/journal/water/special_issues/Urban-Water-Cycle-Modelling-Management)

For citation purposes, cite each article independently as indicated on the article page online and as indicated below:

LastName, A.A.; LastName, B.B.; LastName, C.C. Article Title. *Journal Name* **Year**, *Article Number*, Page Range.

ISBN 978-3-03897-107-8 (Pbk)
ISBN 978-3-03897-108-5 (PDF)

Contents

About the Special Issue Editors

Meenakshi Arora, Dr., is a Senior Lecturer in Environmental Engineering at The University of Melbourne. Dr. Arora has extensive experience in both research and university teaching. Her main research focus is on water resource management, integrated catchment modelling, urban water cycle modelling, the water–energy nexus, water quality, stream health, land and groundwater remediation, and contaminant transport modelling. Dr. Arora was awarded the 2013 'Victoria Fellowship' and has been involved in various projects based on the Integrated Catchment Management approach.

Dr. Arora is the Deputy Director of the Melbourne India Postgraduate Program (MIPP) and winner of the 2017 Award for Excellence in Internationalisation of Research. Dr. Arora chaired a 3-day conference on 'Practical Responses to Climate Change' held in Melbourne in November 2014. She has organized various sessions in national and international conferences and published widely in high impact journals. She serves as a member of editorial boards as well as reviewers for various journals.

Hector Malano graduated in Agricultural Engineering in 1973 at the National University of Cordoba, Argentina. In 1981, he was awarded a Master's degree in Irrigation and Drainage Engineering from Utah State University (USA) for research carried out on the behaviour of infiltration under surge flow hydraulics. He was subsequently awarded a PhD in Irrigation Engineering from the same university in 1985 for his research of two-dimensional numerical modelling of interceptor drains. In 2006, he was appointed the Head of the Department of Civil and Environmental Engineering.

Hector Malano has conducted research on various aspects of water resources at three scales: (i) On-farm modelling of surface irrigation systems; (ii) modelling of irrigation distribution networks; (iii) water allocation between competing uses at the catchment level; (iv) management and modelling of the urban water cycle with special emphasis on complex systems. This research focuses on the design of evidence-based planning policies for achieving fit-for-purpose utilisation of multiple sources and multiple demands of water.

Hector is extensively involved in international professional and research organisations. Recently, he concluded a 3-year term as Vice-President of the International Commission on Irrigation and Drainage. He has consulted for several international organisations including the World Bank, AusAID and Food and Agriculture Organisation of the United Nations.

Hector has authored and co-authored over 150 scientific papers on these topics. He is currently Chief Investigator of several research grants including Allocation Modelling in the Krishna Basin, India, Regional and Economic Benefits through Smarter Irrigation (Hydraulic Modelling). He is the node coordinator of the CRC for Irrigation Futures and is involved in the CRC for eWater.

Preface to "Urban Water Cycle Modelling and Management"

The main aim of this book is to bring together key advances in the integrated management of the urban water cycle. Increasingly, due to concerns arising from reducing emissions associated with climate change to scarcity of water resources for urban populations, the main focus in managing urban water supplies is on the integration of multiple sources and multiples uses of water resources based on fit-for-purpose criteria. Our motivation for preparing this book is to address the key challenges and potential solutions in undertaking the changes needed to achieve integrated urban water resource management goals, and in so doing, assist researchers and practitioners by providing the tools they need to implement these changes.

This book has arisen from the extensive research that the editors have carried out in the field of urban water resource management in the last decade. This experience also assisted us in gaining a greater understanding of the technical, economic and policy challenges facing water managers engaged in this field.

The book includes 12 papers and to assist the reader in navigating through this book, we have grouped the papers into the following five main themes:

- Integrated water supply: papers 1–5
- Urban flood modelling: papers 6–8
- Reservoir operations for urban water supply: papers 9–10
- Science-policy interface: paper 11
- Policy risk: paper 12

We hope that the content of this book contributes and stimulates further discussion and research on these important aspects of urban water management, and also signals possible gaps and directions that future research needs to address in integrated urban water management.

The preparation of this book was only possible because of the contributions from the various authors involved. We are also very grateful to the many reviewers for their quality reviews that greatly assisted us in selecting and improving the quality of these papers. As is usually the case, there are many others who, while not mentioned explicitly as authors, have contributed their time and efforts to carry out the research underpinning these papers. We would also like to acknowledge contributions from MDPI for supporting the publication of this book.

Meenakshi Arora, Hector Malano

Special Issue Editors

water

MDPI

Article

Impact of Hybrid Water Supply on the Centralised Water System

Robert Sitzenfrei [1,*] (ID), Jonatan Zischg [1], Markus Sitzmann [1] and Peter M. Bach [2] (ID)

[1] Unit of Environmental Engineering, University of Innsbruck, Technikerstr. 13, 6020 Innsbruck, Austria; jonatan.zischg@uibk.ac.at (J.Z.); markus.sitzmann@uibk.ac.at (M.S.)
[2] Monash Infrastructure Research Institute, Department of Civil Engineering, Monash University, Clayton VIC 3800, Australia; peter.bach@monash.edu
* Correspondence: robert.sitzenfrei@uibk.ac.at; Tel.: +43-512-5076-2195

Received: 11 September 2017; Accepted: 1 November 2017; Published: 4 November 2017

Abstract: Traditional (technical) concepts to ensure a reliable water supply, a safe handling of wastewater and flood protection are increasingly criticised as outdated and unsustainable. These so-called centralised urban water systems are further maladapted to upcoming challenges because of their long lifespan in combination with their short-sighted planning and design. A combination of (existing) centralised and decentralised infrastructure is expected to be more reliable and sustainable. However, the impact of increasing implementation of decentralised technologies on the local technical performance in sewer or water supply networks and the interaction with the urban form has rarely been addressed in the literature. In this work, an approach which couples the UrbanBEATS model for the planning of decentralised strategies together with a water supply modelling approach is developed and applied to a demonstration case. With this novel approach, critical but also favourable areas for such implementations can be identified. For example, low density areas, which have high potential for rainwater harvesting, can result in local water quality problems in the supply network when further reducing usually low pipe velocities in these areas. On the contrary, in high demand areas (e.g., high density urban forms) there is less effect of rainwater harvesting due to the limited available space. In these high density areas, water efficiency measures result in the highest savings in water volume, but do not cause significant problems in the technical performance of the potable water supply network. For a more generalised and case-independent conclusion, further analyses are performed for semi-virtual benchmark networks to answer the question of an appropriate representation of the water distribution system in a computational model for such an analysis. Inappropriate hydraulic model assumptions and characteristics were identified for the stated problem, which have more impact on the assessments than the decentralised measures.

Keywords: integrated system analysis; rain water harvesting; water quality analysis; UrbanBEATS; urban form

---

1. Introduction

Modern urban water management faces challenges like climate change, urban development and aging infrastructure [1]. Restricted water resources and limited budgets force engineers, researchers and decision makers to rethink the way urban water is managed. Traditional (technical) concepts to ensure a reliable water supply, a safe handling of wastewater and flood protection are increasingly criticised as outdated and unsustainable [2]. These so-called centralised or grey urban water systems—encompassing e.g., piped potable water supply and sewer networks—are, furthermore, maladapted to upcoming challenges because of their long life-span in combination with their short-sighted planning and design. Their design and implementation can result in technological and institutional lock-in effects [3]. A combination of (existing) centralised and decentralised infrastructure is expected to be more reliable and sustainable [4] and can more readily be adapted to upcoming challenges. With regard to stormwater management, new, sustainable water management strategies, such as Sustainable Urban Drainage Systems (SUDS), Green Infrastructure (GI), Water Sensitive Urban Design (WSUD), Low Impact Urban Design and Development (LIUDD), Best Management Practice (BMP), etc., have been gaining increasing interest in recent years, particularly in water scarce regions [5]. It is increasingly recognised that the combination of decentralised and centralised solutions can provide required water services and that dispersed solutions can also provide liveability and sustainability benefits to the local community. Especially for potable water supply, water resources can be used more efficiently through e.g., local reuse or treatment and utilisation of local resources (e.g., greywater reuse or rainwater harvesting). These so-called hybrid water supply systems (a combination of centralised and decentralised technologies) are seen to be more sustainable and resilient, but also introduce complexity into the system by further interlinking drainage and supply.

A modern urban water cycle is a strongly interlinked system. However, in traditional management structures the different sub-disciplines in that cycle are often regarded separately, thus neglecting the complex interactions within such systems [6]. These neglected interactions have usually been of less interest because the interfaces between the sub-systems are more or less well-defined. In new management strategies, the consideration of these interactions is becoming more important and plays a crucial role in gaining confidence in the long-term technical operation of hybrid systems and the entire urban water cycle as part of the city [7,8]. The most important driver for urban water demand is the urban form. Bouziotas, et al. [9] developed a framework for linking the dynamics of the urban growth/form with the spatial distribution of the water demand for testing different water management practices following city evolution. This dynamic approach was further enhanced to also to distribute water-aware technologies [10].

In Sapkota, et al. [11], a conceptual framework was developed to assess the interactions between decentralised water supply systems and existing centralised management practices. In a case study application, it was shown how daily water demands are reduced and concentrations in wastewater flow are altered by implementing decentralised technologies. In that study it was also determined that the wastewater concentrations were increased and changes in peak flow (potable water and waste water) were negligible while the daily volumes were reduced.

In Bach, et al. [12,13] it is shown how the urban form and planning regulations interact with the implementation of decentralised systems. For decision-makers, it is important to understand the implications of different planning regulations on aspects like urban drainage or water supply at a local scale. The software tool UrbanBEATS (Urban Biophysical Environments and Technologies Simulator—www.urbanbeatsmodel.com), which emerged from this study, supports decision makers when planning and implementing such water sensitive strategies. The tool combines and processes spatial and non-spatial data (e.g., land use, population, elevation, rainfall/climate). Multiple benefits arise when following a decentralised rainwater handling strategy, such as the reduction or attenuation of stormwater peak discharges, preserving or at least encouraging a more natural water cycle, on-site treatment (i.e., controlling pollution at the source) or positive effects on the urban microclimate [14]. However, despite the multiple benefits, effects of such distributed infrastructures on the existing (mostly central) water infrastructure also need to be understood and accounted for. For combined urban drainage systems, there is generally a positive effect on hydraulic performance during wet weather events, but it is often neglected that a reduction in wastewater flows due to water reuse or water saving might negatively impact the shear stress performance in sewers, causing increased sedimentation and odour nuisance [15]. In Sitzenfrei and Rauch [16], a spatial sensitivity analysis was developed to quantify the potential impact of a reduction in potable water consumption on the shear stress performance in a combined sewer system and the water quality in the water supply network. Nevertheless, the spatial distribution of possible potable water reduction due to the land use has been neglected in that study. For water distribution systems (WDS), the implementation of decentralised rainwater harvesting or greywater reuse measures could strongly influence the water quality performance of the existing centralised system by reducing the water demand and increasing the travel time and, consequently, the water age in the supply system. However, the impact of increasing the implementation of decentralised technologies on the localised technical performance in potable water supply networks (e.g., stagnation, water quality) has not been analysed.

This paper aims to quantify the impact of 'land use driven' spatial distribution of decentralised technologies (i.e., hybrid water supply systems) on the technical performance of existing water supply networks. With this newly developed approach, the impact of demand reduction scenarios (i.e., rainwater harvesting to substitute private irrigation and water efficiency measures) determined with UrbanBEATS on the water quality can be simulated and are demonstrated in this paper on a case study. The interaction of the low density urban form (low total demand, minimal pipe diameters in the potable supply network) and the high rainwater harvesting potential (a lot of harvesting area in combination with green space for irrigation) was identified as significantly disadvantageous for the WDS, causing water quality problems in localised areas. In contrast, in high demand areas (the high density urban form), there is less potential for e.g., rainwater or stormwater harvesting due to the limited available space and, therefore, also the impact on the technical performance of the potable water supply network in these regions is less significant.

Furthermore, to establish a more generalised and case-independent conclusion, analyses were performed for semi-virtual benchmark networks [17]. In contrast to entirely virtual systems, semi-virtual systems aim to mimic real boundary conditions [15]. Consequently, the question of an appropriate representation in a computational model of a potable water supply system for this kind of analysis is also addressed. We show that the usage of an inappropriately defined hydraulic model can have even more impact than the actual decentralised measures on the water quality.

2. Materials and Methods

2.1. Hybrid Water Supply Estimation with UrbanBEATS

UrbanBEATS combines spatial Geographic Information Systems (GIS) data (e.g., elevation, land use, population and soil type) and non-spatial data (e.g., rainfall) in an integrated model to assist in the planning and management of decentralised water management structures and sustainable urban water strategies (Figure 1). UrbanBEATS was developed within the context of Water Sensitive Urban Design (WSUD) but can be applied/adapted to assist other sustainable water management strategies like LIUDD, GI or BMP.

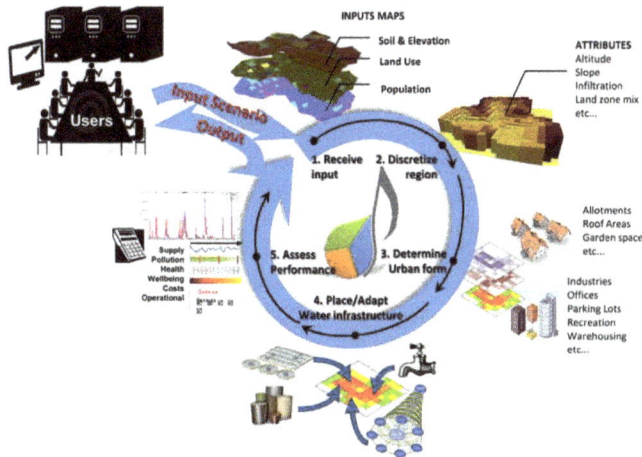

Figure 1. Overview of the UrbanBEATS model (reprinted from [18] with permission from ASCE).

UrbanBEATS models the planning and implementation of decentralised technologies for stormwater/rainwater treatment under various policy, statutory planning and biophysical constraints. Although UrbanBEATS uses a conceptual approach with a grid-based, spatial representation of the data (each grid cell known as a *"Block"*), much of the information about the urban environment (e.g., land use mix, household composition, water use behaviour) is retained in each *Block*. This is necessary, as differentiating between the spatial scale in assessing decentralised options is crucial [19]. The size of a *Block* is determined on a case-by-case basis by the user depending on the purpose of the modelling study. The urban form in the model is abstracted using procedural algorithms that are based on statutory planning regulations and architectural standards [12]. Using input information about land use and population, a collection of algorithms are called to subdivide the land area into allotments and built features (e.g., building area, road, footpaths, garden and other spaces). The concept is illustrated for residential land use in Bach et al. [12] and has since been more extensively developed and tested [20]. Each procedural algorithm is underpinned by various planning ordinances (e.g., [21,22]) and architectural standards [23].

Using the resulting urban form, which is described by a plethora of characteristics (e.g., impervious areas, street widths, building setbacks, garden and public open space), different water infrastructure strategies can be tested and assessed (e.g., installation of rainwater infiltrations measures depending, for example, on the estimated roof areas, available green space and building occupants). Using the conceptual description of the urban form, potential water use reduction strategies can also be assessed. This creates the opportunity to link this spatially explicit information on variable water consumption rates based on the urban form and demographics, to hydraulic water supply models by altering the water demands accordingly or even changing the planning rules in cases of urban renewal. With this approach, it is possible to investigate and quantify how the broad-scale implementation of rainwater harvesting or the impact of water restrictions and behavioural change can impact the centralised water supply system.

In UrbanBEATS, the four aforementioned input spatial maps (10 m × 10 m raster files) of elevation, soil type, land use and population are required, of which the latter two are used to map water demands spatially to the water distribution systems. Water demands are calculated using 'end-use analysis' of typical water use types (e.g., kitchen, toilet, laundry, shower, irrigation) and are then downscaled to sub-daily time steps using seasonal and diurnal scaling patterns. These patterns are stacked and can be varied across different end-use types (which can result in different peak flows at different times of the day). Flow rates are based on typical values for household fixtures taken from the Australian standards AS6400:2016 [24]. Irrigation is applied to garden and public open spaces identified by the model using the spatial input. With this information and the biophysical data in the model, UrbanBEATS can identify suitable layouts of stormwater/rainwater harvesting infrastructure to achieve user-defined demand reduction targets. Alternatively, policy scenarios (e.g., minimum water efficiency compliance or water restrictions) can also be simulated to enact more widespread spatial change. In this study, we specifically explore the latter policy scenarios.

The integration between UrbanBEATS and the water distribution modelling software EPANET 2 [25] is established within UrbanBEATS itself. The model is capable of reading and modifying EPANET 2 input files and considers, explicitly, the spatial variation in water demand at various nodes in the network and across the different diurnal patterns (associated with different end use types). The link between the coarse spatial grid of *Blocks* (containing demand data) and a detailed network is achieved through a geometric operation, which determines the spatial proportion of a *Block* area connected to each node in the water distribution system. As such, an alteration of water demand through a policy scenario or the implementation of decentralised infrastructure in UrbanBEATS can then be propagated to the WDS and the modified EPANET 2 input file can be generated and used for external performance assessment.

Currently, the seasonal demand dynamics as, for example, discussed in [26], are only taken into account in a simplified way. A peak day demand is used for network design, an average day with a diurnal demand pattern is used to assess the impact of the decentralised measures and an assumed low consumption day being one third of the average daily demand is used for water quality assessments in the potable water supply network. Analysis for real demand data show that such low demand days occur on at least one or a few days per year [27]. Future development will implement the proposed approach to better represent seasonal demand dynamics.

2.2. Potable Water Supply Design and Hybrid Supply Systems

The potable water supply system should reliably supply water in sufficient quantity and quality. Although there are specific national requirements on how to design and operate them, traditional (technical) water supply follow first principal technical aspects. One might assume that we still face institutional barriers for novel approaches like hybrid water supply, but as long as a community relies on the essential central services, these first principal technical aspects must be fulfilled to avoid system malfunction under regular conditions and, especially, critical conditions.

A major challenge is the design of the layout and the sizing of pipes within the systems. The layouts of the systems are usually looped networks with redundant capacity, able to provide reliable water supply under critical conditions (pipe breaks, source failures, fire-fighting demand, etc.). Pipe-sizing is based on two conflicting requirements: (1) in cases of high demand, there must be sufficient remaining pressure in the system (high diameters to reduce friction losses and ensure that there is sufficient pressure); and (2) in cases of low demand, the residence time in the water supply system should not be too long to ensure water quality (bacterial growth with increasing water age, chlorine decay, etc.) which results in low diameters.

Water demand can considerably vary over the year [28]. Flow velocities in the systems vary accordingly depending on the pipe diameter and range from a maximum of 2.5 m/s to approximately 0.3 m/s. The variations of high demand (hourly peak demand) and an average day can be a factor of five and, for low demand days (depending on the composition of the supply area), even as high as a factor of 10 [27]. In addition to hourly peak demand design, fire-fighting requirements can also be the driving load case for the design, especially in low-density areas (i.e., low demands). In such a case, a minimal diameter requirement (e.g., a smallest pipe diameter of 80 mm or 100 mm) is usually used to ensure that the required fire-fighting demand can be covered. The implementation of the minimal diameter for fire-fighting requirements can already cause stagnation problems but it is nevertheless mandatory and must be handled with appropriate operational measures (e.g., flushing). Decentralised or alternative technologies can be used instead if they adequately supply fire-fighting requirements. In the context of a progressive installation of decentralised water supply schemes, the average water requirements can be reduced, but a reduction of the peak (design) demands is hardly possible and therefore no reductions in construction costs are foreseen. However, a reduction in operational costs due to a reduced volume for water treatment or pumping is feasible. A reliable supply scheme must ensure its resilience especially under drought conditions when no rainwater is available for a longer period. The highest volume reduction for potable demand from decentralised technologies like rainwater harvesting arises during wet weather with potentially lower outside temperatures. For the potable water supply, these are usually low demand days which are critical for stagnation and water quality problems. As such, alternative decentralised water supply schemes would further intensify these low flow conditions.

Rainwater harvesting is most effective, when there is a large enough catchment area to collect rainwater from in relation to the requested water demand. Therefore, it is important to account for the urban form and potable water supply network when developing a hybrid supply system. In this work, different rainwater harvesting and water efficiency measures determined with UrbanBEATS are propagated to the WDS and the modified potable water supply system is subsequently analysed with EPANET 2 [25] in order to identify such critical areas.

2.3. Case Study Description and Scenarios

As a demonstration case study, we selected the Casey Clyde Growth Area (CCGA) in Melbourne, Australia. The area of 48 km^2, located along Melbourne's south-eastern urban fringe, is expected to grow in population to around 150,000 persons (around 51,700 new households) in the next 50 years. The region is of interest as the local water utility has undertaken an options assessment process of various water servicing strategies for the region. Most of the area is currently still undergoing planning. A hydraulic water distribution model has been developed for the planning process, but a detailed all-pipe model only exists for certain areas. Therefore, the WDS hydraulic model reflects what is anticipated at the planning stage. The newly planned district has an expected hourly design/peak water demand of 1436 L/s. The demand in the water distribution model is aggregated in roughly 1000 demand nodes. In Figure 2, different design demand patterns depending on the urban form are shown. Based on these patterns and the actual land use, the diurnal demand variation in the water distribution model are considered.

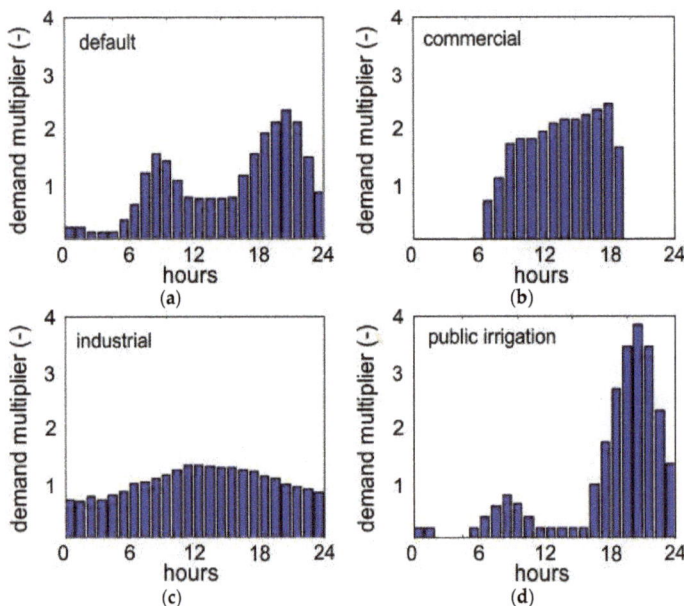

Figure 2. Diurnal design demand patterns used for the Case Clyde model: (**a**) default (domestic) indoor pattern; (**b**) commercial pattern; (**c**) industrial pattern; (**d**) (public and private) outdoor irrigation pattern. In the hydraulic model the average daily demand is used for the demand nodes. In addition these patterns are applied as hourly demand multipliers for the average daily demand.

The WDS model is shown in Figure 3a. The initial model was designed as a distribution grid with a grid length of approximately 400 to 600 m. Pipe diameters in this supply grid vary from 100 mm to 150 mm. In areas where more detailed planning information was available (e.g., in the northern section of the WDS), a higher level of detail could be replicated in the WDS model. In this northern area, the WDS adopts a fine grid structure with most pipe diameters around 100 mm and pipe lengths as low as 50 m. The WDS represents one of the many supply zones in the water utility's network. As such, there are five open connections that connect the CCGA zone to the rest of the network. For this study, these five intakes were modelled as reservoirs that provide the design flow into the CCGA zone. As such, only the relative travel time within our study boundary and not the overall water age from nearest supply tank or reservoir could be calculated. The average pressure in this zone is six bar.

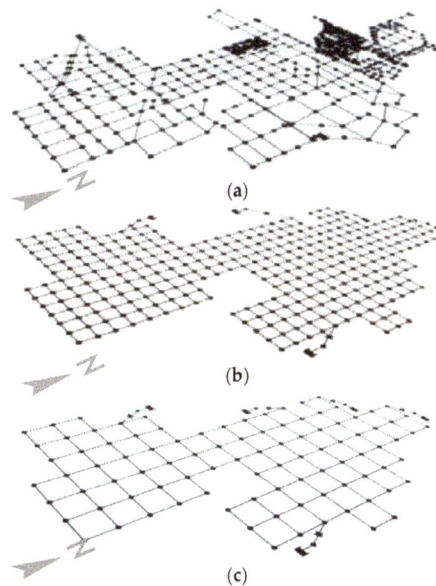

Figure 3. Water distribution models (**a**) CCGA model with different levels of detail due to planning stages; (**b**) semi-virtual model created with the Modular Design System (MDS) for the CCGA zone with a fine grid (MDS1); (**c**) semi virtual model for the CCGA zone with a coarse grid (MDS2).

For minimum pressure analysis, the peak demand is used. For water quality analysis, an extended period simulation of 10 days is used with the last two days modelled as low consumption days (additional demand factor of 0.5). Besides the travel time in the system (i.e., water age simulations), chlorine decay within the supply zone was also investigated. At the five intakes, chlorine booster stations were modelled. The initial chlorine concentrations in these intakes were set to 1 mg/L. Powell, et al. [29] investigated the chlorine decay coefficients in Melbourne. Based on these investigations, a bulk decay coefficient of 0.435 L/d and wall decay coefficient of 0.027 m/d were used for the modelled first order decay in EPANET 2. The minimum required chlorine concentration in the system was constrained to 0.2 mg/L.

The UrbanBEATS model was set up using a 500 m × 500 m grid (see Blocks in Figure 5a,b). The model was calibrated to agree with the determined design demand and specifications by the urban planners and water utility. We obtained information on the independent assessment of the area that informed the CCGA model. Calibration focussed on the total water demand for the region as well as the sub-daily diurnal demand patterns for different end uses. Subsequently, two different scenarios were investigated. The first scenario, referred to as reduced irrigation (RI), emulates a water restriction or drought period where private and open space irrigation demands were lowered from 2.4 ML/ha/year to 1 ML/ha/year [30]. The suggested irrigation value is also reflective of water that would be typically obtainable from rainwater harvesting. The second scenario, referred to as 'water efficiency' (WE), targets widespread demand reduction through forced adoption of six-star water fittings as per Australian standards [24] in all residential households.

2.4. A Case Independent Approach—Network Structure Variations

The CCGA model contains varying levels of network detail due to different levels of detail in spatial planning for the region. To further analyse what impact these planning stages will have on the water quantity and quality simulations and to generalise our findings beyond the CCGA model, we also adopted a case independent approach. We repeated the investigation on different (semi-) virtual water distribution models within the same region. Semi-virtual water distribution models enable us to investigate whether the actual demonstration case study would perform differently if specific characteristics are altered (e.g., topology, level of detail). By analysing these different models, we should gain insight into the questions surrounding model detail, in particular, the granularity of the network and the degree to which it is looped. With the help of the analysed shapes and sizes from the CCGA model, two similar WDS models with different grid sizes and pipe lengths (see Figure 3b,c) were generated using the Modular Design System (MDS) [31]. The MDS is a MATLAB based creation procedure that is freely available (http://www.hydro-it.com/extern/IUT/mds_app/). With this approach, predefined building blocks with a graph-based representation can be used to construct entire water distribution models. These can be entirely virtual with no additional input data needed or semi-virtual, where the information of the supplied area, land use, topography and water demand is included in the model creation [32,33]. Models can be generated with different topological characteristics (e.g., looped/branched layout of the WDS or level of detail in the model) to investigate the impacts of those characteristics on hydraulic or water quality performance. More details on this approach can also be found in [34,35].

We observed a variable grid structure in the CCGA model. The semi-virtual systems have the limitation that only one grid length can be used. Therefore, the two grid sizes of 400 m (Figure 3b) and 800 m (Figure 3c) were used to mimic those, which are most commonly present in the CCGA model.

Water quality degradation frequently takes place in the final sections of the water supply network e.g., in dead end pipes and offtakes to households [36]. Therefore, an additional refinement of the network structure was added to the models MDS1 and MDS2. These additional refinements were investigated in two scenarios: (1) as a looped structure and (2) as a branched structure (see Figure 4). Consequently, a network with a consistent node distribution and level of detail was obtained.

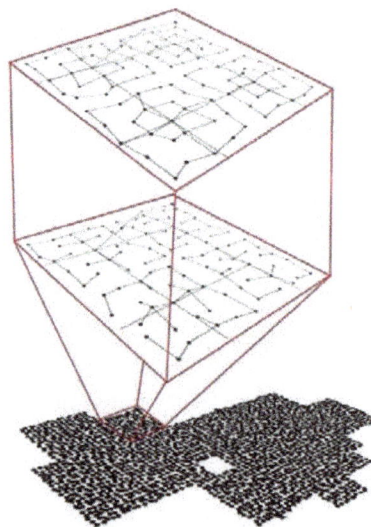

Figure 4. Different refinement strategies of an MDS model with a looped structure (top) or a branched structure (middle).

To use the four created semi-virtual models in our investigation, a network design was necessary. We adopted similar boundary conditions as for the CCGA model (intakes, demands, topography, design pressure). With assistance from the EPANET 2 add-on WaterNetGen [37], the MDS-models were designed. The smallest diameter used to design the coarse models was 100 mm and the largest diameter of 800 mm was chosen. The design load was chosen to obtain similar velocity distributions as in the CCGA model. This would enable us to evaluate the general structural impact (i.e., level of detail) and the topological impact (loop or branched structure).

To summarise, we investigated two water management scenarios (RI and WE) for the CCGA model and four semi-virtual models differing in grid size and network topology. Specifically, we focussed on quantifying the impact these scenarios have on the pressure distribution, water travel time (stagnation) and chlorine decay within the WDS.

3. Results

The UrbanBEATS simulation was set up and the demand categories for the different *Blocks* were calibrated to meet proposed design demand and its characteristics (patterns) in the CCGA model (the same demands and patterns are also used for the MDS models). Subsequently, the spatial distribution of the two demand reduction scenarios RI and WE was determined. As a result, the model not only alters the daily average water demands accordingly across the region, but also the corresponding diurnal patterns. Table 1 summarises the results of these reduction scenarios. For the RI scenario, the hourly peak demand was reduced by 15% and the daily average demand was reduced by 9.3%. For the WE scenario, the hourly peak demand was reduced by 10.8% and the daily average demand by 13.7%. Although the RI scenario has more impact on the hourly peak demand than the WE scenario, the actual daily water saving is higher for the WE scenario (see Table 1). This is due to the fact, that the RI scenario primarily affects the evening peak (see Figure 2d). Note that no reduction of the peak design criteria for pipe-sizing (hourly peak demand for a peak day) occurs, because the simulated day is an average day and not a peak day which is used for WDS design.

Table 1. Summary of the scenarios and simulated the demand reductions.

Scenario	Hourly Peak Demand m^3/h	Peak Demand Reduction (Peak h) %	Daily Average Demand m^3/d	Daily Average Demand Reduction %
Initial situation	5168	-	53,565	-
Reduced irrigation (RI)	4381	15.0	48,591	9.3
Water efficiency (WE)	4601	10.8	46,237	13.7

The spatial impact of both scenarios on water consumption, simulated with UrbanBEATS, is illustrated in Figure 5a for RI and 5b for WE. For areas with more green space available (e.g., single detached dwellings with garden) the RI scenario has a greater impact. Unfavourably for the potable water supply network, these areas are characterised by low building density, where the decisive design criteria is mainly the minimum pipe diameter requirement for fire-fighting. As such, the potable water supply network is over-sized under regular demand conditions, thereby causing low flow velocities and increasing the risk of stagnation (degradation of water quality). The WE scenario has a greater impact in areas with high demand (see Figure 5b—marked area), such as areas with apartment buildings.

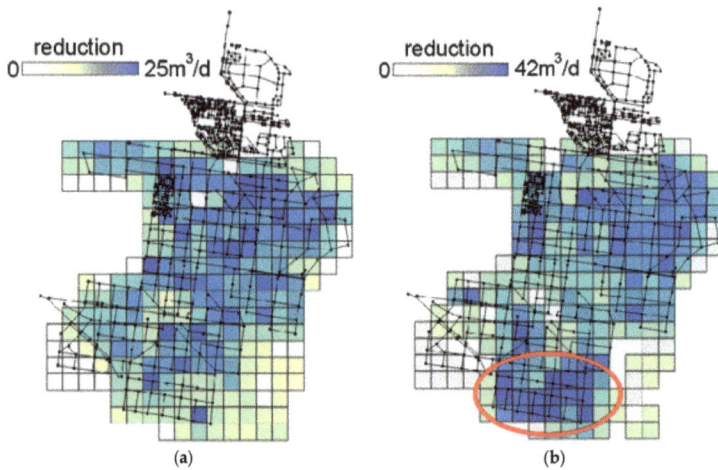

Figure 5. Spatially-distributed demand reductions for the scenarios reduced irrigation (RI) (**a**); and water efficiency (WE) (**b**) (adapted from [18] with permission from ASCE).

In Figure 6, the initial pressure distribution (Figure 6a) and the impact of the two reduction scenarios on the pressure performance (peak hour) are shown (Figure 6b—RI and Figure 6c—WE). It can be observed that with a demand reduction, the pressure increases. A higher pressure in the potable water supply network also increases water losses (losses are approximately proportional to the square root of the pressure head). Nevertheless, in this study the effect of increased water demand due to increasing water losses is neglected.

In Figure 6e,f, the residence time in that supply zone is evaluated for the reduction scenarios RI and WE. These results are based on an extended period simulation over 10 days with eight consecutive average days and the last two days assumed as low consumption days with one third of the volume of the average day. The results for water age are evaluated on the last of the 10-day simulation. Analyses for real demand data show that such low demand days occur on at least one or a few days per year [27].

By visually comparing the residence time distributions after demand reduction with the initial situation shown in Figure 6d, the highest increase in water age is obtained for the scenario WE in Figure 6f, especially for the marked area in Figure 5b. Furthermore, a differentiation for the travel time in the network can be observed for the upper and lower part of the network, in which different levels of detail are implemented in the system.

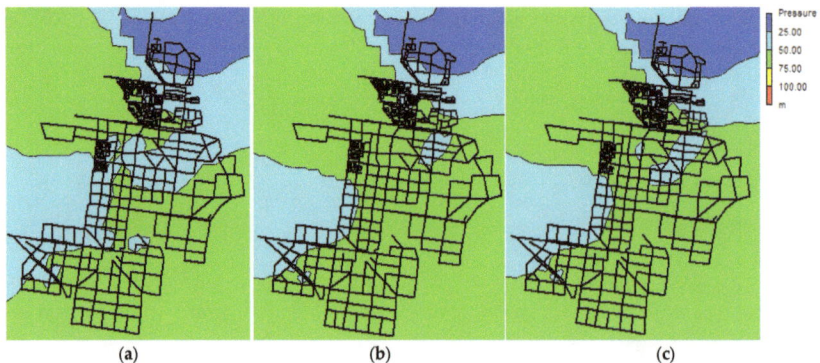

Figure 6. *Cont.*

Figure 6. Pressure and water age distribution for the initial situation and the reduction scenarios: pressure distribution for the (**a**) initial situation; (**b**) RI; (**c**) WE and the residence time in the network (**d**) initial situation; (**e**) RI and (**f**) WE for the last day of a 10-day extended period simulation.

3.1. Results for Water Age and Chlorine Decay for CCGA

In Figure 7, the changes in travel time in the network depending on the water demand reduction scenarios RI (Figure 7a) and WE (Figure 7b) are systematically investigated. On the horizontal axis, the initial travel time without any scenario (initial situation) is shown. On the vertical axes the travel time for the different scenarios is shown. Each dot represents a demand node and gives insight into the change in travel time due to demand reductions, which are indicated by the colour of each dot. No demand reduction (0%) is shown in dark blue and in cases where the demand would be entirely covered by decentralised strategies, a dark red colour is used (see legends in Figure 7). For dots above the first median line (y = x), an increase in the observed parameter is indicated and, below, a decrease. For dots on the first median, no changes occurred and the greater the distance to the first median, the higher the change in the observed parameter.

When comparing Figure 7a,b, higher node-wise demand reductions can be observed for the RI scenario (light blue and even green dots). Nevertheless, for the WE scenario, higher changes in travel time are observed (more points above the first median also with a greater distance to it). In general, there is an increase in travel time, but there are also noticeable nodes where the travel time decreases. This can be explained by the five intakes to the model. When changing the demand in the WDS, flow redistributions occur and the source node from which a demand node gets its water can be changed.

The results for changes in chlorine concentration are shown in Figure 7c,d. For the RI scenario, there is only a slight drop for all the nodes. On average, a reduction of 2.37% (-0.013 mg/L) in chlorine concentration can be observed. For the WE scenario, there is a greater change in concentration. There is also a higher impact of flow redistribution (12.2% of the nodes/points are above the first median). In general, there is a higher quality decline (average reduction of 8.89%, -0.052 mg/L) although the node-wise reductions are lower.

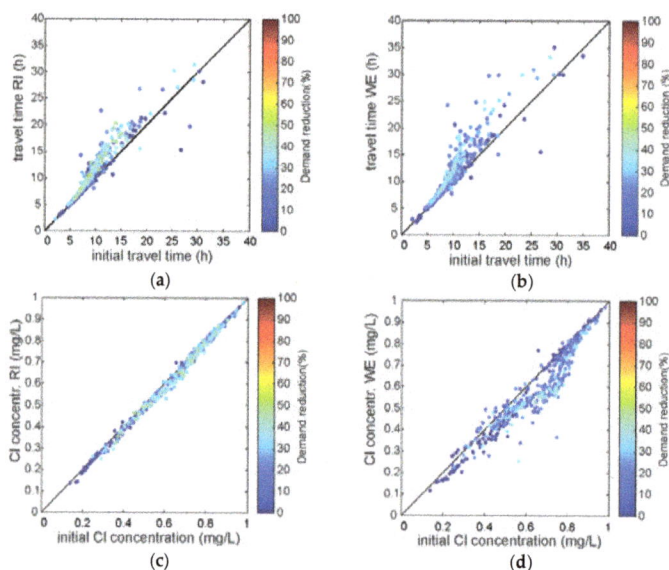

Figure 7. Systematic node-wise comparison of the initial conditions with the scenarios RI and WE with colours indicating the demand reductions for (**a**) and (**b**) travel time in the network; (**c**) and (**d**) chlorine concentrations.

3.2. Sructural Impact (without Scenarios for Demand Reduction)

In this section, the structural impact of the WDS models is analysed. The semi-virtual models MDS1 and MDS2 are compared with the CCGA model (see Figure 3). Additionally, we also investigate the refinements of MDS1 and MDS2 with a looped and branched structure (see Figure 4).

We first compare the results of the different models without applying any demand reduction scenarios. Figure 8 shows the cumulative distribution functions (CDF) of the node values for the different models for the travel time in the potable supply network and the chlorine concentrations. The blue line shows the results for the CCGA model (with partial refinements), which, according to Figure 8a, is between the results for the initial MDS1 model and branched and looped refinements. The same observation is made for MDS2 in Figure 8b. Therefore, it can be concluded, that in terms of travel time distribution in the network, the semi-virtual systems have comparable properties as the actual system. We can therefore reasonably assume that these are suitable to address the question of what would happen if the level of model detail becomes coarser (less details) or finer (more details) than the baseline CCGA model.

For chlorine concentrations, similar behaviour can be observed with the exception, for low concentrations (for MDS1 in Figure 8c below 0.45 mg/L and MDS2 in Figure 8d below 0.55 mg/L) the semi-virtual systems over-value the concentrations. This is due to the fact that the demand in the semi-virtual systems is more equally distributed compared to the CCGA model. Nevertheless, it is assumed that from MDS1 and MDS2 results, reasonable insights into the impact of the network structural behaviour can be obtained.

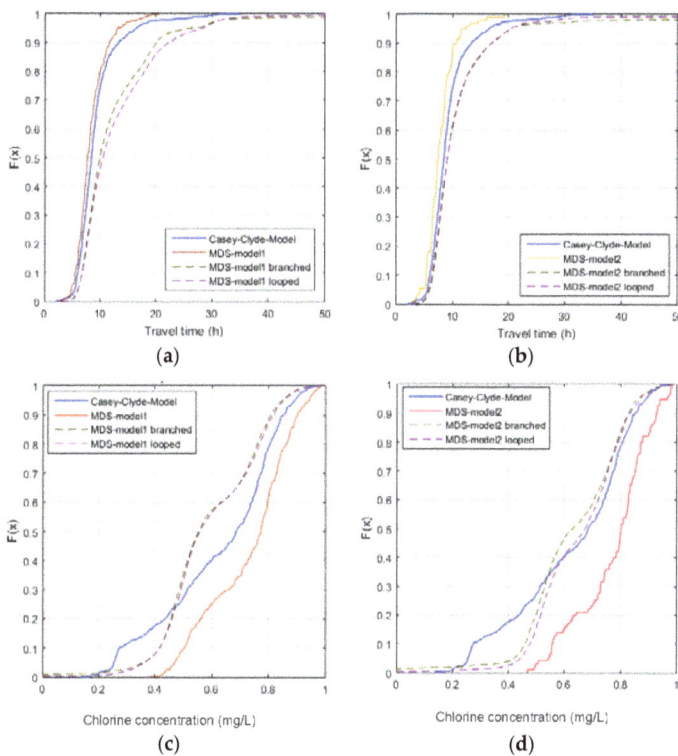

Figure 8. Cumulative distribution function (CDF) of the simulated node values for (**a**) travel time in the CCGA model in comparison to the MDS1 models (coarse, branched and looped refinements); (**b**) travel time in the CCGA model in comparison to the MDS2 models (coarse, branched and looped refinements); (**c**) chlorine concentrations in the CCGA model in comparison to the MDS1 models (coarse, branched and looped refinements); (**d**) chlorine concentration in the CCGA model in comparison to the MDS2 models (coarse, branched and looped refinements).

3.3. Structural Properites and the Effect of Reductions Scenarios on Water Quality

Figure 9 shows the impact of branched and looped refinements of MDS1 for the scenarios RI and WE on the travel time. Similarly, Figure 10 shows the impact on the chlorine concentrations. All analyses were performed for each model and each parameter, but we have focused our discussion in this paper on the most prevalent and significant cases.

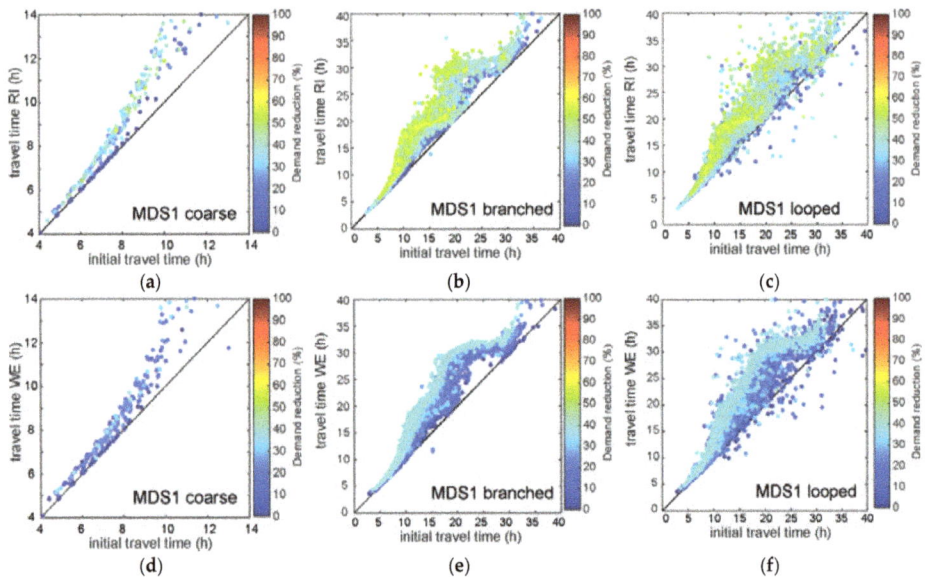

Figure 9. Investigation of travel time and of the impact of the reduction scenarios RI and WE in comparison to model refinements in the model MDS1 (**a**) coarse MDS1 model and RI; (**b**) branched refinement of MDS1 and RI; (**c**) looped refinements of MDS1 and RI; (**d**) coarse MDS1 model and WE; (**e**) branched refinement of MDS1 and WE; (**f**) looped refinements of MDS1 and WE.

In Figure 9a,d the results for the coarse MDS1 model are shown. The coarse model consists of 315 nodes while the refined models consists of 7765 nodes. Without the finer network structure, one can see that the maximum values for the travel time are much lower compared to the finer networks. The average travel time under initial conditions is 8.4 h, for branched refinement it is 14.4 h and for the looped refinement it is 14.2 h. For the RI scenario without refinements it is 9.7 h, for the branched refinement it is 17.1 h and for the looped refinement it is 16.7 h. In the coarse model, an increase of approximately 1.3 h is caused by scenario RI, whereas for the refined models, an approximately 2.6-h increase occurs (slightly less for the looped system).

The topological impact can clearly be observed when comparing the branched refinements with the looped refinements. In the (almost) fully-branched networks, the effect of flow redistributions cannot be observed. Therefore, a demand reduction in these cases mainly causes an increase in travel time (dots over the first median). In the branched system, the highest demand reduction also corresponds with the highest increase in travel time while for the looped systems this effect is weakened by flow redistributions.

In Figure 10, the chlorine concentrations for MDS1 and MDS2 models and their refinements are shown for the WE scenario. The MDS1 models are the same as in Figure 9. The coarse MDS2 model consists of 95 nodes and the refined models of 2250 nodes. In the coarse models, no chlorine concentrations below 0.4 mg/L were observed, while in the refined models there were values even close to 0 mg/L. Evidently, the branched systems in this case (Figure 10b,e) contain more values below the threshold value of 0.2 mg/L compared to the looped refinement (Figure 10c,f). For chlorine concentrations the effect of loops and, consequently, flow redistribution is less pronounced than for travel time.

The average reduction of chlorine concentrations across all models (Figure 10a,f) is between 0.053 and 0.057 mg/L. However, a reduction up to 0.3 mg/L occurred, and even below the required threshold of 0.2 mg/L.

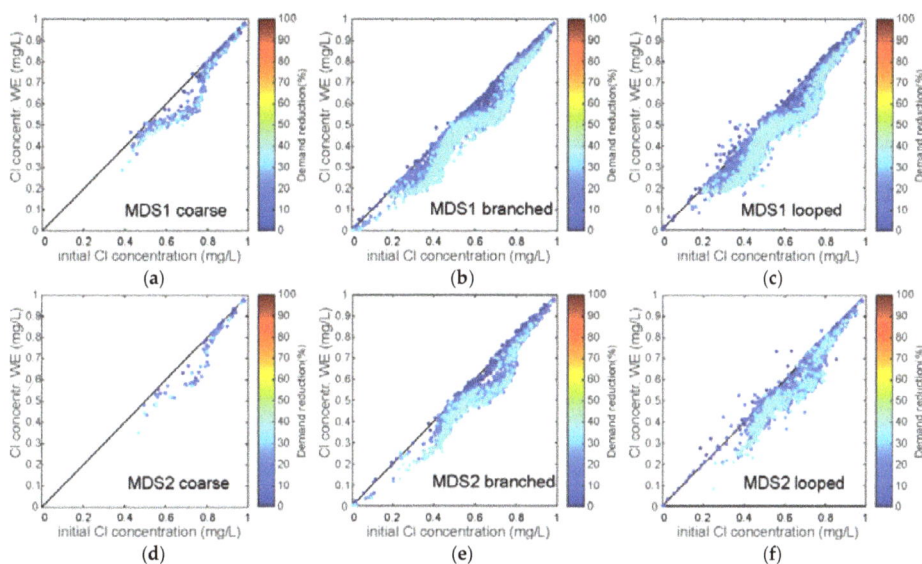

Figure 10. Investigation of chlorine concentrations and of the impact of the reduction scenarios WE in comparison with model refinements in the model MDS1 and MDS2 (**a**) coarse MDS1 model; (**b**) branched refinement of MDS1; (**c**) looped refinements of MDS1; (**d**) coarse MDS2 model (**e**) branched refinement of MDS2; (**f**) looped refinements of MDS2.

4. Discussion

To further support the discussion of results, we statistically evaluated and compared the data from Figures 7, 9 and 10. The outcomes are shown in Figure 11 in the form of boxplots depicting the change in node values for water age and chlorine concentrations. The ratio to the initial conditions (without applying any reductions scenarios) are shown. The boxplots show the median value (red horizontal lines), the interquartile range (IQR) from the 25th percentile to the 75th percentile (blue central boxes), 1.5 range of the IQR as whiskers and all values outside that 1.5 IQR as outliers (red crosses). The notches show the 95% confidence interval of the median value. Outliers in this context do not indicate that they are statistically relevant, but that there is a high variation in the values throughout the investigated systems. For chlorine concentration, these outliers above one can be, for example, the impact of the flow redistribution. A value greater than one means an increase for the scenario (e.g., in general for travel time) and a value below one means a reduction of that parameter when applying the demand reduction scenarios (e.g., in general for the chlorine concentrations).

In Figure 11a, the results are shown for the CCGA model for the scenarios RI and WE for the travel time and the chlorine concentrations. Although the change in travel time distribution is not significant (the median values and their notches are overlapping), notable differences in the outliers can be seen, predominately for increased values >1. For chlorine concentration, there is a general downward trend and the effects of the scenarios RI and WE are more pronounced. The WE scenario causes a much greater variation in changes (greater and smaller than one) than the RI scenario. The flow redistribution and demand reduction throughout the entire day are likely causes of this behaviour.

In Figure 11b, the boxplots of travel time in the network for the model MDS1 and its branched and looped refinements are plotted for the scenarios RI and WE. For the initial system, in both scenarios, there is no decrease in travel time (except for one outlier) while for the refined system there is a decrease. This means that the effect of flow redistribution primarily occurs in the fine structure of the system. It is also plausible that additional reverse flows in the pipes may be occurring, thereby transporting the water back and forth, which can also usually lead to water quality problems. This was, however, not specifically tested in this study. The networks with branched refinements in Figure 11b show a marginally higher travel time increase than the systems with the looped refinements. However, looped systems have a greater impact on the outliers (in both directions, increase and decrease). The general design principal for water supply systems is to consider a looped layout for redundancy and enhanced reliability. Looped structures are, however, also susceptible to larger variations in travel times and other changes, which is inherent to their nature and flexibility.

In Figure 11c the boxplots of chlorine concentrations in the network for the model MDS1 and its branched and looped refinements are compared to the results obtained with MDS2 for the scenario WE. The two coarse MDS models do not show a great difference. However, noticeably, there is no increase in chlorine concentration. Hence, it can be concluded that the MDS2 model is too coarse for flow redistribution to take place when demand reductions are modelled. Interestingly, for branched refinements, reductions in chlorine are higher for the outliers than in the looped systems (the median values and IQR do not deviate significantly).

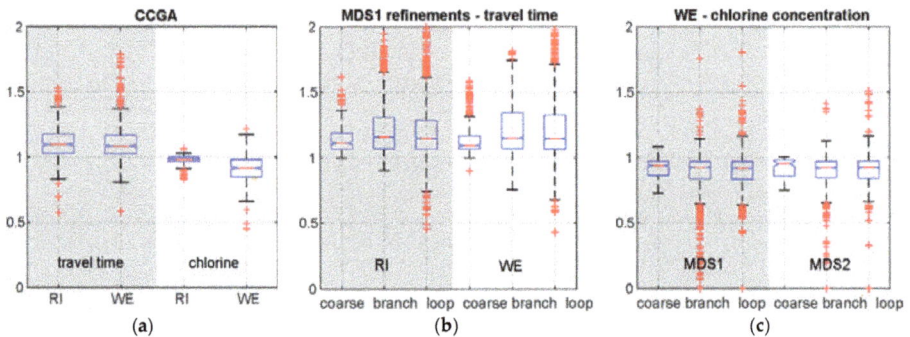

Figure 11. Statistical comparison of the different scenarios (a) CCGA model; (b) the MDS1 model with different refinement strategies (c) structural impact between MDS1 and MDS2.

The application of different semi-virtual networks provided valuable insights into the effect of different levels of network detail that result from planning of future water services such as the one in the CCGA case study. Despite us using the context of Casey-Clyde as a boundary condition, the obtained model results for MDS1 and MDS2 are more case unspecific. The travel time in the water network is usually used a representative indicator for water quality and is, to some extent, also related to the decrease in chlorine concentration. Nevertheless, in the first order chlorine decay equation, the decay coefficient is inversely proportional to the hydraulic radius of the pipe diameter. With different structural layouts (e.g., branched or looped refinements) and, accordingly, different diameter distributions, unexpected behaviours were observed. Mostly, that in the branched network (which usually have smaller maximum travel time) the decrease in chlorine concentrations is highest. In this study, the temperature of the water was kept constant. With an increasing water temperature, the decay rate also increases. In moderate and warm temperature zones, the potable water usually becomes warmer when its location is closer to the consumer. In this instance, the described effect of water quality degradation in the last section of the potable pipe network even amplifies the diameter-dependent decay. Abokifa, Yang, Lo and Biswas [36] showed that in the traditional modelling approach, as shown in this work, the water quality degradation is significantly underestimated and most of the degradation takes place in the very last sections of the network (i.e., closest to the consumer). This reported effect would even further intensify the water quality degradation determined in this work.

For the management of hybrid water supply systems, chlorine management is a crucial issue. The injection usually occurs at booster stations at e.g., the intakes to the system. In this study, the injected concentrations were already at maximum (1 mg/L) and cannot be increased any further. Despite this, many locations in the network resulted in partly insufficient chlorine concentrations (below 0.2 mg/L). It was observed that the water demand reduction throughout the day had the most severe impact on the potable water supply network. Therefore, a re-chlorination at certain nodes might be required in such cases to offset this adverse impact.

Another observed effect of transitioning from a fully centralised water supply towards a hybrid water supply approach is the increase in network pressures due to less flow and hence less friction losses [38]. Revision or development of smarter pressure management strategies for the potable water supply system may be recommended in such a case (e.g., lower pumping heights, pressure reductions). Otherwise, demand reduction from the implementation of hybrid technologies could be partly compensated by an increase in water losses.

Furthermore, with the increasing water efficiency of household devices (e.g., shower, clothes washer, etc.), the volume which can be used for greywater recycling is reduced. To strike an optimal balance in the design and implementation of hybrid water supply systems, a fully integrated picture of the effects and contribution of different water sources needs to be established and system interactions must be fully considered in the infrastructure choices.

Yet another notable issue is the implementation of a minimal diameter for fire-fighting requirements, which can already cause stagnation problems under regular conditions. As long as no other technology is in place (e.g., decentralised storage with an extra pipe network for fire-fighting), it is mandatory and must be handled with operational measures (e.g., flushing), which in itself causes the loss of water.

In the context of a progressive installation of decentralised supply technology, average water demand can be reduced. However, a reduction of the peak (design) demands is hardly possible. For a reliable supply, it has to be ensured that sufficient water is always available, especially when there are longer periods of no rainfall. This is particularly crucial when available decentralised storages are depleted during the peak water demand days/hours due to high temperatures. As such, we cannot rely solely on rainwater harvesting. To fully simulate such a behaviour, a dynamic approach would be required. A limitation of the presented approach at the current stage of development is that long-term system dynamics (e.g., in rainfall and water demands) are not considered for simplification purposes. However, in the conceptualisation of the developed approach, it was ensured that the model can be enhanced for this capability without changing the method for spatial coupling presented in this paper.

In such a dynamic consideration, it is expected that the highest volume reduction from decentralised technologies like rainwater harvesting will arise when there is wet weather with lower outside temperatures. For the potable water supply, these are usually low demand days, which are critical for stagnation and water quality problems. These low flow conditions would even be further intensified by a hybrid solution. Nevertheless, a subset of extreme situations for water supply performance can already be assessed with the presented approach (e.g., an increase in pressure for the design/peak demand or an increase in water age and water quality).

In terms of practical implications, there are several strategies to overcome these issues such as an extensive decentralised storage volume and increasing chlorine booster stations among others. However, the most efficient strategy is expected to be a more location-sensitive approach, which identifies areas where a demand reduction is reasonable and areas where more drawbacks than benefits can be expected. For existing systems that already operate at maximum capacity, to increase the reliability [39] or the performance in failure mode of the existing system [40], hybrid solutions can provide additional water resources, if required, without costly investment in the central water infrastructure. With the presented approach, such adaption strategies for an existing water supply network can be tested to support the infrastructure planning process.

5. Conclusions

A modern urban water cycle is a strongly interlinked system, but the different sub-disciplines in that cycle are still often regarded separately in traditional management structures and interactions in complex systems are often neglected. For the efficient long-term operation of hybrid water supply systems, these interactions are significant. However, the impact of increasing the implementation of decentralised technologies on the local technical performance in sewer systems and potable water supply networks have rarely been analysed in the scientific literature. Urban form and planning regulations frequently interact with the implementation of such decentralised systems. Therefore, for decision-makers it is important to understand the implications of different planning regulations on aspects like urban drainage or water supply not only at the regional, but also at a local scale.

This paper aimed to quantify impact of the spatial distribution of the implementation of decentralised technologies (i.e., hybrid water supply systems) on the technical performance of existing water supply networks. The impact of demand reduction scenarios (i.e., rainwater harvesting to substitute private irrigation and water efficiency measures) determined with a planning-support tool called UrbanBEATS, which combines spatial and non-spatial data, on the water quality in the potable water network are demonstrated. Most importantly, the very disadvantageous interaction of the low density urban form (low total demand, minimal pipe diameters in the potable supply network) and the high rainwater harvesting potential (a large harvestable area in combination with green space for irrigation) was identified as causing local water quality problems in those areas. In contrast, in high demand areas (e.g., the high density urban form) there is less effect from rainwater harvesting due to the limited available space. The impact on the technical performance of the potable water supply network in these regions is therefore less significant. In these high density areas, water efficiency measures result in the highest savings in water volume but do not cause significant problems for the technical performance of the potable water supply network. A limitation of the presented approach at the current stage of development is that the long-term system dynamics (e.g., rainfall and water demand) are not considered for reasons of simplification. However, in the conceptualisation of the developed approach, it was ensured that the model can be enhanced for that capability without changing the method for spatial coupling presented in this paper.

For a more generalised and case-independent conclusion, further analyses were performed for semi-virtual benchmark networks. We questioned the appropriateness of the water distribution model for such an analysis. It was shown that using an inappropriate level of detail for the computational model of a WDS for determining the hydraulics can have a more pronounced impact on the water quality results than the actual decentralised measures. Currently, seasonal demand dynamics are only considered in a simplified way. An average day with a diurnal demand pattern was used to assess the impact of decentralised measures, with an assumed low consumption day being one third of the average daily demand used for water quality assessments in the potable water supply network. Analysis of real demand data showed that such low demand days occur on at least one or a few days per year. Future model development will implement a more detailed approach to considering the seasonal demand dynamics as well as more detailed feedback loops with decentralised systems such as rainwater/stormwater harvesting measures and even wastewater recycling.

Acknowledgments: This research was funded by the Austrian Research Promotion Agency (FFG) within the research project ORONET (project number: 858557) and Monash University's Faculty of Engineering SEED Funding Scheme 2015. We also thank South East Water (Melbourne, Australia) for providing us data and information on the proposed urban development and pipe network infrastructure for the Casey-Clyde case study region as well as valuable feedback on this work.

Author Contributions: Robert Sitzenfrei conceived and designed the experiments; Markus Sitzmann and Jonatan Zischg performed the water supply assessments and analysed the data; Peter M. Bach performed the UrbanBEATS simulations; Robert Sitzenfrei wrote the first draft of the paper which all co-authors revised.

Conflicts of Interest: The authors declare no conflict of interest. The founding sponsors had no role in the design of the study; in the collection, analyses, or interpretation of data; in the writing of the manuscript, and in the decision to publish the results.

References

1. Larsen, T.A.; Hoffmann, S.; Luthi, C.; Truffer, B.; Maurer, M. Emerging solutions to the water challenges of an urbanizing world. *Science* **2016**, *352*, 928–933. [CrossRef] [PubMed]
2. Sharma, A.; Burn, S.; Gardner, T.; Gregory, A. Role of decentralised systems in the transition of urban water systems. *Water Sci. Technol. Water Supply* **2010**, *10*, 577–583. [CrossRef]
3. Marlow, D.R.; Moglia, M.; Cook, S.; Beale, D.J. Towards sustainable urban water management: A critical reassessment. *Water Res.* **2013**, *47*, 7150–7161. [CrossRef] [PubMed]
4. Sapkota, M.; Arora, M.; Malano, H.; Moglia, M.; Sharma, A.; George, B.; Pamminger, F. An overview of hybrid water supply systems in the context of urban water management: Challenges and opportunities. *Water* **2015**, *7*, 153–174. [CrossRef]
5. Fletcher, T.D.; Shuster, W.; Hunt, W.F.; Ashley, R.; Butler, D.; Arthur, S.; Trowsdale, S.; Barraud, S.; Semadeni-Davies, A.; Bertrand-Krajewski, J.-L.; et al. Suds, lid, bmps, wsud and more—The evolution and application of terminology surrounding urban drainage. *Urban Water J.* **2015**, *12*, 525–542. [CrossRef]
6. Sitzenfrei, R. An Integrated View on the Urban Water Cycle: A Shift in the Paradigm? Postdoctoral Thesis, Unit of Environmental Engineering, University of Innsbruck, Innsbruck, Austria, 2015.
7. Rozos, E.; Makropoulos, C. Source to tap urban water cycle modelling. *Environ. Model. Softw.* **2013**, *41*, 139–150. [CrossRef]
8. Sitzenfrei, R.; Rauch, W.; Rogers, B.; Dawson, R.; Kleidorfer, M. Modeling the urban water cycle as part of the city. *Water Sci. Technol.* **2014**, *70*, 1717–1720. [CrossRef] [PubMed]
9. Bouziotas, D.; Rozos, E.; Makropoulos, C. Water and the city: Exploring links between urban growth and water demand management. *J. Hydroinform.* **2015**, *17*, 176–192. [CrossRef]
10. Rozos, E.; Butler, D.; Makropoulos, C. An integrated system dynamics—Cellular automata model for distributed water-infrastructure planning. *Water Sci. Technol. Water Supply* **2016**, *16*, 1519–1527. [CrossRef]
11. Sapkota, M.; Arora, M.; Malano, H.; Moglia, M.; Sharma, A.; George, B.; Pamminger, F. An integrated framework for assessment of hybrid water supply systems. *Water* **2016**, *8*, 4. [CrossRef]
12. Bach, P.M.; Deletic, A.; Urich, C.; Sitzenfrei, R.; Kleidorfer, M.; Rauch, W.; McCarthy, D.T. Modelling interactions between lot-scale decentralised water infrastructure and urban form—A case study on infiltration systems. *Water Resour. Manag.* **2013**, *27*, 4845–4863. [CrossRef]
13. Bach, P.M.; McCarthy, D.T.; Urich, C.; Sitzenfrei, R.; Kleidorfer, M.; Rauch, W.; Deletic, A. A planning algorithm for quantifying decentralised water management opportunities in urban environments. *Water Sci. Technol.* **2013**, *68*, 1857–1865. [CrossRef] [PubMed]
14. Wong, T.; Allen, R.; Brown, R.; Deletić, A.; Gangadharan, L.; Gernjak, W.; Jakob, C.; Johnstone, P.; Reeder, M.; Tapper, N.; et al. *Blueprint2013—Stormwater Management in a Water Sensitive City*; Cooperative Research Centre for Water Sensitive Cities: Melbourne, Australia, 2013.
15. Sitzenfrei, R.; Möderl, M.; Rauch, W. Assessing the impact of transitions from centralised to decentralised water solutions on existing infrastructures—Integrated city-scale analysis with vibe. *Water Res.* **2013**, *47*, 7251–7263. [CrossRef] [PubMed]
16. Sitzenfrei, R.; Rauch, W. Integrated hydraulic modelling of water supply and urban drainage networks for assessment of decentralized options. *Water Sci. Technol.* **2014**, *70*, 1817–1824. [CrossRef] [PubMed]
17. Mair, M.; Zischg, J.; Rauch, W.; Sitzenfrei, R. Where to find water pipes and sewers?—On the correlation of infrastructure networks in the urban environment. *Water* **2017**, *9*, 146. [CrossRef]
18. Sitzenfrei, R.; Zischg, J.; Sitzmann, M.; Rathnayaka, S.; Kodikara, J.; Bach, P.M. Effects of implementing decentralized water supply systems in existing centralized systems. In *World Environmental and Water Resources Congress 2017*; Dunn, C.N., Weele, B.V., Eds.; American Society of Civil Engineers (ASCE): Reston, VA, USA, 2017.
19. Arora, M.; Malano, H.; Davidson, B.; Nelson, R.; George, B. Interactions between centralized and decentralized water systems in urban context: A review. *Wiley Interdiscip. Rev. Water* **2015**, *2*, 623–634. [CrossRef]
20. Bach, P.M. Urbanbeats—A Virtual Urban Water System Tool for Exploring Strategic Planning Scenarios. Ph.D. Thesis, Monash University, Melbourne, Australia, 2014.
21. Telford, T. *Industrial and Commercial Estates–Planning and Site Development*; United Kingdom Development Agencies: London, UK, 1986.

22. Department of Environment, Land, Water and Planning (DELWP). *Victorian Planning Provisions*; DELWP: Melbourne, Australia, 2017.

23. De Chiara, J.; Panero, J.; Zelnik, M. *Time-Saver Standards for Housing and Residential Development*; McGraw-Hill Companies: New York, NY, USA, 1995.

24. Standards Australia. *Water Efficient Products—Rating and Labelling*; AS/NZS 6400:2016; Standards Australia: Sydney, Australia, 2016.

25. Rossman, L.A. *EPANET 2 User's Manual*; U.S. Environmental Protection Agency: Cincinnati, OH, USA, 2000.

26. Rathnayaka, K.; Malano, H.; Maheepala, S.; George, B.; Nawarathna, B.; Arora, M.; Roberts, P. Seasonal demand dynamics of residential water end-uses. *Water* **2015**, *7*, 202–216. [CrossRef]

27. Sitzenfrei, R.; Anawar, H.M.; Strezov, V. Uncertainty in hydropower system: New approach in design of sustainable hydropower system. In *Renewable Energy Systems: Efficiency, Innovation and Sustainability*; Strezov, V., Ed.; Productivity Press Publishing: New York, NY, USA, 2017, in press.

28. Rathnayaka, K.; Maheepala, S.; Nawarathna, B.; George, B.; Malano, H.; Arora, M.; Roberts, P. Factors affecting the variability of household water use in Melbourne, Australia. *Resour. Conserv. Recycl.* **2014**, *92*, 85–94. [CrossRef]

29. Powell, J.C.; Hallam, N.B.; West, J.R.; Forster, C.F.; Simms, J. Factors which control bulk chlorine decay rates. *Water Res.* **2000**, *34*, 117–126. [CrossRef]

30. Wilkenfeld, G. *Water Saving Requirements for New Residential Buildings in Victoria: Options for Flexible Compliance*; Department of Sustainability and Environment: Sydney, Australia, 2007.

31. Möderl, M.; Sitzenfrei, R.; Fetz, T.; Fleischhacker, E.; Rauch, W. Systematic generation of virtual networks for water supply. *Water Resour. Res.* **2011**, *47*. [CrossRef]

32. Sitzenfrei, R.; Möderl, M.; Rauch, W. Automatic generation of water distribution systems based on gis data. *Environ. Model. Softw.* **2013**, *47*, 138–147. [CrossRef] [PubMed]

33. Zischg, J.; Mair, M.; Rauch, W.; Sitzenfrei, R. Enabling efficient and sustainable transitions of water distribution systems under network structure uncertainty. *Water* **2017**, *9*, 715.

34. Sitzenfrei, R. Stochastic Generation of Urban Water Systems for Case Study Analysis. Ph.D. Thesis, Unit of Environmental Engineering, University of Innsbruck, Innsbruck, Austria, 2010.

35. Sitzenfrei, R. A review on network generator algorithms for water supply modelling and application studies. In Proceedings of the World Environmental and Water Resources Congress, West Palm Beach, FL, USA, 22–26 May 2016; pp. 907–916.

36. Abokifa, A.A.; Yang, Y.J.; Lo, C.S.; Biswas, P. Water quality modeling in the dead end sections of drinking water distribution networks. *Water Res.* **2016**, *89*, 107–117. [CrossRef] [PubMed]

37. Muranho, J.; Ferreira, A.; Sousa, J.; Gomes, A.; Marques, A.S. Waternetgen: An EPANET extension for automatic water distribution network models generation and pipe sizing. *Water Sci. Technol. Water Supply* **2012**, *12*, 117–123. [CrossRef]

38. Campisano, A.; Modica, C.; Reitano, S.; Ugarelli, R.; Bagherian, S. Field-oriented methodology for real-time pressure control to reduce leakage in water distribution networks. *J. Water Resour. Plan. Manag.* **2016**, *142*, 04016057. [CrossRef]

39. Pietrucha-Urbanik, K.; Tchórzewska-Cieslak, B. Water supply system operation regarding consumer safety using kohonen neural network. In *Safety, Reliability and Risk Analysis: Beyond the Horizon*; Steenbergen, R.D.J.M., van Gelder, P.H.A.J.M., Miraglia, S., Vrouwenvelder, A.C.W.M., Eds.; Taylor & Francis Group: London, UK, 2014; pp. 1115–1120.

40. Tchórzewska-Cieślak, B.; Pietrucha-Urbanik, K.; Urbanik, M. Analysis of the gas network failure and failure prediction using the Monte Carlo simulation method (Analiza awaryjności sieci gazowych oraz prognozowanie awarii z zastosowaniem symulacyjnej metody Monte Carlo). *Eksploat. Niezawodn.—Maint. Reliab.* **2016**, *18*, 254–259. [CrossRef]

water

MDPI

Article

Simulation of Infrastructure Options for Urban Water Management in Two Urban Catchments in Bogotá, Colombia

Carlos Andrés Peña-Guzmán [1,2,*] (ORCID), Joaquín Melgarejo [2], Inmaculada Lopez-Ortiz [2] and Duvan Javier Mesa [3]

1   Environmental Engineering Program, Universidad Autónoma de Colombia, Bogotá 111711, Colombia
2   Institute of Water and Environmental Sciences, Universidad de Alicante, San Vicente del Raspeig Route, S/n, 03690 Alicante, Spain; jmelgar@ua.es (J.M.); iortiz@ua.es (I.L.-O.)
3   Environmental Engineering Program, Universidad Santo Tomas, Bogotá 110311, Colombia; duvanmesa@usantotomas.edu.co
*   Correspondence: carpeguz@gmail.com; Tel.: +51-318-516-5542

Received: 27 August 2017; Accepted: 1 November 2017; Published: 5 November 2017

Abstract: Urban areas are currently experiencing rapid growth, which brings with it increases in the population, the expansion of impervious surfaces, and an overall jump in the environmental and hydrological impact. To mitigate such an impact, different strategies proposed to tackle this problem often vary; for example, stormwater tanks, the reuse of wastewater and grey water, the installation of equipment to reduce water consumption, and education-based approaches. Consequently, this article presents the simulation and evaluation of implementing infrastructure options (stormwater harvesting, reuse of industrial waters, water-saving technology in residential sectors, and reuse of water from washing machines) for managing urban water in two urban catchments (Fucha and Tunjuelo) in Bogotá, Colombia, over three periods: baseline, 10 years, and 20 years. The simulation was performed using the software Urban Volume Quality (UVQ) and revealed a possible reduction in drinking water consumption of up to 47% for the Fucha Catchment and 40% for the Tunjuelo Catchment; with respect to wastewater, the reduction was up to 20% for the Fucha Catchment and 25% for the Tunjuelo Catchment. Lastly, two scenarios were evaluated in terms of potential savings related to water supply and sewage fees. The implementation of strategies 3 and 6 insofar as these two strategies impacted the hydric resources. Therefore, there would be a significant reduction in contaminant loads and notable economic benefits attributable to implementing these strategies.

Keywords: urban water cycle; urban water management; simulation; UVQ; cost-benefit analysis

---

1. Introduction

The world is witnessing fast-paced urban development. At present, urban areas account for more than half of the global population, and more than 500 cities are already home to one million people or more [1,2]. It is estimated that 60% of world's population will inhabit such areas by 2025 [3].

In urban areas, water's importance is hard to overstate; it is an essential resource for human and environmental well-being in a city. It should come as no surprise that Integrated Urban Water Management (IUWM) has taken on prominence internationally [4,5] given that it enables the definition of problems, identifies and determines solutions, and facilitates the implementation of these solutions [6]. Most efforts in this vein focus on new sources of water [7], wastewater reduction [8], the control of flooding caused by stormwater [9], effective treatment systems [10], and active participation by the different actors in the urban water cycle (users, government, etc.) [11].

Several strategies for urban water management have been proposed, including structural and nonstructural actions. These actions have been applied to different scales, from regions, cities, towns, and neighborhoods to houses, apartments, and independent properties, all of which contribute—to varying degrees—to the IUWM [12]. Some examples include: (i) harvesting stormwater; (ii) reusing greywater and wastewater; (iii) treating stormwater, wastewater, and greywater; (iv) runoff and contaminant reduction in stormwater via structural work (filters, artificial wetlands, porous pavement, etc.); (v) water-saving technology; and (vi) environmental education, among other proposals [13–20]. To evaluate the implementation or the selection of these options, software and mathematical models have been used, offering appropriate technical support to plan cities' growth and sustainability.

Among the most frequently employed infrastructure options for smaller-scale alternatives are the use of rainwater, reuse of greywater, and water-saving devices. The collection of rainwater is a clear and direct strategy for the sustainability of the urban water cycle. It offers myriad benefits, including: reduced water demand (30–70%), less hydric stress, minimized contaminant loads, and flooding mitigation [21–25].

The use of greywater offers notable potential in light of the constant supply of this source, i.e., a high population density in cities leads to large volumes of greywater [26]. Water-saving devices, for their part, have been gaining traction due to their ease of implementation and affordability. These strategies have been studied across the globe with promising reductions in the consumption of drinking water, namely from 20% to 60% [27–30]. Combining these two strategies (greywater reuse and domestic water-saving systems) enhances the positive impact: Baskaran et al. [31] found savings of up to 77% relative to normal consumption. It is important to mention that these strategies entail benefits from an economic standpoint for cities, regions, and countries, as well as individuals.

Bogotá, Colombia's capital and largest urban center, has more than eight million inhabitants and essentially requires the implementation of such strategies. The treatment of water for human consumption in the city is led by a public company whose primary sources are rivers, dams, and gullies; the company has achieved full (100%) coverage. The drinking water supply comprises three systems (Tibitoc, Chingaza, and La Regadera), which have five drinking water treatment plants (DWTP) and present an installed capacity of 27.5 m^3/s, of which 53% is used. Groundwater represents an insignificant source and is mostly used by industrial and commercial sectors. The city's public sewage system has achieved coverage exceeding 90%. This system comprises four sanitary catchments, as can be observed in Figure 1; these catchments bear the names of the four principal rivers in Bogotá: Torca, Salitre, Fucha, and Tunjuelo. Within these catchments, there are a total of 65 subcatchments, which are subdivided into 49 sanitary subcatchments and 16 rain subcatchments. Most of the sanitary systems are separated, with only a small portion in the Fucha and Salitre Catchments [32]. Bogotá currently has only one wastewater treatment plant (El Salitre), which is located at the confluence of the Salitre River with the Bogotá River. At present, the El Salitre plant only runs primary treatment and has a treatment capacity of 4 m^3/s. It receives wastewater from approximately 2,200,000 people in the north of the city, the area circumscribed by the Salitre Catchment.

It is important to mention that in Colombia, as is the case for many countries around the world, clear policies for urban water management have not been set forth; for that reason, in a previous study, the authors of this paper underscored the urgent need to carry out activities aimed at conserving water to sustainably supply Bogotá's population. The previous study highlighted alternative sources of water supply, and the authors demonstrated that renewed emphasis should be placed on managing the city's drainage systems and treatment systems, primarily to reduce wastewater, contamination, and runoff. These three aspects should be considered holistically, that is, in tandem, for the design of policies for the management of urban water in line with the principles of IUWM [33].

In this article, the simulated implementation of four strategies (stormwater harvesting, reuse of industrial waters, water-saving technology in residential sectors, and reuse of water from washing machines) for managing urban water in two urban catchments (Fucha and Tunjuelo) are evaluated at three periods: baseline, 10 years, and 20 years. This simulation was done using Urban Volume Quality (UVQ) software. In addition, two scenarios are assessed and compared in terms of potential savings related to water supply and sewage fees.

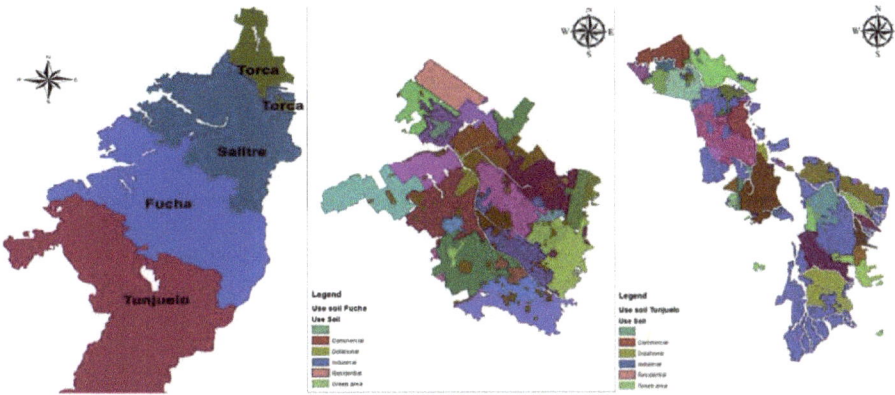

Figure 1. Land use of Fucha and Tunjuelo Catchments.

2. Materials and Methods

2.1. Location

Only the Fucha and Tunjuelo catchments were evaluated, for they account for 50% of Bogotá's population, which is a sufficient sample size for the purposes of the present article. The Fucha Catchment has a total area of 14,024.61 ha for wastewater and 3824.67 ha for stormwater. This catchment has the highest area of industrial development in the city, including dry cleaners, tanneries, food-related businesses, metal-mechanic companies, and chemical companies. Also, this catchment's eastern end has a combined sewage system, with a total coverage of 40.5 km^2. The Tunjuelo Catchment is predominately rural (agriculture use), thus serving mostly residences, with some industrial and commercial sectors. Furthermore, this is the only completely separate sanitary catchment in Bogotá. Its total coverage area is 21,956.69 ha for wastewater and 62,606.85 ha for stormwater. Figure 1 shows the dominant land use of the Fucha and Tunjuelo Catchments (these uses will be employed as inputs for the software on the movement of water flows).

2.2. Urban Volume and Quality (UVQ)

The UVQ software program is an extension of Aquacycle, another software program; UVQ was initially developed as support for the evaluation of alternatives in the management of the urban water cycle in Australia within the feasibility stage of the CSIRO urban water program, which began in the year 1999 [34]. Subsequently, it became part of the fifth European-Australian Assessing and Improving Sustainability of Urban Water Resources and Systems (AISUWRS) project. The objective of this project was to evaluate and improve the sustainability of the urban groundwater resources by means of computer tools [35,36]. UVQ was selected for use in four cities globally, with detailed case studies available: Mt. Gambier (Australia), Doncaster (UK), Rastatt (Germany), and Ljubljana (Slovenia). As a management tool, UVQ has thus far generated a positive impact on urban water management [34,37–39].

This software sequentially simulates the water balances in the provision of drinking water, as well as the hydrological (precipitation, runoff, and evapotranspiration) and wastewater processes within the study areas. These balances are given as loops that traverse the entire system at a daily timescale. Spatially, UVQ relies on three different scales: single dwelling, neighborhood, and study area. Additionally, it has the capacity to simulate contaminants through the already-obtained water balances. Note that the model assumes no degradation or conversion of the contaminants during the different steps [34,40,41]. UVQ was chosen in light of the fact that it is a free tool, encompasses all components of the urban water cycle, and is suitable for expansive urban areas such as Bogotá.

2.3. Input Data

The input variables and their source are presented in Table 1.

Climate data were obtained from the Institute of Hydrology, Meteorology, and Environmental Studies (IDEAM per its name in Spanish). For this research, two stations were selected (one for each catchment). The stations selected had data for at least the last seven years. Model calibration was performed by trial and error for runoff, wastewater, and consumption values reported by the Aqueduct, Sewage, and Sanitation Company (EAAB per its name in Spanish), which was in charge of providing these services in the city of Bogotá for these two catchments in the year 2014. As a reference for calibration, the data with the best fit between simulated and reported values for each water flow were chosen. The model was validated with data from the second half of 2015 and the first half of 2016.

Table 1. Input variables and their source.

Variable	Type	Unit	Source			Observation
			Measured	Local Literature	International Literature	
Total area	N.A.	ha	X			Measured using ArcGis
Road area	N.A.	ha	X			Measured using ArcGis
Open space	N.A.	ha	X			Measured using ArcGis
Exfiltration of wastewater	N.A.	ratio		X		The rate was estimated using a value of 7.67 L/d/m, as proposed by [42]
Number of dwellings in a neighborhood	N.A.	#	X			Measured in the field by a pilot study and using ArcGis
Dwelling area	N.A.	m2	X			Measured using ArcGis
People per dwelling	N.A.	#	X			Measured in the field by a pilot study
Dwelling garden area	N.A.	m2	X			Measured using ArcGis
Dwelling roof area	N.A.	m2	X			Measured using ArcGis
Dwelling paved area	N.A.	m2	X			Measured using ArcGis
Consumption of water—kitchen, bathroom, toilet and laundry	N.A.	L/d/c		X		[43]
Contaminant load—kitchen, bathroom, toilet and laundry	BOD	m/c/d			X	[44]
	COD	m/c/d			X	[44]
	TN	m/c/d			X	[44]
	TF	m/c/d			X	[44]
	TSS	m/c/d			X	[44]
Precipitation	N.A.	mm	X			Weather stations
Temperature	N.A.	°C	X			Weather stations
Actual evaporation	N.A.	mm	X			Weather stations

BOD: Biochemical oxygen demand, COD: Chemical oxygen demand, TSS: Total suspended solids, TN: Total nitrogen, TP: Total phosphorus.

2.4. Scenarios

For the simulation, six scenarios were developed, as shown in Table 2. Each was run at three adopted points in time. The first time point reflects baseline conditions (2015–2016). The second represents forecast population growth after 10 years. The third forecasts population growth after 20 years. For the forecasts, historical rates of population growth for these catchments were used; these rates were calculated using a census administered by Bogotá's mayoral office.

Table 2. Scenario Descriptions.

Scenario	Strategy	Description
Scenario 1	Stormwater use	1-m³ stormwater tanks for dwellings and 10-m³ tanks for industries, big businesses commercial, and business or office centers.
Scenario 2	Stormwater use	2-m³ stormwater tanks for dwellings and 20-m³ tanks for industries, big businesses, and business or office centers.
Scenario 3	Stormwater use, wastewater treatment for the industrial and commercial sector, and reuse of wastewater in industries	1-m³ stormwater tanks for dwellings and 10-m³ tanks for industries, big businesses commercial, and business or office centers. Wastewater treatment with removal of 40% TN, 40% TP, 40% BOD, 40% COD, and 60% TSS. Water tanks to reuse industrial waters after treatment with the 10-m³ tanks.
Scenario 4	Greywater use	0.1-m³ tank for dwellings with water from washing machines.
Scenario 5	Water-saving technology	Installation of water-saving equipment in dwellings, sinks, dishwashers, toilets, and showers.
Scenario 6	Combination of scenarios 4 and 5	0.1-m³ tank for dwellings with water from washing machines and installation of water-saving equipment in dwellings for sinks, dishwashers, toilets, and showers.

Rainwater was used given its immense potential as an alternative source in city with a climate such as Bogotá's. Wastewater reuse was chosen as an alternative source because of the general need for treatment systems to control the contamination of this water. Lastly, greywater reuse and water-saving technology were chosen due to the fact that they represent affordable strategies for the residential sector.

The stormwater tank volumes for residential areas were calculated by looking at water use for toilets, dishwashers, sinks washbasins, and showers, and by accounting for infrastructure capacity, especially as pertains to areas for household and apartment building storage tanks. For industrial and commercial areas, tank volumes for stormwater and wastewater reuse were calculated based on water demand. These tank volumes represent average consumption, so a single volume was determined for each sector. For this purpose, the following equation, developed by Santos and Taveira-Pinto [45], was used:

$$V = \text{Min(Annual non} - \text{potable demand)} * 0.06 \tag{1}$$

2.5. Statistical Analysis

To evaluate the differences in water flows between the base scenario and the proposed scenarios, a nonparametric Wilcoxon statistical test was used. With this test, it was possible to assess the statistical significance, or lack thereof, of the observed (forecast) reductions. The test was performed using all values from each simulated neighborhood. For all statistical testing, significance was established using a 95% confidence interval ($\alpha = 0.05$) [46].

2.6. Economic Evaluation

To conduct the economic evaluation, two strategies that would impact drinking water consumption and wastewater generation, respectively, were employed. The average consumption per user (residential, industrial, and commercial) for the year 2015 was determined. For the same period, fixed and base fees for water supply and sewage were obtained to determine the average costs of each. Subsequently, the forecast results for the implementation of either of the strategies were compared to average consumption. It is important to mention that, during this evaluation, only the cost of service was taken into account (i.e., the cost of infrastructure was excluded).

3. Results

Although results were obtained at a daily scale, they are presented here in m m^3/s. The results show that the Fucha Catchment presented higher levels for all water flows relative to those obtained for the Tunjuelo Catchment. With respect to the consumption of drinking water, approximately 2 m^3/s was the biggest difference between these catchments. For the generation of wastewater, this value was approximately 1.5 m^3/s, which may be associated with the fact that the Fucha Catchment has more users, mainly residential, industrial, and commercial, than the Tunjuelo Catchment. In contrast, the stormwater flows were also greater for the former catchment, given that the Tunjuelo Catchment has more pervious surfaces, e.g., green areas, parks, etc., allowing for a higher infiltration rate. In the following paragraphs, the results are broken down by catchment.

3.1. Fucha Catchment

3.1.1. Flow of Drinking Water, Wastewater, and Stormwater

In terms of drinking-water consumption, scenarios 1, 2, and 3 were projected to bring about 42 to 47% reductions relative to the baseline scenario without any modifications/improvements. Scenarios 5 and 6 led to a reduction in drinking-water consumption of 29%. The results revealed that scenarios 1, 2, 3, 5, and 6 produced statistically significant differences. However, scenario 4's reduction was 4.9%. For wastewater, the biggest drop was observed for scenario 6—at 20%—with statistical significance. For scenarios 1, 2, 3, and 5, decreases between 11% and 13% were projected. Scenario 4 presented a drop of 7%. Finally, for stormwater values, the first three scenarios provided 21% to 23% reductions, with statistical significance confirmed, whereas scenarios 4, 5, and 6 presented a reduction though not to the level of significance. The decrease in the all flows can be seen in Figure 2.

For the forecast water flows (10 and 20 years), an approximate increase of 49% in water consumption is expected for the 10-year period and 89% for the 20-year period (relative to the current values). With respect to wastewater, increases of approximately 24% and 48% are expected for the 10- and 20-year periods, respectively (relative to the current values). The infiltration expected for the stormwater sewage system from the sanitation systems from the drinking-water piping is 19% over the next 10 years and 21% over the next 20 years, with these percentages referring to increases relative to current values. Infiltration is caused by faulty wastewate-system connections and exfiltration by piping for drinking water and wastewater.

To compare the future values with current ones, a red line was included in Figure 3. This line clearly indicates the current values for drinking-water consumption, wastewater generation, and stormwater flows. As shown in Figure 2, the expected drinking-water consumption when implementing strategies 1, 2, and 3 for the 10-year period is less than current consumption levels, representing a reduction of 7%, 12.2%, and 8%, respectively, when employing these three strategies, which indicates that beginning to implement any of these measures for this period would be insufficient if aiming to maintain current consumption levels. Moreover, the use of stormwater as a water source allows for these same scenarios to help reduce stormwater flows. As for wastewater, scenario 6 for the 10-year period resulted in values that are minimally below the current levels, with a difference of only 1.8%. Finally, for the 20-year period, all wastewater values were greater than the current base state.

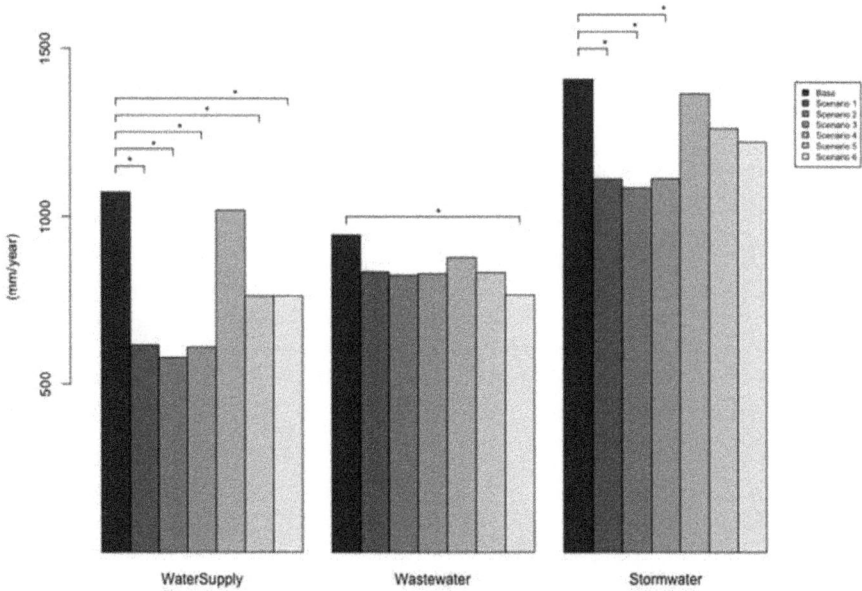

Figure 2. Flow of Drinking Water, Wastewater, and Stormwater in the Fucha Catchment.

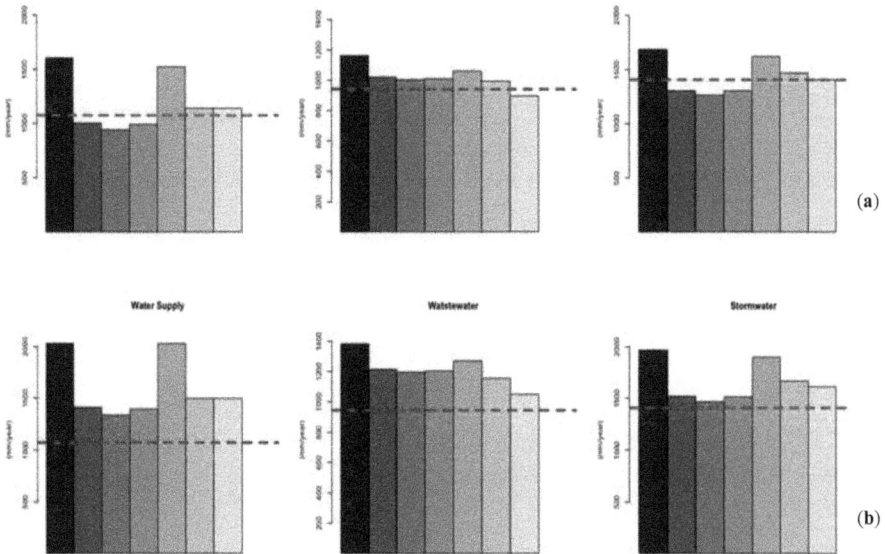

Figure 3. (**a**) Flow of Drinking Water, Wastewater, and Stormwater for the 10-year period and (**b**) Flow of Drinking Water, Wastewater, and Stormwater for the 20-year future cases in Fucha Catchment.

3.1.2. Contaminant Loads

Scenario 6, which displayed the greatest reduction in terms of wastewater flow, also produced the greatest reduction in terms of contaminant loads. If implementing this scenario, when compared to the current state, decreases would be: 13% for total phosphorus (TP), 12% for total nitrogen (TN), 26% for biological oxygen demand (BOD), 27% for chemical oxygen demand (COD), and 30% for total suspended solids (TSS). For scenarios 1, 2, 3, and 5, reductions were between 7% and 9% for TP and TN; for BOD and COD, these reductions were between 18% and 20%. Finally, TSS levels dropped between 20% and 25%. Despite the fact that industrial wastewater treatment was simulated for scenario 3, the reduction observed was not significant because industrial commercial users do not represent more than 10% of the Fucha Catchment's users.

3.2. Tunjuelo Catchment

3.2.1. Flow of Drinking Water, Wastewater, and Stormwater

In terms of drinking-water consumption, scenarios 1, 2, and 3 presented reductions of 35% to 40%. For scenarios 5 and 6, reductions in drinking-water consumption were 24.5%. These five scenarios led to statistically significant decreases relative to the baseline. For scenario 4, the drinking-water reduction was 10%. For wastewater, a statistically the significant statistics reduction stemmed from scenario 6 (19%). For scenarios 1, 2, 3, and 5, drops were between 11% and 13%; for scenario 4, this value was 7.2%. For the stormwater flow, the first three scenarios led to a drop of 22% to 25%, though scenarios 4, 5, and 6 presented a drop, albeit an insignificant one. The results can be seen in Figure 4.

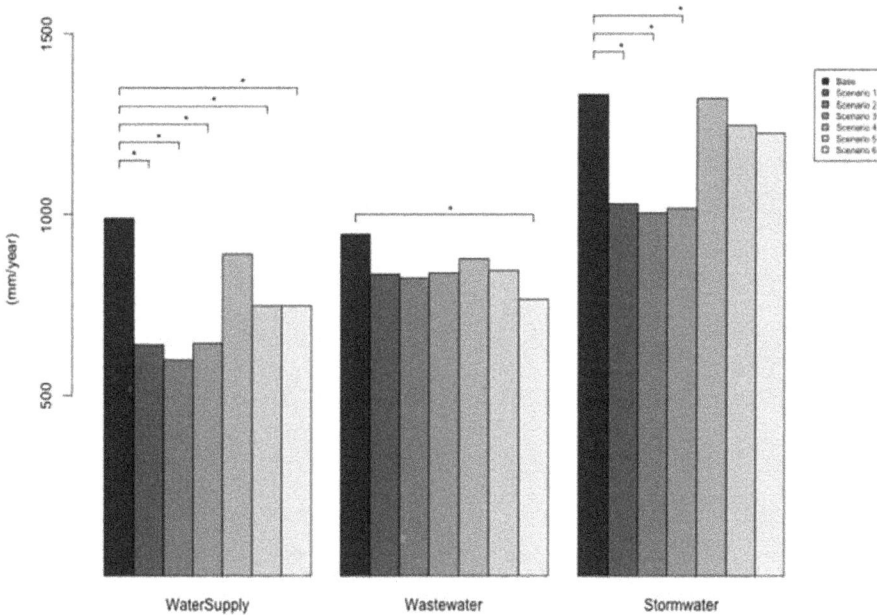

Figure 4. Flow of Drinking Water, Wastewater, and Stormwater in Tunjuelo Catchment.

Looking at the projected scenarios, increases of 38% in water consumption for the 10-year period and 66% for the 20-year period were forecast; for wastewater, increases of approximately 30% and 42% for the 10-year and 20-year periods, respectively, were forecast. The infiltration expected for stormwater drainage from the sanitation system and drinking-water pipelines is 11% and 23% higher for the 10-year and 20-year periods relative to the current scenario.

Comparing the projected flow values to the current ones, strikingly similar behavior was observed for the Fucha Catchment. For drinking water, in scenarios 1 and 3 for the 10-year period, the values were almost the same as the current drinking-water consumption (differences of 4% and 6%, respectively, for each scenario). For scenario 2, the difference between the 10-year projection and the baseline was 180 mm/year in consumption, equivalent to an 18% increase. The flow of wastewater in scenario 6 for the 10-year period was lower than that of the current state (though only by 3%); although this value is relatively small, its impact may be high, and for the catchment growth, it would be 23%. Finally, for the 20-year period, all values were above those of the baseline state (see Figure 5).

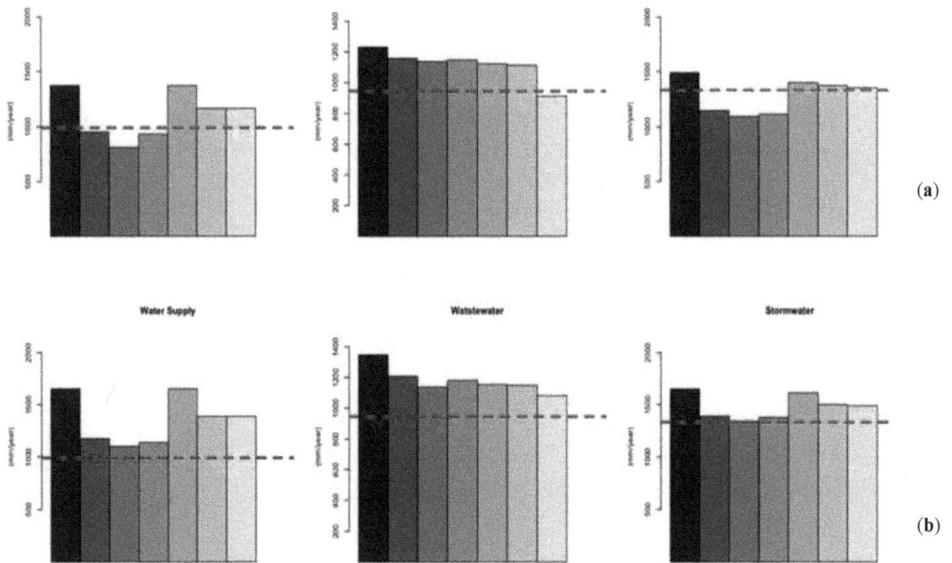

Figure 5. (a) Flow of Drinking Water, Wastewater, and Stormwater for the 10-year period and (b) Flow of Drinking Water, Wastewater, and Stormwater for the 20-year future cases in Tunjuelo Catchment.

3.2.2. Contaminant Loads

The simulation results show that scenario 6 generated a decrease of 6% for TP, 4% for TN, 12% for BOD and COD, and, 15% for TSS. For scenarios 1, 2, 3, and 5, the values for TP and TN were 2%. For BOD and COD, these values were 5% and 7%, respectively; for TSS, they were 8%.

3.3. Comparing Scenarios 3 and 6

The (hypothetical) implementation of different scenarios pointed to the importance of strategies 3 and 6 insofar as these two strategies impacted the hydric resources. Therefore, there would be a significant reduction in contaminant loads and notable economic benefits attributable to implementing these strategies. Scenario 3 was selected because of the low costs associated with 1-m³ tanks, and scenario 6 was chosen primarily because of the practicality of its implementation in residential sectors.

The first step in comparing these strategies consisted of determining the reduction in the water volume captured under maximum conditions for each scenario. For the Fucha Catchment, simulation results showed that each drinking-water treatment plant supplying this catchment (Tibitoc and Wiesner) reduced its capture by 1 m³/s for scenario 3 and 0.6 m³/s for scenario 6. The reduction for the

Tunjuelo Catchment was 0.55 m^3/s for each plant (South treatment plant and Wiesner) for scenario 3 and 0.22 m^3/s for scenario 6.

When running scenario 3, the treatment plants delivering water to the Fucha Catchment would capture approximately 5.26 GL/month less water. For the Tunjuelo Catchment, scenario 3 would reduce water capture by 2.23 GL/month. In other words, the two catchments would capture 7.8 GL/month less. For scenario 6, the reduction in drinking-water consumption for each plant supplying the Fucha Catchment would be 3.15 × GL/month 1 and 5.78 × 10^5 m^3/month for the Tunjuelo Catchment; thus, a total of 3.73 GL/month in water would no longer be captured if implementing scenario 6. Therefore, there was a difference in flow captured of 52% between scenarios 3 and 6, with the former providing greater benefits. However, both strategies would positively impact the ecosystem, environment, society, and economy at city and national levels.

In light of the fact that neither catchment has a wastewater treatment system, the reduction of this water and the concomitant reduction in contaminant loads represent an optimal—perhaps even necessary—strategy for controlling contamination in the city's water bodies. On balance, scenario 3 would lead to a reduction of 32% for nutrients, 20% for organic material, and 26% for solids for both catchments. For scenario 6, the contaminant improvements would be 36%, 38%, and 39% for nutrients, organic material, and solids, respectively, revealing that an improvement in all parameters associated with this scenario present the highest reduction in wastewater flow.

Lastly, savings on water supply and sewage fees were evaluated for both proposed scenarios. In Table 3, the fixed and base fees for sanitation and sewage, in addition to the average values of drinking-water consumption and sewage, for both catchments are displayed.

As shown in Figure 6 scenario 3 for the Fucha Catchment was identified as undergoing the greatest drop in total cost (water supply and sanitation) per month for all land uses, which is primarily attributable to the fact that this strategy allows for the greatest reductions in drinking-water consumption (the highest-cost service). Furthermore, this scenario showed how the industrial and commercial sectors could cut total costs between 40% and 47%compared to the baseline; in residential sectors, this drop could be between 10% and 20%.

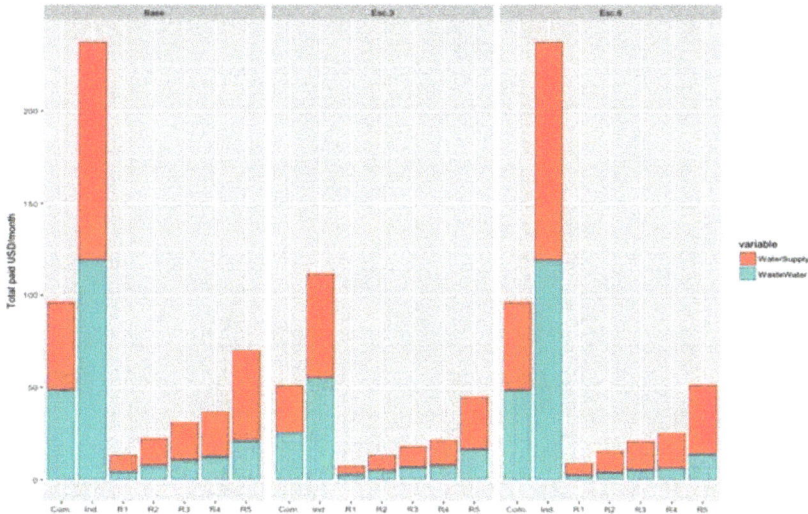

Figure 6. Comparison of the cost for services in the baseline and scenarios 3 and 6 for the Fucha Catchment.

Table 3. Cost to pay for services and values of drinking-water consumption and sewage for Fucha and Tunjuelo catchments.

Land Use	Fixed Fee—Water Supply (USD)	Base Fee—Water Supply (USD)	Average Consumption—Water Supply Catchment Fucha (m³ month⁻¹)	Average Consumption—Water Supply Catchment Tunjuelo (m³ month⁻¹)	Fixed Fee—Sanitation (USD)	Base Fee—Sanitation (USD)	Average Consumption—Sanitation Catchment Fucha (m³ month⁻¹)	Average Consumption—Sanitation Catchment Tunjuelo (m³ month⁻¹)
Residential Stratus 1	1.58	0.27	27.6	22.6	0.8	0.16	22.08	15.03
Residential Stratus 2	3.16	0.54	22.34	19.12	1.61	0.33	17.46	12.71
Residential Stratus 3	4.47	0.76	21	19.2	2.28	0.47	17.46	12.77
Residential Stratus 4	5.26	0.89	21.5	18.5	2.68	0.55	17.46	12.3
Residential Stratus 5	11.78	1.38	26.82	N.A.	6.67	0.83	17.46	N.A.
Industrial	6.84	1.23	91	46.2	6.84	1.23	91	30.72
Commercial	7.89	1.34	30.1	30	7.89	1.34	30.1	19.95

Scenario 3, once more, presented the lowest total costs per month for all land uses. In contrast to what was observed for the Fucha Catchment in the industrial and commercial sectors, the reduction was between 18% and 27% of the total payment for the residences served by the Fucha Catchment; for residential sectors, reductions were between 5% and 9.5%. However, the difference between scenarios 3 and 6 with respect to the residential sector for this catchment did not exceed 4%, showing that the implementation of either of these two strategies would essentially generate the same reduction in expenses (see Figure 7).

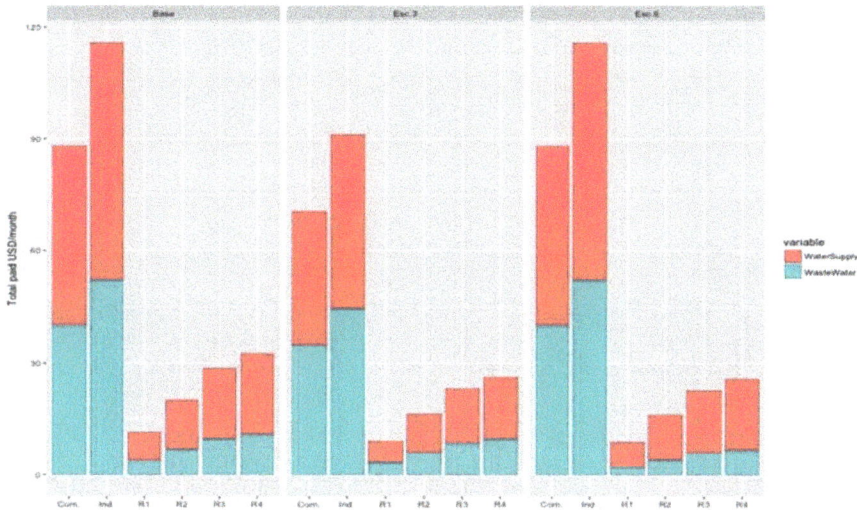

Figure 7. Comparison of the pay for services in baseline and the sceneries 3 and 6 for Tunjuelo Catchment.

It is worth mentioning that the annual savings on water supply and wastewater treatment is significant in Colombian pesos (COP), such that implementing either of these two strategies would engender a major positive impact on Bogota residents' wallets. Moreover, by estimating the costs of materials and construction labor per building in scenario 3, it is expected that dwellings could recoup their investment in 1 to 2.5 years. Turning to the industrial and commercial sectors, recouping the investment would take approximately 2.5 to 3.5 years. In fact, one scenario 6 should leave policy makers optimistic: In the residential sector, investments (infrastructure) would be offset in less than one year. However, to reap any of these benefits, the public sector must develop a model of financing or subsidization, as recommended by Furlong et al. [47], for doing so would facilitate the mass implementation of these strategies.

4. Conclusions

The simulation results obtained using UVQ software, which allow for the simulation of all components of the urban water cycle, revealed a reduction between both catchments of nearly 2 m^3/s in drinking-water consumption and 1.5 m^3/s in wastewater volume treated at when implementing.

Scenarios 1 (1-m^3 stormwater tanks for dwellings and 10-m3 tanks for industries, big businesses commercial, and business or office centers), 2 (2-m^3 stormwater tanks for dwellings and 20-m^3 tanks for industries, big businesses commercial, and business or office centers), and 3 (1-m^3 stormwater tanks for dwellings and 10-m^3 tanks for other uses. Wastewater treatment with removal of 40% TN, 40% TP, 40% BOD, 40% COD, and 60% TSS) for the Fucha Catchment presented the highest reductions in drinking-water consumption, with values ranging from 42% to 47%, which were statistically significant.

Yet, for scenarios 5 (Installation of water-saving equipment in dwellings, sinks, dishwashers, toilets, and showers) and 6 (0.1-m³ tank for dwellings with water from washing machines and installation of water-saving equipment in dwellings for sinks, dishwashers, toilets, and showers), the reductions were not as impressive, though they were significant. Regarding wastewater, scenario 6 presented the greatest reduction at 20%, with statistical significance. Finally, for stormwater, the first three scenarios were projected to offer reductions in the range of 21 to 23%. Turning to the Tunjuelo Catchment, scenarios 1, 2, and 3 presented reductions of 35% to 40% for drinking-water consumption. The biggest drop in wastewater flow was observed for scenario 6 (19%). For stormwater flows, the first three scenarios resulted in reductions of 22% to 25%. Reductions in contaminant loads in organic material for the Fucha Catchment exceeded 20%, though only 12% for the Tunjuelo Catchment. However, either case would notably help the Fucha and Tunjuelo rivers, which do not have treatment systems, as well as the Bogotá River, for the Fucha and Tunjuelo are tributaries of the Bogotá River.

Evaluating scenarios 3 and 6 showed that the two catchments, in conjunction, would stop capturing 7.8 GL/month if implementing strategy 3 and 3.73 GL/month if implementing strategy 6—a difference of 52% between these scenarios, favoring the former. Scenario 3 would reduce contaminant loads by 32% for nutrients, 20% for organic material, and 26% for solids for both catchments. For scenario 6, decreases of 36%, 38%, and 39% for nutrients, organic material, and solids, respectively, were projected.

In terms of the costs associated with users' sewage and sanitation services, scenario 3 presented the greatest reductions for both catchments, mainly benefiting the industrial sector for the Fucha catchment, though the implementation strategy 3 would offset the investment in residential sectors in less time.

Author Contributions: Carlos Peña-Guzman worked on model construction and the development of scenarios. All authors contributed to analysis of results.

Conflicts of Interest: The authors declare no conflicts of interest.

References

1. Fletcher, T.D.; Andrieu, H.; Hamel, P. Understanding, management and modelling of urban hydrology and its consequences for receiving waters: A state of the art. *Adv. Water Resour.* **2013**, *51*, 261–279. [CrossRef]
2. Lee, T.R. Urban water management for better urban life in Latin America. *Urban Water* **2000**, *2*, 71–78. [CrossRef]
3. McIntyre, N.E.; Knowles-Yánez, K.; Hope, D. Urban Ecology as an Interdisciplinary Field: Differences in the use of "Urban" Between the Social and Natural Sciences. In *Urban Ecology*; Marzluff, J.M., Shulenberger, E., Endlicher, W., Alberti, M., Bradley, G., Ryan, C., Simon, U., ZumBrunnen, C., Eds.; Springer: New York, NY, USA, 2008; pp. 49–65. ISBN 978-0-387-73411-8.
4. Wen, B.; van der Zouwen, M.; Horlings, E.; van der Meulen, B.; van Vierssen, W. Transitions in urban water management and patterns of international, interdisciplinary and intersectoral collaboration in urban water science. *Environ. Innov. Soc. Transit.* **2015**, *15*, 123–139. [CrossRef]
5. Makropoulos, C.K.; Natsis, K.; Liu, S.; Mittas, K.; Butler, D. Decision support for sustainable option selection in integrated urban water management. *Environ. Model. Softw.* **2008**, *23*, 1448–1460. [CrossRef]
6. Furlong, C.; Brotchie, R.; Considine, R.; Finlayson, G.; Guthrie, L. Key concepts for Integrated Urban Water Management infrastructure planning: Lessons from Melbourne. *Util. Policy* **2017**, *45*, 84–96. [CrossRef]
7. Mitchell, V.G. Applying Integrated Urban Water Management Concepts: A Review of Australian Experience. *Environ. Manag.* **2006**, *37*, 589–605. [CrossRef] [PubMed]
8. Brown, R.R.; Keath, N.; Wong, T.H.F. Urban water management in cities: Historical, current and future regimes. *Water Sci. Technol.* **2009**, *59*, 847–855. [CrossRef] [PubMed]
9. Larsen, T.A.; Gujer, W. The concept of sustainable Urban Water Management. *Water Sci. Technol.* **1997**, *35*, 3–10. [CrossRef]
10. Price, R.K.; Vojinović, Z. *Urban Hydroinformatics: Data, Models, and Decision Support for Integrated Urban Water Management*; IWA Publishing: London, UK, 2011; ISBN 978-1-84339-274-3.

11. Furlong, C.; De Silva, S.; Guthrie, L.; Considine, R. Developing a water infrastructure planning framework for the complex modern planning environment. *Util. Policy* **2016**, *38*, 1–10. [CrossRef]
12. Furlong, C.; Silva, S.D.; Guthrie, L. Planning scales and approval processes for IUWM projects; lessons from Melbourne, Australia. *Water Policy* **2016**, *18*, 783–802. [CrossRef]
13. Asano, T.; Levine, A.D. Wastewater reclamation, recycling and reuse: Past, present, and future. *Water Sci. Technol.* **1996**, *33*, 1–14. [CrossRef]
14. Brodie, I. Stormwater harvesting and WSUD frequent flow management: A compatibility analysis. *Water Sci. Technol.* **2012**, *66*, 612–619. [CrossRef] [PubMed]
15. Christova-Boal, D.; Eden, R.E.; McFarlane, S. An investigation into greywater reuse for urban residential properties. *Desalination* **1996**, *106*, 391–397. [CrossRef]
16. Fletcher, T.D.; Mitchell, V.G.; Deletic, A.; Ladson, T.R.; Séven, A. Is stormwater harvesting beneficial to urban waterway environmental flows? *Water Sci. Technol.* **2007**, *55*, 265–272. [CrossRef] [PubMed]
17. Geller, E.S.; Erickson, J.B.; Buttram, B.A. Attempts to promote residential water conservation with educational, behavioral and engineering strategies. *Popul. Environ.* **1983**, *6*, 96–112. [CrossRef]
18. McArdle, P.; Gleeson, J.; Hammond, T.; Heslop, E.; Holden, R.; Kuczera, G. Centralised urban stormwater harvesting for potable reuse. *Water Sci. Technol.* **2011**, *63*, 16–24. [CrossRef] [PubMed]
19. Mitchell, V.G.; Deletic, A.; Fletcher, T.D.; Hatt, B.E.; McCarthy, D.T. Achieving multiple benefits from stormwater harvesting. *Water Sci. Technol.* **2007**, *55*, 135–144. [CrossRef] [PubMed]
20. Strecker, E.; Quigley, M.; Urbonas, B.; Jones, J.; Clary, J. Determining urban storm water BMP effectiveness. *J. Water Resour. Plan. Manag.* **2001**, *127*, 144–149. [CrossRef]
21. Eroksuz, E.; Rahman, A. Rainwater tanks in multi-unit buildings: A case study for three Australian cities. *Resour. Conserv. Recycl.* **2010**, *54*, 1449–1452. [CrossRef]
22. Gikas, G.D.; Tsihrintzis, V.A. Assessment of water quality of first-flush roof runoff and harvested rainwater. *J. Hydrol.* **2012**, *466*, 115–126. [CrossRef]
23. Farreny, R.; Gabarrell, X.; Rieradevall, J. Cost-efficiency of rainwater harvesting strategies in dense Mediterranean neighbourhoods. *Resour. Conserv. Recycl.* **2011**, *55*, 686–694. [CrossRef]
24. Aladenola, O.O.; Adeboye, O.B. Assessing the Potential for Rainwater Harvesting. *Water Resour. Manag.* **2010**, *24*, 2129–2137. [CrossRef]
25. Cook, S.; Sharma, A.; Chong, M. Performance Analysis of a Communal Residential Rainwater System for Potable Supply: A Case Study in Brisbane, Australia. *Water Resour. Manag.* **2013**, *27*, 4865–4876. [CrossRef]
26. Zhang, Y.; Grant, A.; Sharma, A.; Donghui, C.; Liang, C. Assessment of rainwater use and greywater reuse in high-rise buildings in a brownfield site. *Water Sci. Technol.* **2009**, *60*, 575–581. [CrossRef] [PubMed]
27. Ghisi, E.; Mengotti de Oliveira, S. Potential for potable water savings by combining the use of rainwater and greywater in houses in southern Brazil. *Build. Environ.* **2007**, *42*, 1731–1742. [CrossRef]
28. Zhang, D.; Gersberg, R.M.; Wilhelm, C.; Voigt, M. Decentralized water management: Rainwater harvesting and greywater reuse in an urban area of Beijing, China. *Urban Water J.* **2009**, *6*, 375–385. [CrossRef]
29. Liu, S.; Butler, D.; Memon, F.A.; Makropoulos, C.; Avery, L.; Jefferson, B. Impacts of residence time during storage on potential of water saving for grey water recycling system. *Water Res.* **2010**, *44*, 267–277. [CrossRef] [PubMed]
30. Penn, R.; Hadari, M.; Friedler, E. Evaluation of the effects of greywater reuse on domestic wastewater quality and quantity. *Urban Water J.* **2012**, *9*, 137–148. [CrossRef]
31. Muthukumaran, S.; Baskaran, K.; Sexton, N. Quantification of potable water savings by residential water conservation and reuse—A case study. *Resour. Conserv. Recycl.* **2011**, *55*, 945–952. [CrossRef]
32. Rodríguez, J.; Díaz-Granados, M.; Camacho, L.; Raciny, I.; Maksimovic, C.; McIntyre, N. Bogotá's urban drainage system: Context, research activities and perspectives. In Proceedings of the 10th National Hydrology Symposium, British Hydrological Society, Exeter, UK, 15–17 September 2008.
33. Peña-Guzmán, C.; Melgarejo, J.; Prats, D. El ciclo urbano del agua en Bogotá, Colombia: Estado actual y desafíos para la sostenibilidad. *Tecnol. Cienc. Agua* **2016**, *7*, 57–71.
34. Mitchell, V.G.; Diaper, C. Simulating the urban water and contaminant cycle. *Environ. Model. Softw.* **2006**, *21*, 129–134. [CrossRef]
35. Wolf, L.; Morris, B.L.; Burn, S.; Hötzl, H. The AISUWRS approach. In *Urban Water Resources Toolbox—Integrating Groundwater into Urban Water Management*; Wolf, L., Morris, B., Burn, S., Eds.; International Water Association: London, UK, 2008.

36. Mitchell, V.G.; Diaper, C. UVQ: A tool for assessing the water and contaminant balance impacts of urban development scenarios. *Water Sci. Technol.* **2005**, *52*, 91–98. [PubMed]
37. Wolf, L.; Klinger, J.; Held, I.; Hötzl, H. Integrating groundwater into urban water management. *Water Sci. Technol.* **2006**, *54*, 395–403. [CrossRef] [PubMed]
38. Mitchell, V.G.; Diaper, C.; Gray, S.R.; Rahilly, M. UVQ: Modelling the Movement of Water and Contaminants through the Total Urban Water Cycle. In Proceedings of the 28th International Hydrology and Water Resources Symposium: About Water, Barton, Australia, 10–13 November 2003.
39. Rueedi, J.; Cronin, A.A.; Moon, B.; Wolf, L.; Hoetzl, H. Effect of different water management strategies on water and contaminant fluxes in Doncaster, United Kingdom. *Water Sci. Technol.* **2005**, *52*, 115–123. [PubMed]
40. Burn, S.; DeSilva, D.; Ambrose, M.; Meddings, S.; Diaper, C.; Correll, R.; Miller, R.; Wolf, L. A decision support system for urban groundwater resource sustainability. *Water Pract. Technol.* **2006**, *1*, wpt2006010. [CrossRef]
41. Marleni, N.; Gray, S.; Sharma, A.; Burn, S.; Muttil, N. Impact of water management practice scenarios on wastewater flow and contaminant concentration. *J. Environ. Manag.* **2015**, *151*, 461–471. [CrossRef] [PubMed]
42. Martinez, S.E.; Escolero, O.; Wolf, L. Total Urban Water Cycle Models in Semiarid Environments—Quantitative Scenario Analysis at the Area of San Luis Potosi, Mexico. *Water Resour. Manag.* **2011**, *25*, 239–263. [CrossRef]
43. Niño, D.; Martinez, C. *Estudio de las Aguas Grises Domésticas en Tres niveles Socioeconómicos de la Ciudad de Bogotá*; Pontificia Universidad Javeriana: Bogotá, Colombia, 2013.
44. Sharma, A.; Grant, A.; Tjandraatmadja, G.; Gray, S. *Sustainability of Alternative Sewerage and Water Servicing Options-Yarra Valley Water, Stage 2—Backlog Areas*; CISRO: Canberra, Australia, 2006.
45. Santos, C.; Taveira-Pinto, F. Analysis of different criteria to size rainwater storage tanks using detailed methods. *Resour. Conserv. Recycl.* **2013**, *71*, 1–6. [CrossRef]
46. Kruskal, W.H. Historical Notes on the Wilcoxon Unpaired Two-Sample Test. *J. Am. Stat. Assoc.* **1957**, *52*, 356–360. [CrossRef]
47. Furlong, C.; De Silva, S.; Gan, K.; Guthrie, L.; Considine, R. Risk management, financial evaluation and funding for wastewater and stormwater reuse projects. *J. Environ. Manag.* **2017**, *191*, 83–95. [CrossRef] [PubMed]

water

MDPI

Article

Sponge City Construction in China: A Survey of the Challenges and Opportunities

Hui Li [1,2], Liuqian Ding [1,2], Minglei Ren [1,2] , Changzhi Li [1,2] and Hong Wang [1,2,*]

1   State Key Laboratory of Simulation and Regulation of Water Cycle in River Basin, China Institute of Water Resources and Hydropower Research, Beijing 100038, China; lihui@wihr.com (H.L.); dinglq@iwhr.com (L.D.); mingleiren@163.com (M.R.); lichangzhi@iwhr.com (C.L.)
2   Research Center on Flood and Drought Disaster Reduction of the Ministry of Water Resources, Beijing 100038, China
*   Correspondence: wanghong@iwhr.com; Tel.: +86-10-6878-1593

Received: 15 June 2017; Accepted: 7 August 2017; Published: 28 August 2017

Abstract: Rapid urbanization in China has caused severe water and environmental problems in recent years. To resolve the issues, the Chinese government launched a sponge city construction program in 2015. While the sponge city construction initiative is drawing attention and is spreading fast nationwide, some challenges and risks remain. This study surveyed progress of all 30 pilot sponge cities and identified a broad array of challenges from technical, physical, regulatory, and financial, to community and institutional. The most dominant challenges involve uncertainties and risks. To resolve the issues, this study also identified various opportunities to improve China's sponge city construction program. Based on the results, recommendations are proposed including urging local governments to adopt sponge city regulations and permits to alleviate water quality and urban pluvial flooding issues, fully measuring and accounting for economic and environmental benefits, embracing regional flexibility and results-oriented approaches, and focusing on a wider range of funding resources to finance the sponge city program. Coordination among other government agencies is critical, and this is true at all level of governments. Only through greater coordination, education, and broader funding could the sponge city program be advanced meaningfully and sustainably.

Keywords: green infrastructure; low impact development; public–private-partnership; urban stormwater management; urban flood

1. Introduction

China's large scale urbanization started around the 1980's with a rapid rise in the urban population from 36.22% in 2000 to 54.77% in 2014 (Figure 1) [1]. Consequently, more cities are facing challenges associated with urban sustainability and urban water issues such as aging/outdated water and wastewater infrastructures, urban flooding, combined sewer overflow, water quality deterioration, water scarcity, and a high frequency of extreme weather [2–4]. Among these, urban flooding is one of the most frequent and hazardous disasters that can cause enormous impacts on the economy, environment, city infrastructure and human society [5–7]. Recent survey shows that 62% of Chinese cities experienced floods, and direct economic losses amounted to up to $100 billion between 2011 and 2014 [1]. Other research indicates increasing trends to both urban flood disasters as well as economic and human life losses. To address these challenges, the Chinese government had been searching for viable options and launched pilot sponge city construction programs.

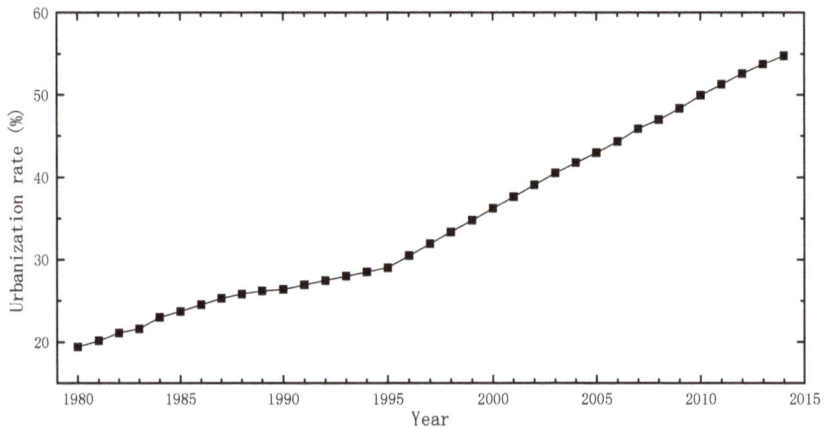

Figure 1. China's urbanization rate from 1980 to 2014.

A sponge city refers to sustainable urban development including flood control, water conservation, water quality improvement and natural eco-system protection. It envisions a city with a water system which operates like a sponge to absorb, store, infiltrate and purify rainwater and release it for reuse when needed [8]. The sponge city program takes inspiration from the low impact development (LID) and green infrastructure in the US [9,10] and Canada [11,12], sustainable drainage systems (SusDrain) in the UK [13,14] and other European countries [15,16], and water sensitive urban design (WSUD) in Australia [17] and New Zealand [18]. It promotes natural and semi-natural measures in managing urban stormwater and wastewater as well as other water cycles. The primary goals for China's sponge city construction are: retaining 70–90% of average annual rain water onsite by applying the green infrastructure concept and using LID measures, eliminating water logging and preventing urban flooding, improving urban water quality, mitigating impacts on natural eco-systems, and alleviating urban heat island impacts [19]. The sponge city program will also create investment opportunities in infrastructure upgrading, engineering products and new green technologies.

The sponge city program was launched at the end of 2014, under the direct guidance and support of the Ministry of Housing and Rural-Urban Development (UHURD), Ministry of Finance (MOF) and Ministry of Water Resources (MWR). These three ministries are responsible for reviewing, evaluating and selecting candidate cities recommended by their respective provincial governments, based on a series of criteria concerning the rationality and feasibility of pilot goals, financing mechanisms and the effectiveness of supporting measures from local governments. The three ministries are also responsible for the assessment of pilot city performance. In April 2015, the first group of 16 cities was selected as the pilot sponge cities; one year later in April 2016, the pilot program was expanded to another 14 cities. Figure 2 illustrates the locations of these pilot cities. The central government allocated to each pilot city between 400 and 600 million Chinese Yuan (CNY) each year for three consecutive years, and pilot cities are encouraged to raise matching funds through public–private-partnership (PPP) and other financial ventures. The money will be used to implement innovative water and wastewater management measures that would transform these cities into sponge cities [20].

With the strong support of the central government along with the enthusiastic participation of local governments and private sectors, the sponge city program is gaining momentum; however, the obstacles and challenges should not be overlooked, associated risks should not be ignored, and future opportunities should be fully recognized.

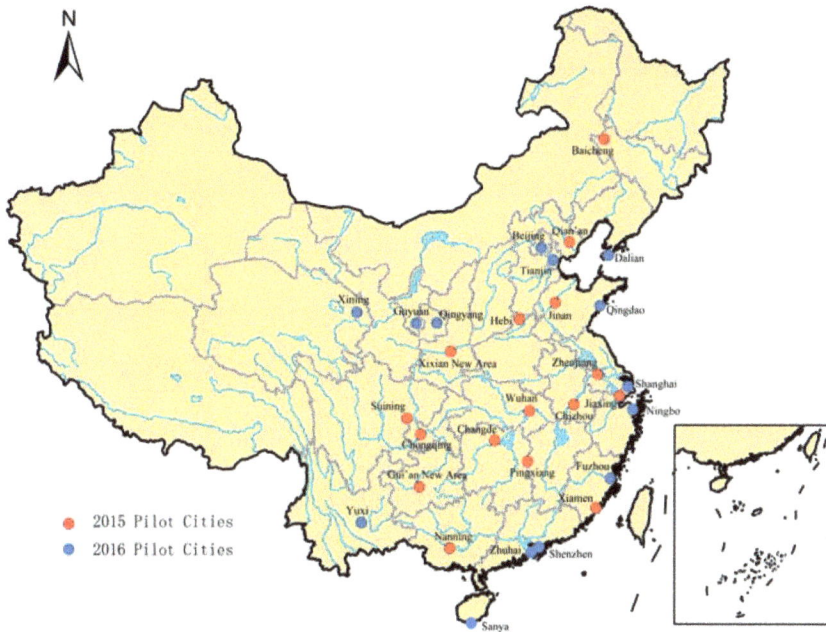

Figure 2. Location of pilot sponge cities.

2. General Descriptions of Pilot Cities

China is a vast country with great physical diversity. The climate of China is extremely diverse, ranging from tropical in the far south to subarctic in the far north and alpine in the higher elevations of the Tibetan plateau in the southwest. Precipitation is unevenly distributed in time and space. Temporally, it is almost invariably concentrated in the warmer months; and spatially, it increases from the northwest inland to the southeast coast, with the annual totals range from less than 20 mm in northwestern regions to easily exceeding 2000 mm in southern coast of the country (Figure 3). As sponge city construction involves green planning and LID measures, impacts of annual precipitation and temperature were considered along with other regional factors when selecting pilot cities. Owing to the nature and potential benefits of green infrastructures, most pilot cities are located in the central and southeastern regions, where annual precipitation ranges from 410 to 1830 mm and annual average temperature from 4.6 to 25.5 °C. In order to gain diverse experiences, a few pilot cities were also selected in northeastern cold area and in arid areas near the central-north of China.

Similar to other cities in China, pilot cities are facing various urban sustainability and urban water issues [21]. Ahead of others, these cities are actively seeking solutions in recent years; most are exposed or engaged in the early stages of green infrastructure planning and LID practices. In addition, these cities also have a fairly good governance and support basis both technically and financially. In order to focus on innovation and new approaches, as well as safeguarding success, each pilot city was designated a pilot area (not smaller than 15 km^2). Most LID measures are planned inside the designated pilot areas, and other green infrastructure planning and gray infrastructure improvements may expand outside the pilot areas.

Table A1 in Appendix A presents the regional characteristics of and general information on the 30 pilot cities including precipitation and climate, sizes of the designated pilot areas, existing conditions of stormwater management systems, proposed goals of the sponge city construction, and the total capital investment in the three-year period. As indicated in Table A1, all 30 cities set up clear goals on runoff volume control, urban drainage standards as well as waterlogging and flood prevention standards. However, the first group of 16 cities had less clear standards on rainwater resourcing, stormwater runoff pollution control, and wastewater recycling, comparing to the second group of 14 cities. Note that Table A1 only listed the primary goals of sponge cities. There are other secondary goals set up by sponge cities based on the specific conditions of each city.

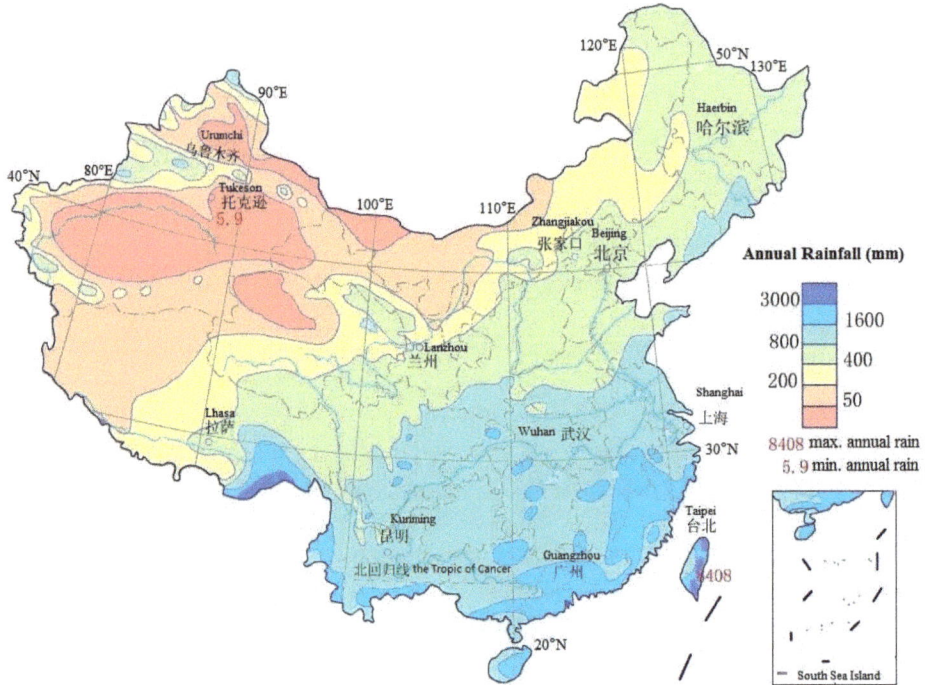

Figure 3. Annual precipitation distribution in China.

3. Materials and Methods

3.1. Study Design

This study utilized a descriptive design to survey pilot sponge cities. The survey consists of a literature review, information collection and review, field visits to pilot cities, and interviews with the public. All 30 pilot cities were included in the survey. Information collected in this survey relates to current implementation efforts and challenges to the further spreading of the sponge city program. The survey was organized by four challenge categories: technical/physical, legal/regulatory, financial, and community/institutional challenges. It was designed to collect both quantitative and qualitative information regarding the benefits of sponge city program, the type of challenges that it may encounter, and if/how these challenges can be overcome; however, more qualitative information was gained due to the nature of this study.

The goal of the survey was to identify challenges to sponge cities at the local, provincial and national levels of government, and provide concrete and provocative recommendations on how these challenges can be overcome. The flow chart of our survey is shown in Figure 4.

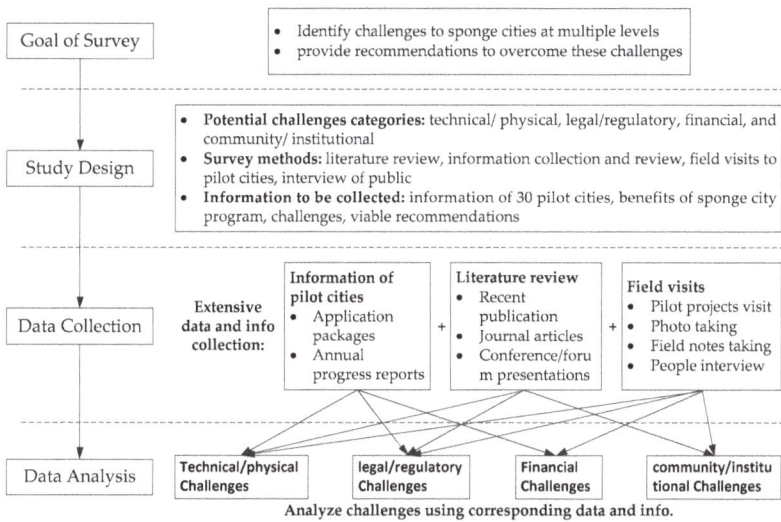

Figure 4. Flow chart of the survey work.

3.2. Data Collection Procedures

Extensive data and information were collected primarily from three sources (1) reviewing application packages and annual progress reports submitted by each pilot sponge city; (2) literature review; (3) field visits to pilot cities.

To compete for pilot sponge cities, each candidate city submitted an application package for review, evaluation and selection. The packages generally include a three-year sponge city implementation plan and other supporting documents such as long-term sponge city planning, future regulatory frameworks, and related research. The implementation plans are important documents, which contain variety of valuable information such as physical conditions; existing issues; proposed scopes and goals; proposed projects including green, gray, and non-structural measures; schedule of projects; responsible organizations; and financial arrangements. Annual progress reports are other significant resources for evaluating the progress as well as assessing and analyzing the success of the sponge city program.

The literature review was focused on recent publications, journal articles, and conference/forum presentations in China. Using the PRISMA (Preferred Reporting Items for Systematic Reviews and Meta-Analyses) methodology of literature review, 90 articles and presentations were retrieved from the CNKI [22], WANFANG [23], CSCD [24], and VIP [25] databases, as well as several conference sites in China including the 2016 International Low Impact Development Conference, in Beijing, China, 2016 International Conference on Green Infrastructure and Resilient City in Shenzhen, China, and the International Conference on Sustainable Infrastructure 2016 in Shenzhen, China. A total of 70 articles/presentations were included in the initial review, and 20 were excluded from the review after reading the whole content because they did not match the objectives of the literature review and the inclusion criteria. The key words used to search the materials were: urban flooding, urban water safety, green infrastructure, low impact development, sponge city construction, sustainable urban planning, rainwater resourcing, wastewater recycling, storm runoff pollution, socio-economic impact of urbanization, urban eco-system. The literature search was for publications starting from the year 2002 to the year 2016.

In addition, the research team also conducted field trips to 22 sponge cities from July 2015 to April 2017. Some field trips were accomplished during sponge city annual reviews and inspection or were organized by conferences, while others were conducted with expert consultation activities. Total of 156 pilot projects were visited, numerous photos and field notes were taken, and over 200 people were surveyed including scholars, practitioners, government officials, urban dwellers, and local stakeholders. The outline of the survey questions is presented in Table 1.

Table 1. The outline of the survey questions.

Technical:
• Design and construction codes and standards
• Performance and sustainability of sponge city measures
• Technology and materials
• Monitoring techniques and standards
• Education and training
• Operation and maintenance
Legal:
• Local, provincial and national rules, ordinance, policies, regulations, laws and guidelines
• Municipal structure for maintenance and ownership
• Opportunities
Financial:
• Full life cycle and maintenance costs of sponge city measures
• Social, economic and environmental benefits of sponge city measures
• Financial sources
• Private sector's interests
• Incentives
Community/Institutional:
• Public knowledge, interests, and involvement of sponge city construction
• Community education
• Aesthetics
• Cooperation between agencies and communities
• Available information

3.3. Data Analysis

Data and information were organized based on four challenge categories identified in the beginning. The data from application packages and annual progress reports were primarily used for analyzing technical/physical challenges along with legal/regulatory and financial challenges. The information retrieved from the literature review contains a fair number of expert and practitioner opinions and was largely used for analyzing the technical/physical, legal/regulatory, and community/institutional challenges. The data obtained from field visits contained first-hand knowledge and was used for analyzing all four categories.

4. Major Challenges

The survey results show that despite the promising benefits that the sponge city program could provide to urban environment, many challenges were present, which may inhibit wide-scale implementation and the long-term success of the sponge city program. The following sections present the primary challenges that were identified in this survey.

4.1. Technical Challenges

(1) Ambitious goals without sound research basis

The original goal of the sponge city construction was defined as runoff-volume-focused LID to retain 60–90% runoff on sites. One year later in 2016, it was expanded to a full array of urban sustainability goals by adding restoration of eco-systems, improving deteriorated urban water bodies, reducing urban heat island impacts, and building smart urban water cycle. Although the concept and practices of LID were introduced into China more than a decade ago [26], and recent research has been carried out on sustainable urban stormwater management [27,28], the research foundation for sponge city construction on such a large scale is rather weak. The rapid implementation of sponge city measures with such ambitious goals is largely based on very little research domestically and locally. A sponge city is an integrated approach that involves a broad range of concepts such as multi-scale conservation and water system management, multi-function of ecological systems, urban hydrology and runoff control frameworks, and impacts of urbanization and human activities on the natural environment. Lacking a sound research foundation can unnecessarily restrict the potential positive effects of this new urban water cycle management approach. To successfully implement sponge cities, appropriate definition of goals and adequate research to understand this new approach, along with sufficient knowledge are necessary.

(2) One model to fit every part of the country

Although the basic theories and primary concepts of the sponge city approach are largely applicable to any climate, geographical/geological, hydrological and soil conditions; the implementation strategies and the selection of specific measures should be considered with local conditions. The current practices, however, exhibits a pattern of using one model for every part of the country. Table 2 presents LID strategies proposed by four pilot sponge cities.

Table 2. LID strategies for various cities.

Data	Qian'an	Baicheng	Shenzhen	Yuxi
Annual Temp. (°C)	11.5	4.6	22.4	19.2
Annual Rainfall (mm)	672	410	1837	909
Annual Evap. (mm)	1100	1840	1675	1801
Runoff Control (%)	80	85	70	82
Runoff Control (mm)	28.0	25	31.3	23.9
Rain Resourcing (%)	7	6	≥ 8	10
Green Roof (%)	6	5	≥ 30	12.6
Depressed Green (%)	35	36.4	≥ 10	35.4
Permeable Paver (%)	46.5	46.2	≥ 70	38.2

As illustrated in Table 2, although the natural weather and geographical/hydrological conditions for these four cities vary drastically, the LID strategies proposed by these cities remain similar. For example, Baicheng City is located in a cold and arid region (as shown in Figure 2) with an annual temperature of 4.6 °C (Celsius), annual rainfall and annual evaporation of 410 and 1840 mm, respectively. The urgent issues for this city are water quality deterioration, water body shrinking, and water resources shortage. However, the LID strategies this city proposed will not alleviate these problems; instead, green roof, depressed green space and permeable pavement all have a potential to increase evaporation/evapotranspiration, and consequently, worsen the situation. Shenzhen, on the contrary, is a lowland coastal city located in a tropical region. It suffers from a high groundwater table, poorly drained soil, sea water intrusion, land salinization and heavy seasonal storms. Depressed green space and permeable pavements are clearly not suitable for it, and 30% of green roofs may not be economically viable. The success of a sponge city approach depends on local conditions and cannot be transferred in a standardized way to another context, as it might not prove as successful as local conditions change. The sponge city strategies should be developed based on a careful assessment of local conditions and potentials along with special issues, and mitigate these problems by leveraging the local potential and regional resources.

(3) In need of guidance and education/training

Up to recently, complete local, provincial and national guidance, design standards and codes were not available. A national level guideline was published at the end of 2014, and a few city-level guidelines were completed through 2015 to 2016 by some pilot sponge cities. These guidelines, however, are rather simple and general. They are, by large, merely the translation and combination of similar guidelines widely used in the US and do not consider variations in regional and local conditions such as soil, climate and topography variances. Lack of design standards and codes has limited the ability of local communities to implement sponge city projects based on local conditions.

In addition, the most common technical challenge is perhaps an overall lack of education/training, knowledge, and experience of sponge city planning, design, construction, maintenance, and maximizing benefits at all levels of government and all related professions across the board. In addition to local practicing staff and management personnel, the development and consulting industries also lack sufficient knowledge of sponge city concepts and practices, resulting in an industry culture that is either skeptical of the sponge city approach or one that produces poor planning and design.

(4)   Inappropriate strategies causing further problems

The lack of knowledge and guidance, design standards and codes, as well as appropriate education and training may result in the poor planning and implementation of certain sponge city measures in some incidences. For example, an existing drainage ditch, which used to run along the foot of a hill, was covered with rain gardens and concrete grates (a few meters of rain garden and a few meters of grate alternation); the small openings on the grate are around 3 cm in diameter. The design concept is to filter the mud and debris from hill-flow; however, it neglected the sudden and greater damaging nature of the hill-flow. The new design severely restricted the runoff collecting capacity of the original open ditch; consequently, it may cause additional damage to the roadway traffic, bypassers, and even damage to the newly installed rain gardens.

Other frequent problems involve excessively using green measures or constructing green measures at inappropriate or unnecessary locations. For example, some sponge parks are largely in undeveloped natural areas with over 75% of land covered by trees, plants and greens, and another over 10% of open water. Despite the lightly-developed nature, these parks are fully loaded with various LID measures including pervious pavement, rain gardens, dry creeks, depressed green spaces, infiltration swales, and even underground retention tanks. In another incidence, a 10 km long, two-lane roadway features large rain gardens on each side of the road with retention chambers underneath each rain garden, pervious pavers line the sidewalks, and depressed green spaces are planned beside the sidewalks. It appears that many LID measures are built in undeveloped/underdeveloped areas such as parks and large open spaces, while in urban and old town centers, where retrofit and restoration are really needed, sponge city interventions are largely avoided.

In other incidences, infiltration-related LID measures have been constructed on mountain sides with no consideration of mud-slides; an infiltration reservoir was built on top of karst bedrock with no protection from groundwater; green roofs were planned in arid areas with no concern for water scarcity; and rain gardens and depressed green spaces were designed in coastal areas with high groundwater tables and sea water intrusion problems with no knowledge of the suitability of the plants.

(5)   Unavailable green products and materials

In recent decades, the green infrastructure industry has been booming in the US and Europe, and many products and materials have been developed to assist green infrastructure implementation. These range from LID products including rain garden systems, tree planting systems, green/blue roof systems, tree root protection modular, landscape/sports/playground solution systems; blue products including vortex flow control device and inlet filter device; and different monitoring equipment. Many of these pre-made, ready-to-install products are designed and tested to meet industrial standards. Besides easy to design and install and easy to operate and maintain, these products also achieve stable/consistent performance and, in certain cases, provide standard monitoring components. In China currently, sponge city products and materials are mainly imported or introduced from abroad such as computer software and design and implementation techniques. Counterparts of the ready-to-install systems/modular/devices are largely unavailable. Table 3 exhibits the availability of various green infrastructure systems/modular/devices. Absence of similar green infrastructure products and materials may greatly hinder the progress of sponge city construction, or reduces the effectiveness of the sponge city program.

Table 3. Availability of various green infrastructure systems/modular/devices.

Products	AVLB	LMTD	UNAV	Remarks
Cistern/Rain Barrel		√		Simple small-size above ground barrels available without control apparatus
Rain Garden System			√	
Tree Planter System			√	
Green/Blue Roof System			√	Include green and blue roof modular
Infiltration Planter System			√	
Pervious Pavement	√			Various products available, but quality and durability are uncertain
Underground Infiltration			√	
Underground Detention		√		Very limited small-scale, small-sized products available.
Water Quality Control			√	refers to vortex flow control device, inlet filter device, etc.
Monitoring Equipment		√		Mostly copy versions of international products, and generally in poor quality and poor accuracy

Notes: AVLB = Available; LMTD = Limited; UNAV = Unavailable.

(6) Insufficient performance data

As a new approach, long-term performance data for sponge city measures is not available in most regions of China. Local communities are uncertain when implementing sponge city measures as part of the development process. In addition, the information on life cycle costs and operation/maintenance requirements under different flow regimes, soil types and climates is also unavailable. As a remedy, computer models have been used to predict the unavailable information, which appears difficult to use to convince public and local governments. Currently, there is not enough information on how well some sponge city measures will perform in long-term service and what it may take to maintain their various functions.

(7) Unaddressed operation and maintenance difficulties

Compared to traditional stormwater management systems, sponge city measures may require more frequent, periodic maintenance. Maintenance requirements vary depending on the specific measures, their functions and local conditions. These tasks may be as simple as weeding a vegetated swale and removing debris from curb cuts, or as complex as maintaining a large-scale wetland or an underground storage tunnel with multiple functions. One unique maintenance challenge posed by sponge city measures is that they are often scattered around a large area, and some are located on private property, making it difficult for public agencies to ensure that proper maintenance is carried out. Sometimes sponge city projects may be filled in or removed during other projects by private owners. This, in turn, presents great difficulties for the sustainability of sponge city construction.

4.2. Physical Challenges

(1) Geographical location

Some sponge city measures may not be suitable in certain locations due to the physical characteristics of the land, climate, or other conditions. For example, infiltration-related practices should not be used in areas where infiltration is not desirable, such as poorly drained soil, high groundwater tables, steep slopes, landslide hazard areas, floodplains, contaminated soils, and wellhead protection areas, unless special measures are employed. In arid and semi-arid areas, certain practices that increase evaporation are also not desirable. These restrictions create challenges to sponge city construction.

(2) Land scarcity in urban areas

China is a densely-populated country. Land is highly valuable, especially in developed urban areas. Whereas traditional systems in urban areas convey stormwater via underground pipes, sponge city practices, which allow stormwater to infiltrate into the ground or be stored on-site, may require additional land space; this may present a challenge when designing new developments or retrofitting existing urban areas. It is in developers' financial interests to maximize the amount of buildable land, while minimizing the costs. Setting aside space for sponge city measures sometimes may conflict with other development goals. Space limitations can also present a challenge when installing sponge city measures in the right-of-way along public streets. There are multiple demands for space in the right of way, including stormwater treatment, bicycle lanes, sidewalks, utilities, parking and traffic lanes.

(3) Climate

It is known that certain sponge city measures are not effective in managing stormwater in cold, hot or arid climates. In some regions of China, where the ground is frozen half of the year or permafrost exists, the potential for water to infiltrate into the ground is reduced. This limits year-round functioning and presents challenges to local governments.

(4) Soil conditions

One very specific challenge is the lack of understanding of soil characteristics, soil conservation/restoration, and plant–soil–water relations. Clay soils are a substantial impediment to LID because the full effect is hard to achieve. This problem has not been addressed in China and remains poorly understood. To compensate, projects in regions with poorly drained soil must be designed with underdrains, thereby reducing the benefits of the systems. Sponge city measures in clay soils need further investigation, standard development and championing to tout the benefits.

4.3. Financial Challenges

(1) Uncertainty of life cycle costs and benefits

Although the design and construction costs for sponge city projects are quite clear, the life cycle costs including operation and maintenance are unknown. In certain cases, even the life-spans of certain sponge city projects are uncertain. In current practices, sufficient funding for initial construction is allocated, however, funding needs for future operation and maintenance are not addressed. In addition, due to the uncertainty of the life-span and life-cycle performance, the life-cycle benefits—including environmental, ecological and social benefits—cannot be assessed appropriately. For an investment without a clear picture of the future benefits, the financial risk is rather high for both public and private entities.

(2) Challenges in public–private partnership

As the demand for infrastructure investment climbs around the globe, public–private partnerships (PPPs) are increasingly playing a crucial role in bridging the gap. In Western countries, these partnerships—in which the private sector builds, controls, and operates infrastructure projects subject to strict laws, regulations and government oversight—tap private sources of financing and expertise to deliver large infrastructure improvements. When managed effectively, PPPs not only provide much needed new sources of capital, but also bring significant discipline to project selection, construction, and operation. Successfully forming and managing PPPs, however, is no small feat.

As a new approach, sponge city construction is pushing its funding resources through PPPs, another new venue in China. Due to the nature of sponge city projects, the PPPs are mostly performance based. This survey identified following challenges: (1) The regulatory environment. There are no specific laws governing PPPs, and there is no independent PPP regulating agency in China. To better regulate PPPs and attract more private funding, a more robust regulatory environment, with clear laws and an independent regulating agency are essential; (2) Lack of information. The PPP program lacks a comprehensive database regarding the projects/studies to be awarded under PPPs. An online database, consisting of all the project documents including feasibility reports, concession agreements and the status of various clearances and land acquisitions would be helpful to all bidders; (3) Project development. The project development activities, such as detailed feasibility studies, land acquisition, and environmental/forest/floodplain clearances, are not given adequate importance by the concession authorities. The absence of adequate project development by authorities leads to misunderstood interests by the private sector, mispricing, and many times delays at the time of execution; (4) Lack of institutional capacity. The limited institutional capacity to undertake large and complex projects at all levels of government, especially at the local level, hinder the translation of targets into projects.

Other issues involve risk transfer, financial implications, contractual matters, politics, management and accountability. For example, some well financed pilot cities are capable of completing all sponge city projects with their own funding; for some political reasons, they are forced into using PPPs against their will and public interests. It defies the purpose of PPPs and can induce unnecessary costs to the public and increase managing difficulties for local governments.

4.4. Legal and Regulatory Challenges

Locally, challenges include local ordinances; building codes; plumbing and health codes; restrictions involving street width, drainage codes, and parking spaces; and restrictions on the use of reclaimed stormwater. Municipal codes and ordinances often favor gray over green infrastructure. A challenge that may be both technical and legal is that green infrastructure is often located on private property, and public agencies face difficulties ensuring that proper maintenance is occurring and will continue long-term. At the provincial level, water and land-use policies and property rights can be complicating factors. For example, downstream water rights may be impacted if upstream water management practices reduce the quantity of water to downstream users. The lack of national guidelines and performance standards are a complicating factor.

4.5. Public Acceptance Challenges

Contrary to traditional stormwater management systems, which are generally buried underground, sponge city systems are mostly built above ground and scattered in large regions; some located in private land interfering with public life. Therefore, public opinion and acceptance of sponge city construction can easily hinder its success. Considering the importance of public acceptance, the education efforts in China are deficient both in quality and in scale. More educational efforts are needed for a broad array of groups including political leaders, administrators, agency staff, planning and design professionals, developers, builders, landscapers, and the public. To achieve public outreach goals and shift public perceptions, a complete education program involving the technical training of municipal staff and lessons in sponge city concepts for the public are in demand. These lessons must be incorporated into formal and informal education programs for institutions and communities to fully understand the sponge city concepts.

4.6. Inter-Agency Cooperation and Data Sharing Challenges

Since sponge city construction involves a broad field of knowledge including stormwater, water quality, the eco-system, transportation, neighborhood retrofitting, and energy management, inter-agency and community cooperation is critical. While the partnerships and cooperation between agencies leverage efficiencies and economic benefits, they require significant patience and finesse. So far, the inter-agency cooperation and working across functions has not always been easy in China due to the difficulty of working across divisions, agencies, and political boundaries with diverse groups and diverse interests. It seems that some agencies compete to be a dominant party and reluctant to cooperate, whereas others view it as someone else's responsibility. Consequently, holistic efforts are hard to coordinate, focus and keep moving forward.

Lack of inter-agency cooperation also leads to difficulties in data and information sharing. It hinders research and innovations. In some situations, repetitive efforts were directed to collect the same data or information, while in others, research funding was awarded to organizations solely because they owned the critical data that was needed for the research.

5. Future Opportunities

As a new approach for urban water cycle management, the success of China's sponge city construction relies on the identification of challenges and adaption of effective improvements. To achieve a bright future, the following critical improvement opportunities are identified.

5.1. Taking an Integrated, Watershed Scale Approach

Sponge city construction aims at resolving various problems associated with urbanization at multiple scales, and ultimately at establishing greener and more holistic urban environments. Some earlier efforts defined the sponge city as runoff-volume-focused LID measures, and set up volume control criteria as the sole control parameter. It narrowed the sponge city concept; thus, some pilot cities focused on discretized LID measures at the source level and lessened the importance of connectivity at multiple scales. Taking an integrated, watershed-scale approach and focusing on the connectivity of the source–community–region–watershed scales can prevent the segmentation and isolation of the system and focus on the full benefits of sponge city approach—such as natural conservation, flood reduction, eco-service enhancement, and water resource protection—and ultimately promote a healthier watershed [29].

5.2. Enhancing Guidance and Design Standards for Local Conditions

The success of sponge city approach relies on the understanding of local issues, conditions, and potentials. It is essential to carefully assess specific problems in the city and resolve these by leveraging the local potential and regional resources. Currently, the lack of guidance and design standards from national, provincial and local levels make the following elements of sponge city projects difficult: assessment, planning, design, construction, operation/maintenance, and monitoring/evaluation. To ensure the success of sponge city construction, it is urgent that national and provincial guidelines be completed to help local governments develop local sponge city construction manuals/design standards. These documents should be based on a careful assessment of local conditions and potentials with input from local developers, planners, and engineers. Various education programs should also be established to provide training to government officials, public works employees, planning and design professionals, and the public. In addition, there is a strong demand for performance, costs and life-cycle data from pilot demonstration projects/practices in various natural conditions.

5.3. Promoting Government Leadership and Inter-Agency Cooperation

Leadership at all levels of government along with inter-agency cooperation is another key element essential for the effective implementation of sponge cities. Local leadership and knowledge of local conditions, as well as the potential benefits of sponge cities, need to grow. Agencies should work together to identify the needs for changing current municipal building codes, street/transportation/ parking ordinances, conflicting agency policies, and other uniquely local management constraints. Provincial leadership is necessary to clarify sponge city definitions and water rights implications, and to integrate and reconcile multiple local and provincial agency policies that impact sponge city practices.

National leadership can take many forms without creating a one size fits all approach that stifles provincial or local flexibility. Flexible performance criteria can help promote the performance of this new approach [30]. Standard-setting, permitting and enforcement agencies need to recognize that the sponge city approach often demands more time and different performance milestones than traditional approaches.

5.4. Establishing Locally Based Legislation Framework

Another important element to China's sponge city construction is locally based legislation that is formulated based on local conditions. These laws and regulations should consider the following:

(1) Establish decision-making processes surrounding land development activities that protect the integrity of the watershed and preserve the health of water resources.
(2) Require that new developments, redevelopments and all land conversion activities maintain the natural hydrologic characteristics of the land to reduce flooding, stream bank erosion, siltation, nonpoint source pollution, property damage, and to maintain the integrity of stream channels and aquatic habitats.
(3) Establish minimum post-development LID management standards and design criteria and control of stormwater runoff quantity and quality; establish minimum design criteria for the protection of groundwater resources; establish minimum design criteria for measures to minimize nonpoint source pollution from stormwater runoff.
(4) Establish design and application criteria for the construction and use of structural stormwater control facilities that can be used to meet the minimum post-development LID management standards.
(5) Encourage the use of LID practices such as reducing impervious cover and the preservation of green space and other natural areas to the maximum extent practicable.
(6) Establish provisions for the long-term responsibility for and maintenance of sponge city facilities to ensure that they continue to function as designed and pose no threat to public safety;
(7) Establish provisions to ensure that there is an adequate funding mechanism including guarantee for the proper review, inspection and long-term maintenance of the sponge city facilities implemented.
(8) Establish administrative procedures and fees for the submission, review, approval or disapproval of sponge city plans, and for the inspection of approved active projects, and long-term follow up.

5.5. Finding Innovative Ways to Create More Funding Options

While PPP is gaining interest in both developed and developing countries, it alone may not be able to raise enough funding to support China's large-scale sponge city construction. Additional and more innovative funding opportunities and mechanisms at all levels are in demand, including better integration between national agencies to cost-share national funds for local sponge city projects. China should adopt a model which takes into consideration the various local conditions. Different cities should select different economic strategies concerning their varied natural conditions and economic situations. Other funding sources such as tax-increments, development charges, value-capture taxes, loans, and bonds should be explored based on local conditions. Meanwhile, PPPs should be utilized as a way of developing local private-sector capabilities through joint ventures with large domestic and/or international firms, as well as providing sub-contracting opportunities for local firms. Ultimately, the economic strategy of a city should be beneficial to the local public and local economic growth.

Furthermore, incentives, both financial and non-financial, are in strong need at the provincial and local levels to encourage the adaptation of the sponge city approach. The incentives can range from instituting tax incentives, utility rate reductions, and/or regulatory credits. Non-monetary incentives that can encourage sponge city implementation include development incentives such as streamlined permitting and transfer of development rights, regulatory credits, and watershed trading for sponge city projects.

5.6. Continuous Research

In-depth research into sponge city concepts and practices is needed in various areas. At the national level, the following research should be conducted: developing computer models and tools to assist planning, design, and monitoring and performance evaluation purposes; assessing urban soils across the country for the suitability of sponge city practices; completing design, operation, maintenance and decision support guidelines; and establishing a national database for sponge-city-related data, information, technology, and demonstration. At the provincial and local scales, the following research should be carried out: assessing sponge city impacts on watersheds; developing incentives to encourage the sponge city; conducting cost of service studies and fiscal impact analyses to determine the impact on the fiscal health and viability of the community; conducting triple bottom-line (social, economic, and environment) analysis to identify means for saving and/or funding sponge city practices as opposed to gray infrastructure. Additionally, many other studies are also in demand.

6. Conclusions

As a new urban water planning and management approach, China's sponge city construction initiative is entering the third year and is quickly taking root in cities across the country. This research surveyed 30 pilot sponge cities and identified a wide array of challenges that may hinder the progress of the sponge city program. These challenges are classified into four categories: technical and physical, legal and regulatory, financial, as well community and institutional. The results show that these challenges come in various shapes and sizes depending on the local context; however, risks and uncertainties appear in each pilot sponge city, especially uncertainty about the outcomes, standards, techniques, and procedures. While significant challenges remain, important opportunities are opening for safer, greener, more holistic urban environments. Based on this study, the following conclusions are reached. (1) broad and diverse coalitions are necessary for discovering the benefits, exploring the possibilities, piloting the projects and probing system-wide changes; (2) increased research efforts into the techniques, levels of performance, range of multiple benefits, life cycle analysis of costs, and other key areas of sponge city implementation are needed; (3) greater coordination is needed among agencies and at all levels; communication among stakeholders, government officials and staff, and practitioners is also in need of improvement; (4) similarly to all new things, this new approach will require investment, coordination and patience. The development of green solutions that are acceptable in modern cities will take time. Time is needed for professional training and public education; time is needed for stepwise learning including learning from previous experiences. With the appropriate guidance and adjustments, it obstacles can be overcome, resulting in fewer technical, legal, financial, and cultural barriers. As understanding grows and methods improve, risks will be reduced.

Acknowledgments: This research was supported by the IWHR Research & Development Support Programs (No. JZ0145B322016; No. JZ0145B042017). We are grateful for the efforts of editors and reviewers and believe that the valuable comments reviewers provided are beneficial to this paper.

Author Contributions: Hong Wang served as lead and corresponding author, and designed the proposal; Liuqian Ding perfected the thoughts; Hui Li, Minglei Ren and Changzhi Li collected the data; Hui Li and Hong Wang analyzed the data; Hui Li wrote the paper; Hong Wang and Liuqian Ding provided editorial improvements to the paper.

Conflicts of Interest: The authors declare no conflict of interest.

Appendix A

Table A1. Regional characteristics and general information of pilot sponge cities.

No	Pilot Cities	Annual Average Rainfall (mm)	Annual Average Evapor. (mm)	Annual Average	Average High/Low	Pilot Area (km²)	Existing Drainage Capacity (a)	Ex. Flood Control Capacity (a)	Average Annual Runoff Contl (%)	Rain Water Resourcing (%)	Water Quality Control SS (%)	Wastewater Recycling (%)	Drainage Standard (a)	Pluvial Flood Standard (a)	Fluvial Flood Standard (a)	Investment (Billion-RMB)
1	Qian'an	672	1100	11.5	26/−5	21	0.5–1	20–50	80	7	–	30	2	20	50	4.493
2	Baicheng	410	1840	4.6	38/−32	21	1–3	10/20	85	6	60	25	3–5	20	50	4.230
3	Zhenjiang	1063	1277	16.1	29/3	22.0	2–5	20–50	75.0	8	60	25	2–5	30	100	3.060
4	Jiaxing	1194	1313	17.2	29/5	18.4	0.5–1	50	75.0	–	40	25	2–5	30	100	1.948
5	Chizhou	1483	1444	12.7	24/1	18.5	1–2	10–20	80.0	–	–	20	2–5	20–30	50–100	4.045
6	Xiamen	1530	1651	21.3	29/14	45.5	2–5	50	75.0	5	–	–	2–5	50	50	6.474
7	Pingxiang	1600	–	18.1	30/6	28.8	2–3	–	80.0	–	–	–	2–3	30	50	4.600
8	Jinan	665	1526	14.8	28/−1	39	1–5	<100	75	–	–	–	2–10	30–50	50	7.600
9	Hebi	665	2016	14.1	28/−1	29.8	1	50	70	–	–	–	2–5	30	100	3.476
10	Wuhan	1257	950	17.2	30/4	38.0	10	50–100	75.0	8	50	–	5–10	50	200	10.278
11	Changde	1366	–	17.5	29/5	41.2	2–5	50	80.0	–	75	–	2–5	30	100	17.350
12	Nanning	1298	1367	22.6	29/14	60.2	2	20–50	75.0	–	50	20	2–5	20	100	9.519
13	Chongqing	1107	1193	18.0	7/35	18.7	2–5	50–100	80.0	5	50	–	3–5	50	100	7.047
14	Suining	928	950	17.8	28/7	25.0	1–3	20–50	80.0	–	–	–	2–5	30	50	5.760
15	Gui'an	1158	1200	15.3	24/5	19.1	–	–	85.0	10	56	–	2–5	30	100	4.760
16	Xixian	520	1481	14.3	27/1	17.8	–	–	80	–	>60	30	2–5	50	50–200	3.123
17	Fuzhou	1360	970	19.7	28.8/10.6	36.9	1–2	20	75.0	2	45	2	3–5	20–50	20–200	7.800
18	Zhuhai	1766	1469	23.0	32.2/−3	52.0	1–3	20–50	70.0	10	50	15	3–5	30–50	100	10.656
19	Ningbo	1517	830	17.2	29/6	31.0	1–3	20–100	80.0	22	60	40	3–10	50	100–200	6.042
20	Yuxi	909	1801	19.2	22/10	20.9	–	20	82.0	10	50	20	3–5	30	100	4.873
21	Dalian	736	1551	9.1	22/−8.1	21.8	1–3	50	75.0	5	50	25	>2	20	50	2.898
22	Shenzhen	1837	1675	22.4	29/16	24.9	1–5	20–50	70.0	≥8	60	30	3–5	50	200	3.529
23	Shanghai	1191	1420	15.7	29/5	79.0	2–5	200	80.0	8	80	20	5	100	200	8.560
24	Qingyang	510	1425	9.5	23/−8.4	29.6	1	20	90.0	5	60	–	2–5	30	100	4.735
25	Xining	460	1364	6.2	14.9/−0.3	21.6	–	50–100	88.0	2	60	50	2–5	50	100	6.375
26	Sanya	1392	2361	25.5	28.8/21.6	20.3	–	–	70.0	8	–	20	2–5	30	100	4.040
27	Qingdao	776	1401	12.2	25.1/−1.2	25.2	2–3	50	75.0	5	65	30	2–5	30	100	4.870
28	Guyuan	458	1099	6.1	24.7/−14.3	23.0	–	–	85.0	10	40	30	2	30	50	3.654
29	Tianjin	511	1639	13.5	27.2/−2.4	39.5	1–3	50	80.0	5	65	60	3–5	20–50	50–200	7.490
30	Beijing	573	1164	11.7	26/−4.7	19.4	3–5	50	84.4	3	42	75	2–10	50	100	3.937

Note: No. 1–16 are the regional characteristics and general information of the first group of pilot sponge cities and No. 17–30 are those of the second group of pilot cities.

References

1. National Bureau of Statistics of China. *China Statistical Yearbook 2015*; China Statistics Press: Beijing, China, 2015.

2. Research Group of Control and Countermeasure of Flood (RGCCF). Control and countermeasure of flood in China. *China Flood Drought Manag.* **2014**, *3*, 46–48.

3. Lv, Z.; Zhao, P. First report about urban flood in China: 170 cities unprotected and 340 cities down-to-standard. *Zhongzhou Constr.* **2013**, *15*, 56–57.

4. Chen, Z.; Lu, M.; Ni, P. *Urbanization and Rural Development in the People's Republic of China*; ADBI Working Paper 596; Asian Development Bank Institute: Tokyo, Japan, 2016; Available online: https://www.adb.org/publications/urbanization-and-rural-development-peoples-republic-china/ (accessed on 11 November 2016).

5. Zevenbergen, C.; Veerbeek, W.; Gersonius, B.; Van Herk, S. Challenges in urban flood management: Travelling across spatial and temporal scales. *J. Flood Risk Manag.* **2008**, *1*, 81–88. [CrossRef]

6. Zhou, Q.; Mikkelsen, P.S.; Halsnaes, K.; Arnbjerg-Nielsen, K. Framework for economic pluvial flood risk assessment considering climate change effects and adaptation benefits. *J. Hydrol.* **2012**, *414–415*, 539–549. [CrossRef]

7. Chang, H.K.; Tan, Y.C.; Lai, J.S.; Pan, T.Y.; Liu, T.M.; Tung, C.P. Improvement of a drainage system for flood management with assessment of the potential effects of climate change. *Hydrol. Sci. J.* **2013**, *58*, 1581–1597. [CrossRef]

8. Ministry of Housing and Urban-Rural Development (MHURD). Technical Guide for Sponge Cities—Water System Construction of Low Impact Development. Available online: http://www.mohurd.gov.cn/zcfg/jsbwj_0/jsbwjcsjs/201411/W020141102041225.pdf (accessed on 22 October 2016).

9. United States Environmental Protection Agency (US EPA). *Low-Impact Development Design Strategies: An Integrated Design Approach*; EPA 841-B-00003; US EPA: Washington, DC, USA, 1999.

10. Benedict, M.; Mcmahon, E. Green infrastructure: Smart conservation for the 21st century. *Renew. Resour. J.* **2002**, *20*, 12–17.

11. British Columbia Ministry of Environment (BCME). Stormwater Planning: A Guidebook for British Columbia. Available online: http://www.env.gov.bc.ca/epd/mun-waste/waste-liquid/stormwater/ (accessed on 4 September 2016).

12. Olewiler, N. *The Value of Natural Capital in Settled Areas of Canada*; Ducks Unlimited Canada and The Nature Conservancy of Canada: Toronto, ON, Canada, 2004.

13. Alexander, D.; Tomalty, R. Smart growth and sustainable development: Challenges, solutions and policy directions. *Local Environ.* **2002**, *7*, 397–409. [CrossRef]

14. Lehmann, S. UNESCO Chair in Sustainable Urban Development. In *The Principles of Green Urbanism*; Earthscan: London, UK, 2010.

15. Beatly, T. *Green Urbanism: Learning from European Cities*; Island Press: Washington, DC, USA, 1999.

16. Barthod, C.; Deshayes, M. Trame Verte et Bleue, the French Green and Blue Infrastructure. Ministère de l'Écologie, de l'Énergie du Développement durable et de l'Aménagement du Territoire. Available online: http://www.developpement-durable.gouv.fr (accessed on 3 September 2016). (In French)

17. Sharma, A.K.; Pezzaniti, D.; Myers, B.; Cook, S.; Tjandraatmadja, G.; Chacko, P.; Chavoshi, S.; Kemp, D.; Leonard, R.; Koth, B.; et al. Water Sensitive Urban Design: An Investigation of Current Systems, Implementation Drivers, Community Perceptions and Potential to Supplement Urban Water Services. *Water* **2016**, *8*, 272. [CrossRef]

18. Jenkins, S. Towards Regenerative Development. Available online: www.planning.nz (accessed on 10 September 2016).

19. General Office of the State Council (GOSC). Guideline to Promote Building Sponge Cities. Available online: http://www.gov.cn/zhengce/content/2015-10/16/content_10228.htm (accessed on 16 October 2015).

20. Ministry of Finance of the People's Republic of China (MOF). Notice on the Implementation of the Central Financial Support to the Construction of Pilot Sponge Cities. 2014. Available online: http://jjs.mof.gov.cn/zhengwuxinxi/tongzhigonggao/201501/t20150115_1180280.html (accessed on 31 December 2014).

21. Office of State Flood Control and Drought Relief Headquarters; Disaster Reduction Committee of Chinese Hydraulic Engineering Society. *China Urban Flood Control*; China Water & Power Press: Beijing, China, 2008.

22. China National Knowledge Infrastructure (CNKI). Available online: http://nvsm.cnki.net/KNS/ (accessed on 10 February 2017).
23. WANFANG DATA (WANFANG). Available online: http://www.wanfangdata.com/ (accessed on 13 February 2017).
24. Chinese Science Citation Database (CSCD). Available online: http://sciencechina.cn/search_sou.jsp (accessed on 6 February 2017).
25. VIP JOURNAL INTEGRATION PLATFORM (VIP). Available online: http://lib.cqvip.com/ (accessed on 20 February 2017).
26. Wu, C.; Li, Z. The current situation and future trend of urban rain water harvesting. *Water Wastewater Eng.* **2002**, *28*, 12–14.
27. Wang, H.; Cheng, X.; Li, C. Quantitative Analysis of Stormwater Management Strategies in the Process of Watershed Urbanization. *J. Hydraul. Eng.* **2015**, *46*, 19–27.
28. Wang, H.; Li, C.; Li, N.; Yu, Q. Green infrastructure design principles and integration of gray and green infrastructures. *Water Wastewater Eng.* **2016**, *42*, 51–56.
29. Wang, H.; Li, C.; Zhang, W.; Jiang, X. Framework for the planning of urban stormwater infrastructures. *Urban Plan. Int.* **2015**, *30*, 72–77.
30. Wang, H.; Ding, L.; Cheng, X.; Li, N. Hydrologic control criteria framework in the United States and its referential significance to China. *J. Hydraul.* **2015**, *46*, 1261–1271.

water

MDPI

Article

Mitigation Options for Future Water Scarcity: A Case Study in Santa Cruz Island (Galapagos Archipelago)

Maria Fernanda Reyes [1,*], Nemanja Trifunović [1], Saroj Sharma [1], Kourosh Behzadian [2] iD, Zoran Kapelan [3] and Maria D. Kennedy [1,4]

1   Department of Environmental Engineering and Water Technology, UNESCO-IHE Institute for Water Education, P.O. Box 3015, 2601 DA Delft, The Netherlands; n.trifunovic@un-ihe.org (N.T.) s.sharma@un-ihe.org (S.S.); m.kennedy@un-ihe.org (M.D.K.)
2   Centre for Water Systems, School of Engineering, University of West London, London W5 5RF, UK; kourosh.behzadian@uwl.ac.uk
3   College of Engineering, Mathematics and Physical Sciences, University of Exeter, Exeter EX4 4QF, UK; z.kapelan@exeter.ac.uk
4   Faculty of Civil Engineering and Geosciences, Delft University of Technology, P.O. Box 5048, 2600 GA Delft, The Netherlands
*   Correspondence: fersireyes@hotmail.com; Tel.: +31-(0)1-5215-1715

Received: 1 May 2017; Accepted: 9 August 2017; Published: 12 August 2017

Abstract: Santa Cruz Island (Galápagos Archipelago), like many other tourist islands, is currently experiencing an exponential increase in tourism and local population growth, jeopardizing current and future water supply. An accurate assessment of the future water supply/demand balance is crucial to capital investment for water infrastructure. This paper aims to present five intervention strategies, which are suggested to solve the future water crisis. The strategies combined include environmentally sustainable options such as rainwater harvesting, greywater recycling and water demand management, as well as desalination. These strategies were evaluated under four population growth scenarios (very fast, fast, moderate and slow growths) by using several Key Performance Indicators (KPI's) including water demand, leakage levels, total costs, energy consumption, rainwater delivered and greywater recycled. Moreover, it also aims to develop a methodology for similar islands, using the WaterMet^2 modelling approach, a tool for integrated of sustainable-based performance of urban water systems. The results obtained show that by 2044 only a small portion of the future water demand can be covered assuming business as usual. Therefore, desalination seems to be the most viable option in order to mitigate the lack of water at the end of the planning period considering the growth trends. However, strategies comprising more environmentally friendly alternatives may be sufficient, but only under slow population growth scenarios.

Keywords: intervention strategy; key performance indicators; water demand prediction; water scarcity; WaterMet^2

1. Introduction

Santa Cruz is the main tourist island in the Galápagos Archipelago, located around the Equator on the Pacific Ocean, holding more than 60% of total local population and visitors [1]. The island has two main towns of Puerto Ayora and Bellavista with 12,000 and 2500 inhabitants respectively [2]. Furthermore, the number of tourist arrivals in 2013 was approximately 204,000 [3]. As a consequence of the exponential and unsustainable growth rates of tourism and local population, the demand for public services has increased exponentially, especially in Puerto Ayora, the main centre of tourist activities. Due to the fast economic development of the island, priority is given to tourism without considering the related environmental impacts [4]. The demand for natural resources such as potable water has abruptly increased, resulting in deficient water supply services and contributing to fast ecological degradation.

Due to the uncontrolled expansion of tourism [5], touristic premises have increased with annual growth rates between 8% and 11% [6]. The fragile ecosystem is further endangered by the increasing number of immigrants coming from the mainland, contributing to annual population growth rates of almost 5% in 2010, (from 2.5% in 1960). Consequently, local authorities have been facing serious challenges to cope with this and the adequate provision of basic services due to limited financial resources, limited water resources and lack of required infrastructure. The main reasons for these limitations can be found in some factors such as weak and unstable governance, lack of policies considering the ecosystem, unplanned urbanisation and fast economic development [2,7]. Moreover, land-based tourism has predominated, requiring greater urban development and supporting infrastructure, more severe depletion of drinking water resources, consumption of energy, etc. [8].

Currently, the water supply on the island is intermittent, with an average supply of three hours per day. Also, several studies have assessed water losses as high as 70% [9] or more recently, 35% of system input volume as Non-Revenue Water (NRW) [7]; this is caused by ageing networks and the lack of proper maintenance. In addition, there is excessive water loss within premises in the form of leaks and overflows from tanks, which is likely to be the consequence of fixed water tariff structures in Puerto Ayora [10].

Estimation of water demand in the Santa Cruz Island is difficult due to the lack of water meters for customers in Puerto Ayora. Next to the municipal supply, there are numerous extractions from crevices by individuals and/or institutions (contributing to the unknown demand due to the lack of records), and bottled-desalinated water is distributed by small private companies. However, there have been some studies about water demand estimations for the island. The water demand based on data from Puerto Ayora, considering 13 installed water meters, suggested that the specific water demand ranged from 92 to 1567 L/cap/day, suggesting that some domestic premises are also (informal) tourist accommodations [11]. A recent study by [12] estimated the water demand per category of users, considering all sources of water as shown in Table 1. Based on this analysis, the specific water demand in Puerto Ayora was estimated to be between 163 and 177 L/cap/day where the former figure corresponds to municipal demand and the latter one to total domestic water demand.

Table 1. Total water demand quantification for Puerto Ayora for different categories [12].

Category	Municipal Supply (m³/day)	Bottled Water (m³/day)	Water Trucks (m³/day)	Total Demand (m³/day)
Domestic	1951	20	158	2129
Hotels	1107	21	1789	2917
Restaurants	69	8	51	128
Laundries	29	0	20	49
TOTAL	3156	48	2018	5222

Although the above-mentioned studies have intensively estimated existing water demand, none of them have tried to forecast future water demand in relation to water availability, developing a water balance for the next 30 years. This may seriously endanger the future of the island's development with respect to the highly increasing water demands, affecting not only local population and tourism, but the fragile ecosystem as well.

Based on the above, the current study aims to develop a water balance model for Puerto Ayora, in order to compare the baseline conditions (business as usual) with a number of possible intervention strategies to meet future demand, under different scenarios of population growth rates. This should enable decision makers to investigate the impact of population growth on the level of water services on Santa Cruz Island. The model considers a 30-year period during which four different population/tourist growth scenarios are analysed using the WaterMet² model (University of Exeter, UK)) [13].

First, a brief literature review of water demand forecasting using Urban Water Systems (UWS) models and further details of the WaterMet^2 model are presented in the next section. Later, the methodology and assumptions applied to the analysed case study are introduced, followed by the description of the modelling approach. This methodology is then applied to the case study of Santa Cruz. Subsequently, the results are presented and discussed and, at the end, several conclusions are drawn.

2. Literature Review

2.1. Urban Water System Modelling Approach

Owing to the increase in computational power and computer technologies, the modelling of UWS has shifted towards more holistic approaches rather than viewing each process separately. Processes are considered as components of a whole, complete and integrated water cycle [14]. According to [15], the primary objective of the UWS analyses should be first to balance out demand with supply. Therefore, the water demand forecast has been developed mainly to understand spatial and temporal patterns of water use in the future [16], as well as for better management of water resources [17]. Water demand forecasting has encountered many problems due to the nature and quality of available data, numerous variables that are hypothesized to affect demand, and the variety of forecast horizons [18].

There are several types of urban water cycle models: (1) detailed models (Infoworks, SMURF, Hydro Planner), which are characterized by their limited scope and high data requirements; (2) catchment scale models (Water StrategyMan, Aquatool (Mitchell et al., 2010) [19], Systems Modelling RioGrande), characterized by wide system boundaries, but do not provide sufficient analysis of the urban water system at sub-city scale; (3) urban water scoping programs (UWOT-University of Exeter [20], UK, Aquacycle [19], UVQ-CSIRO, Australia, WaterCress, CWB-University of Birmingham, UK [21], WaterMet^2 [13], which model city scale dynamics including all of the important processes within the urban water cycle (but in less detail than more focused models).

The latter ones are suitable to be used as sustainability assessment tools and for strategic planning. They forecast water demand and include the main components of the urban water cycle in a holistic and integrated approach. These models strive to include water supply, stormwater, wastewater and groundwater by considering both aspects of water quantity and quality simulation.

WaterMet^2 is a metabolism-based model which quantifies a number of flows/fluxes (e.g., water and energy) in urban water systems (UWS) [22], and can be used for the assessment of sustainability-based performance of the analyzed water system, including quantifying the likely impact of different intervention strategies. Therefore, WaterMet^2 was found to be the most suitable in this study with respect to the data availability. Also, the aspects of modelling the water cycle in the UWS, as different and separate components made it appropriate for this case study, since there is no sewage system and only the supply and demand component could be modelled. Moreover, it concurrently forecasted water demand under various scenarios and evaluated the performance of several strategic solutions for a water supply problem [23]. Also, this model is able to represent the daily, seasonal, and annual (future) dynamics of the water demand (i.e., caused by demography or changes in consumption pattern). Furthermore, it has a wide range of output indicators resulting from the simulation of wide range of fluxes: water flow, energy flow, greenhouse gas emission, among other, which offers several sustainability indicators for assessment of the intervention strategies.

Unlike the previously mentioned models, data input requirements were extent and focused on the modelling of other aspects but future water demand coverage. Furthermore, due to its conceptual nature, modelling the UWS in WaterMet^2 can be achieved with relatively little input data. This has been an important attribute and specifically a significant constraint for many cases, especially in developing countries. The lack of substantial and historic data has limited the development of different strategies that allow improvement in water demand and supply management. Also, none of the other models is considered as a holistic systemic perspective for the analysis of resource flows and their impacts on the future performance of UWS. Furthermore, this model allows to examine selected intervention strategies for a long-term planning. These two characteristics are addressed through a metabolism-based approach, which refer to the different fluxes and conversion processes related to all water flows, materials and energy in a UWS. With this approach, decision makers are able to identify critical components that have more or less impact in sustainability, and it allows the minimization of negative environmental impact by applying intervention strategies that save water, as well as energy, chemicals and costs. Thus, it presents a generic modeling approach based on WaterMet^2 that can be used to address the long-term supply demand balance under data scarcity conditions.

WaterMet^2 is a quantitative UWS performance model, which simulates mass balances throughout the whole urban system and calculates the principal water-related flows [13]. It also models future urban water balance and cycle, and specifies key performance indicators which can be used for evaluation of alternative intervention strategies under a number of different scenarios. The urban water cycle is simulated within four main subsystems referred to as water supply, water demand, wastewater, and water recovery (Figure 1).

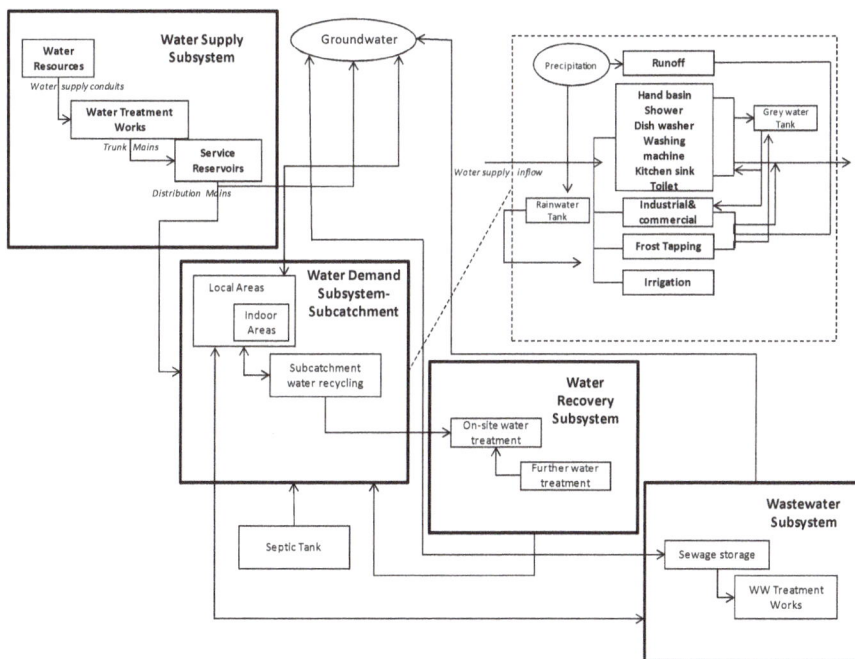

Figure 1. Components used for modelling different spatial levels in WaterMet^2.

Modelling of UWS using WaterMet^2 adopts specific spatial and temporal limits. There are four spatial scales represented by the model: (1) indoor area (e.g., residential, industrial, commercial, and public properties); (2) local area as a group of similar properties and same characteristics; (3) subcatchment area as a group of neighbouring local areas and (4) city area. There are different temporal scales for the simulation of UWS performance over a long-term planning horizon such as annual, monthly and daily variations.

2.2. Population and Tourism Growth Scenarios

Population projection is very important in these types of studies because it shows the interaction of an UWS in the future. Generally, these types of projections gather three factors such as mortality, fertility and/or migration. In order to project the size of a population in the future, many assumptions have to be made regarding the different factors that may influence the tendency of growth and how they will change over the selected period of time. In the case of the WaterMet^2 software, the growth needs to be specified for each year included in the planning horizon.

2.3. Alternatives and Intervention Strategies

The WaterMet^2 model allows the proposition of several types of intervention strategies, which might help improve the UWS performance when dealing with increasing water demands in the future. Among them are leakage reduction levels, pipe rehabilitation, water demand management, water meter installation, rainwater harvesting (RWH) and greywater recycling (GWR). The impact of these alternatives can be evaluated through different KPIs on the software. The KPIs can be further used as measurements for the evaluation of specific criteria, especially when comparing different scenarios or introducing different options in the model.

3. Methodology

In this study, the WaterMet^2 model was used to forecast the urban water balance for a future 30-year period. First, four different growth scenarios were chosen to predict and assess future water demand based on previous studies. Afterwards, after sufficient literature review, the model was built for a Puerto Ayora case study and the baseline condition (business as usual) was analysed in order to develop relevant strategies that will solve the future water deficit. Based on this, six water supply and demand management alternatives (individual strategies) were analysed in the Puerto Ayora model. The impact of each of these alternatives to meet future water demand was assessed by analysing the percentage of coverage of water demand with supply at the end of the planning horizon (KPI used was fraction of water demand delivered).

Due to the low fractions of water demand delivered at the end of the planning horizon, these individual alternatives were further combined in order to improve the future coverage of water demand with supply, developing five more complex intervention strategies. These new strategies combined also several sustainable options, recurring to desalination as the last option. These analysed strategies were compared using a number of KPIs in order to analyse the impact of the selected growth rates on different aspects of water demand and water supply over the period selected. The KPIs used include the ratio of water delivered to consumers, total costs (i.e., capital and Operations & Management) and total energy use (i.e., direct and embodied). With these indicators, each strategy was assessed in order to find the most sustainable and most optimal solution for this case study, considering the fragility of the ecosystem, which will be addressed in the discussion.

Finally, conclusions were drawn, assessing all the KPI's used and the extension to which each strategy would comply at the end of the planning horizon. The intention was to portray to local authorities and stakeholders the limitations and benefits of each strategy included in this study.

4. Case Study

4.1. Description

The case study analysed here is the water supply system in the main urban settlement (Puerto Ayora) on Santa Cruz Island, considered in the model as the 'baseline condition'. The schematic representation of the water supply system of Puerto Ayora is shown in Figure 2. Here, the water is abstracted from crevice 'La Camiseta' by two pumps of 25 hp, leading to average supply of 3024 m^3/day. There is no water treatment; therefore the groundwater withdrawn, which is slightly brackish, is directly distributed to the households. The water is further conveyed over a distance of 2800 m to two reservoirs with the volume of 600 m^3 and 800 m^3, respectively, through two PVC pipes of 315 mm. The water is then distributed to consumers by gravity, one supplying the northern part of the town and the other the southern part. The distribution network has 2156 connections (registered up until December 2013). There are no individual water meters installed for consumers and a fixed monthly tariff is applied based on the category of customers (domestic, commercial, touristic, etc.). The distribution network is approximately 30 years old and consists of PVC pipes of diameter ranging from 63 to 250 mm. The total NRW is estimated to be ±35%, based on some surveys on the consumption of domestic households, hotels, restaurants and laundries [10]. NRW is equal to water losses plus unbilled authorised consumption [24]. However, due to negligible unbilled authorised consumption in the case study, it is assumed that water losses are equal to non-unbilled authorised consumption. As a result, an estimated value of 35% NRW is assumed to be total water losses in the case study. However, for model building purposes, we have chosen 28% of leakage level, instead of 35% as initially intended, mainly due to calibration purposes of the model.

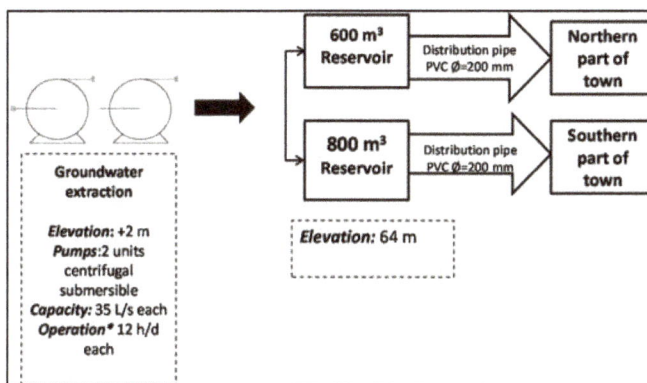

Figure 2. Current layout of Puerto Ayora water supply system; * Operation is only 12 h per day due to lack of sufficient power capacity for the three existing pumps to operate full time. As a result, only two operate at the same time for the specified period.

4.2. WaterMet² Model Building

The main input data to model the UWS in WaterMet² are shown in Figure 3 and are divided into three primary categories: 'Water Supply', 'Subcatchment' and 'Water Resource Recovery'. This figure differs from Figure 1 because it describes how the model is built based on the input data requirements for each component/subsystem.

The 'Water Supply' specifications of the water supply conduits, trunk mains and distribution mains include storage capacity, initial volume as well as energy, chemicals and cost used per unit volume of water. Also, the transmission component of the water supply system, which is the connecting flow routes for conveying water between storage components in the WaterMet² model, is specified here as well. Their general specifications include transmission capacity, leakage, energy and Operations &Management costs per unit volume of transmitted water. The detailed input data of water supply components used in the WaterMet² model for Puerto Ayora is presented in Table 2. The energy equation shown below is used to calculate the energy requirements of pumps as one of the requirements of the WaterMet² in energy consumption of components in the water supply subsystem. Therefore, the theoretical energy consumption was calculated based on the equation [25]:

$$N = \frac{\rho g Q h_p}{\eta_p} \tag{1}$$

where ρ is density (kg/m³), g is gravity constant (m/s²), Q refers to the flow (m³/s), h_p refers to the pumping head (m) and η_p is the efficiency of the pump.

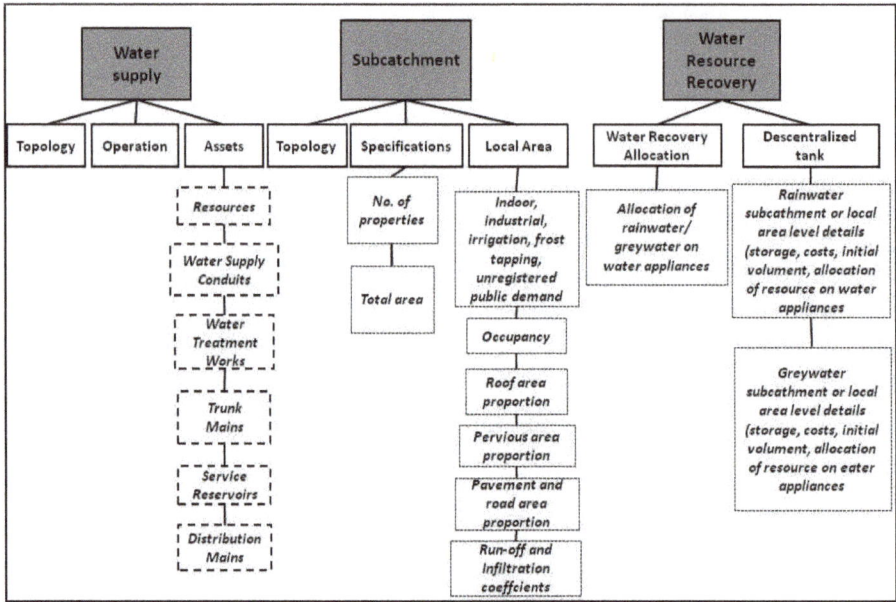

Figure 3. Main Urban Water Systems (UWS) components used in the WaterMet^2 model.

The sub-catchment and local area components were used in WaterMet^2 to define water demand categories and rainfall-run off characteristics of the model. Each settlement was represented as a single sub-catchment with a single local area. In Puerto Ayora, two water demand categories were defined: (1) 'indoor' water representing domestic water use; and (2) 'industrial' representing water demands of restaurants, hotels and laundries. The percentage share of water use for appliances and fittings in domestic water consumption in both cases were assumed to be 7% for hand basin, 20% for kitchen sink, 24% for showers and 49% for toilet flushing, based on a study made by [26]. Furthermore, rainfall-run off simulation was modeled considering run off from roofs only (local area level). For the software calculations, roof area proportion and pervious areas proportions were determined. A summary of the input data used for modelling the sub-catchment components in WaterMet^2 is given in Table 3. The water demand variations of local areas can be defined in WaterMet^2 for different temporal scales (i.e., annually and monthly). The annual variations were analysed under four selected population growth scenarios while the monthly variations were adjusted during the model calibration process, which will be discussed later. Despite the fact that in many touristic islands, there is a significant seasonal variation regarding summer and winter, there is no significant change in the case study of the Galapagos Islands due to: (1) the average temperatures of the Islands, which have no considerable variations. Therefore, tourists are present in the Islands throughout the entire year and there is no significant difference for holiday seasons (e.g., Christmas or summer) and hence an average value of water demand was used here. The daily variations and temperature influence on these variations was ignored in this step due to the lack of daily consumption registration.

Table 2. Input information for the water supply component.

Water Resources Form		
Component Name	**Unit**	**Puerto Ayora**
Type	-	Groundwater
Energy consumption (electricity and fossil fuel)	kWh/m^3 (for electricity) L/m^3 (for fossil fuel)	0.66 kWh/m^3 and 0.3 L/m^3
Fixed annual operation costs	EUR/ year * (cost of elec: 0.17 EUR/kWh Cost of fuel: 0.22 EUR/L)	219,120
Water Loss	%	Assumed there are no water losses at the point of extraction.
Water Supply Conduits		
Component name	Unit	Puerto Ayora
Transmission capacity ***	m^3/day	3024
Leakage **	%	8
Pumping system	m^3	N/A
Energy consumption	kWh/m^3 (for electricity) L/m^3 (for fossil fuel)	0.66 and 0.5
Fixed annual operation costs	EUR/year * (cost of elec: 0.17 EUR/kWh Cost of fuel: 0.22 EUR/L)	4980
Service Reservoirs		
Storage Capacity ***	m^3/day	3024
Initial volume	m^3	1500
Operational cost	EUR/year *	2490
Distribution Mains		
Transmission Capacity ***	m^3/day	3024
Leakage **	%	20
Operational Costs	EUR/year *	58,100

Note: There is no individual water treatment works with specifications defined in WaterMet^2 for the case study. * All the costs are based on information from the municipality ** Leakage figure of 28% was used and not 35% as previously identified as Non-Revenue Water (NRW), due to calibration purposes. *** Transmission components in WaterMet^2 (e.g., water supply conduits and distribution mains) are defined based on a daily transmission capacity expressed in m^3/day while storage components (e.g., service reservoirs) are defined based on storage capacity expressed in m^3. The definition for transmission components is for a conceptual model used in WaterMet^2 and is estimated based on the average hydraulic capacity of the components.

Table 3. Input information for the WM^2 model sub-catchment component.

Component Name	**Unit**	**Puerto Ayora**
Topology	Defined as only one sub-catchment area and one local area	
Number of properties	-	1996 (domestic)
Total area	Ha	163
Current indoor water demand	L/cap/day	160
Current Industrial/Commercial water demand	m^3/day	1200
Average occupancy per property	Inhabitants/household	4
Roof area proportion	(%)	40
Pervious area proportion *	(%)	30
Pavement & road area proportion *	(%)	30
Run-off coefficient *	(0-1)	0.85
Infiltration coefficient *	(0–1)	0.9

* Values calculated based on literature review [27].

"Water Resource Recovery" refers to RWH and GWR schemes. In this study, rainwater was assumed to be collected only from roof runoff and provide water for toilet flushing, showers, sinks, indoor irrigation and commercial uses. GWR collected from hand basins and showers was allocated only for toilet and indoor irrigation. The associated costs and energy was also considered for treatment and purification of RWH and GWR, since the existing supply system does not have any rainwater and/or greywater infrastructure; thus, these are considered as new alternatives. Other input information includes unit costs (of electricity and diesel fuel, water meter installation cost, and inflation rate), climate constants (elevation and geographical location to be used in rainfall-runoff modelling for calculation of evaporation), coefficients for all water demand categories, including percentage of conversion from water to wastewater (assumed here as 95%), percentages of domestic water appliances and their possible energy consumptions, based on personal communications with personnel from the Municipality of Santa Cruz. The historic time series of weather data (e.g., precipitation, temperature and etc.) for the past 30 years were used here in the WaterMet^2 model assuming that the same trend will take place in the future planning horizon. Time series data were obtained from the National Institute of Meteorology and Hydrology of Ecuador (INAMHI). It is relevant to mention that the annual average rainfall over the last years in Puerto Ayora is 380 mm, but in other settlements located higher, such as in Bellavista the average is 1100 mm or even higher were the annual average can reach 2500 mm [28] having significant higher precipitation rates on the hot "invierno" season, than in the cold "garua" season, characterized by big and strong events of rain. On the other hand, the evapotranspiration average on both seasons is around 400 mm.

The input data of 'Distribution Network Pipelines' about pipe materials, diameter and lengths were obtained from the municipality. These data were used for rehabilitation and leakage reduction in the water network.

4.3. WaterMet^2 Model Calibration

The WaterMet^2 model calibration in this study was done based on historical data of monthly water abstraction from the crevices serving as the water source of the UWS. The calibration parameters related to the capacity of the water resources in the WaterMet^2 model were adjusted by using the monthly records on water abstraction available at the municipal water department. The calibration was performed with historical groundwater abstraction rates (m^3/day), which were divided into two periods: year 2012 for calibration and year 2013 for validation. Figure 4a shows a graphical comparison of the performance of the model by plotting simulated versus observed figures. Later, they were likewise validated with the following period (year). The statistical correlation coefficient (R^2) of 0.886 (Figure 4b) represents an acceptable value for this particular case study, regarding the lack of consistent data on the water pumping for supply, where daily records of water extractions for several days were missing. The value of the correlation obtained is significant based on the acceptable ranges suggested by other similar research works [22]. The model accuracy can be improved by increasing the amount of measured data used in calibration.

Based on the calibration and validation processes, the monthly coefficients of water demand profiles for Puerto Ayora applied in WaterMet^2 were calculated (Table 4). These coefficients were calculated based on the supply average for that particular month and the total daily average supply average for the years chosen for validation and calibration (2012 and 2013). These were applied for the entire planning horizon and used for both domestic and industrial water demand.

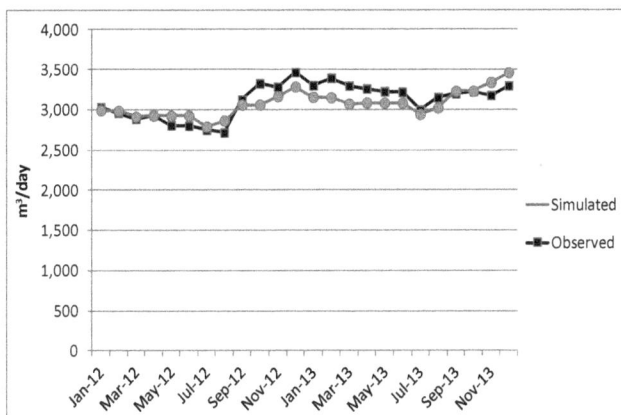

(a)

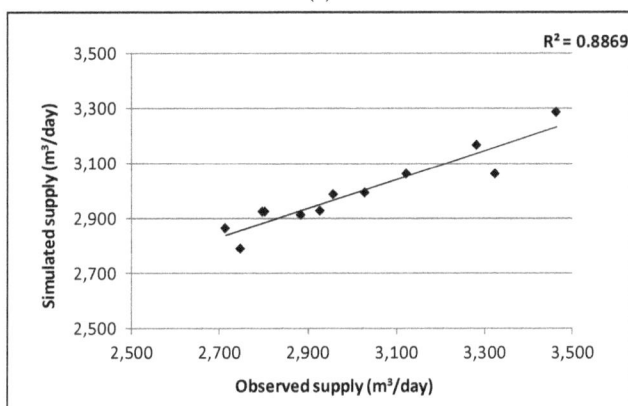

(b)

Figure 4. Comparison between simulated and recorded supply as (**a**) a time series of supply and (**b**) scatter plot.

Table 4. Selected monthly coefficients for Puerto Ayora.

Month	Puerto Ayora
January	1.027
February	1.019
March	0.991
April	0.993
May	0.967
June	0.965
July	0.922
August	0.941
September	1.014
October	1.050
November	1.036
December	1.080

4.4. Population and Tourist Growth Scenarios for Puerto Ayora

The population growth scenarios were chosen based on the suggestions made by Mena et al., 2014 [6], according to historical growth and government planning. This was conducted by deriving relationship between the number of tourists and local residents in Galápagos, representing a demographic model with projections until 2033. Here, an ordinary least squares (OLS) linear regression was used to model the relationship based on population censuses in 1982, 1990, 2001 and 2010. This resulted in determining the number of residents in each year based on the corresponding number of tourist arrivals. Only land-based tourists were considered (excluding tourist cruise/ships). The growth scenarios were developed depending mainly on the migration rate since it is the primary demographic parameter due to the strong ties between the future growth of local residents and tourism growth. More specifically, the growth scenarios developed, based on Mena *et al.*, 2014 [6] were defined as follows:

Slow growth: The tourist arrivals were 180,000 in 2012, therefore that number is considered to be the average per year. This scenario is suggested and preferred by environmentalists, NGOs and the Galápagos National Park.

Moderate growth: The tourist arrivals maintain an average annual increase of 7066 visitors (4%). This figure was estimated from the recorded tourism growth in the last 20 years.

Fast growth: The tourist arrivals increase exponentially by an annual rate of 7%. This scenario is preferred by the central government following their objective to increase tourism revenues in the whole country.

Very fast growth: The number of tourist arrivals would be eight times greater than the number of residents at the end of the planning horizon (i.e., in year 2044), suggesting a rate of annual growth of 9%.

A summary of growth rates used in the four scenarios is shown in Table 5.

Table 5. Annual population and tourist growth scenarios used for water demand forecast.

Growth Scenario	Local Population Increase	Tourist Visitors Increase *
Very Fast	7%	9%
Fast	5%	7%
Moderate	3%	4%
Slow	1%	1%

* The historic average growth per year is 7%.

4.5. Alternatives and Intervention Strategies

Six potential alternatives were developed in this study. These alternatives vary from those aiming to increase water supply (e.g., desalination plant construction or RWH and/or GWR) to those aiming to reduce demand (e.g., leakage reduction, water meter installation or any other form of general water demand management). All of these alternatives have been identified with the aim to balance the long-term water demand. The detailed description of all alternatives is shown in Table 6. Leakage reduction and water meter installation are alternatives that have been proposed already over the last years; nevertheless, they have not yet been implemented [26]. With this study, we aim to quantify the impact of them in the short- and long-term. Furthermore, desalination with reverse osmosis has also been suggested due to popularity within authorities, since water quality issues (salinity) would also be improved. RWH has been proposed as a more sustainable option and due to the attractiveness in the smaller town of Bellavista. Moreover, GWR has also been proposed as a sustainable option as well and with the aim of reducing wastewater disposal, since the submerged membrane bioreactor (MBR) offers a low-footprint with a high quality effluent for recycling domestic water [29]. Finally, water demand management was also included, since the specific demand is currently considered high compared to other domestic demand in water scarce areas. Each alternative was analysed separately in order to assess the impact on the UWS. All of the previous proposed alternatives were discussed with the Department of Water and Sanitation of Santa Cruz. Some of them have already been proposed and some are new suggested alternatives.

The proposed alternatives were first simulated individually in the WaterMet^2 model over the planning horizon (2014–2044) in order to analyze their impact to balance future water demand under different population/tourist growth scenarios. Table 7 shows the results of respective simulation model runs in terms of the fraction of the water demand delivered (i.e., covered) at the end of the planning horizon (year 2044), including the corresponding baseline at the start of the planning horizon (year 2014). This fraction is calculated as the ratio of the total water supplied to the total water demand over the long-term planning horizon. Due to unsatisfactory (i.e., low) fractions of water demand covered by supply for most of the individual alternatives (Table 7), these were combined to form five more complex intervention strategies (Figure 5).

Generally, the selected intervention strategies can start at any year during the planning horizon period. This study assumed that all the alternatives will be implemented starting at year 3, in order to give time to the municipality to implement and construct the different infrastructure needed for each of the proposed alternatives. The combination of strategies aimed to complement each other, and to improve the fraction of water demand delivered at the end of the planning horizon. Intervention Strategy 5 is considered a combination of all sustainable alternatives, except the option with desalination.

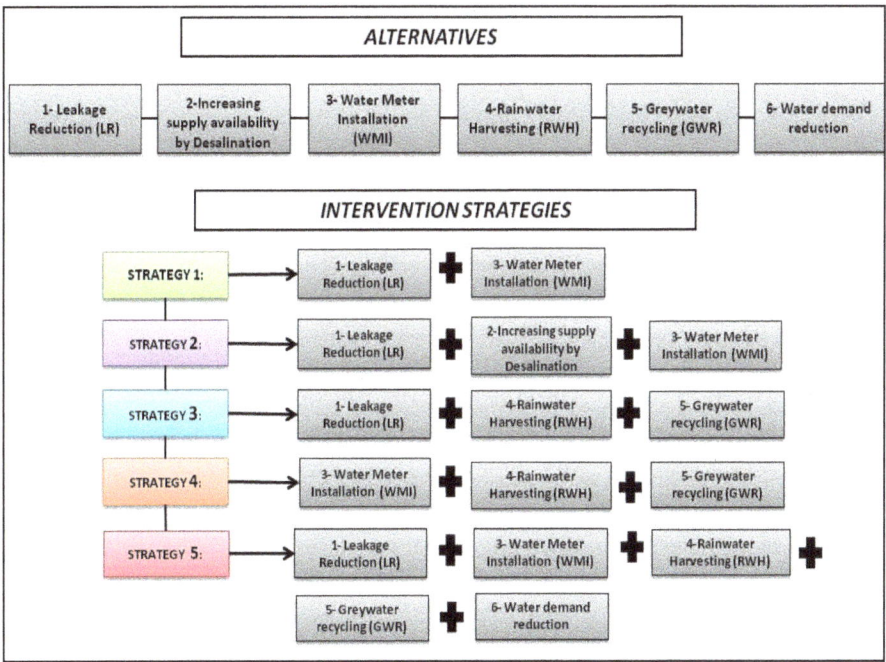

Figure 5. Intervention strategies applied to the UWS model.

Table 6. Suggested alternatives for improvement of the UWS.

Alternative	Description	Input Values	Assumptions	Total Costs [b] (EUR/m³)	Reference
(1) Leakage Reduction	Reduction from 28% [a] to 13% (1% annually).	Energy consumption: 0.66 KWh/m³ (current use of energy). The same values for all four growth scenarios	Installation of automatic and computerized leakage and control system (e.g., pressure and flow monitoring). Replacement of old pipes (17,800 m of PVC pipes).	0.66	Municipality of Santa Cruz and local providers
(2) Desalination Plant	Installation of a new SWRO desalination plant (BWRO was not considered to avoid extra pressure on the basal aquifer and increase of salinity) with energy recovery system. Open seawater intake (35,000 ppm). 55% recovery rate, 99% salt rejection.	(1) small growth (9000 m³/day) (2) moderate growth (16,000 m³/day) (3) fast growth (28,000 m³/day) (4) very fast growth (50,000 m³/day) Energy consumption [c]: 3 KWh/m³	Cost includes plant, land, civil works and amortization costs, chemicals for pre and post water treatment, energy requirement, brine dissolution and discharge, cooling towers(including electricity and steam), spares and maintenance (including membrane replacement every 5 years), and labour.	(1) 1.27, (2) 1.25, (3) 1.23, (4) 1.22	[31–33]
(3) Water Meter Installation	Installation of water meters per premise with a rate of 10% annually.	140 EUR/unit (including installation and maintenance) The same unit cost for all growth scenarios	Installation of Flodis-single jet turbine device)	0.04	Municipality of Santa Cruz
(4) Rainwater Harvesting	Installation of a household rainwater harvesting tank for indoor and/or outdoor use (2 m³)	Capacity calculated as 4000 m³ (approx. 2000 households) Energy consumption: 2 Kwh/m³	Water collected from roofs only [e]. The collected rainwater used for toilet flushing, hand and kitchen basin, showers and outdoor use. The cost includes purchase cost of tank, pumping, delivery and installation, household plumbing, and mains water switching devices, energy consumption, maintenance and pump replacement (every ten years).	0.21	[34–36]
(5) Greywater Recycling	Installation of single house on-site and decentralized greywater treatment using a submerged membrane (MBR), including disinfection unit	Based on household greywater treatment capacity of 350 L capacity and 2000 households; 5 inhabitants per household and 163 Lpcpd [d]. Flow capacity of 200 L/population equivalent	Greywater collected from kitchen, hand basins and showers, which account to approximately 48% of total water demand). Household treatment assumed with membrane bioreactor plant (biological treatment, aeration, and membrane filtration. Treated greywater used on-site for toilet flushing and outdoor use.	1.08	[37–41]
(6) Water Demand Reduction [f]	Reduction of specific demand of municipal water	Reduction from 163 lpcpd [d] to 120 lpcpd [d] (assuming 1% annual reduction on water demand starting in year 3, in order to complete the reduction at the end of the planning horizon	Assumed the change of "water tariff" structure to reduce the average specific demand	-	-

[a] This value was considered for calibration purposes [b] Total costs include investment costs, operations and managements costs, interest rate and extra costs and the municipality will assume all of them; [c] The cost of energy, as observed in the literature, ranges widely from 2 to 12 kwh/m³; however, since this would be a brand new plant, we have selected a value towards the lower side; [d] lpcpd corresponds to litres per capita per day; [e] Only runoff from roof was considered since, pavement/road run-off will incur in a significant extra cost for water treatment; [f] It is assumed that adjustment for water tariff as an intervention by the water utility can lead to water demand reduction, based on previous studies and policy from the municipality. Only capital investment and Operations & Management costs are analysed.

Table 7. Fraction of the water demand delivered.

Population Growth	Baseline	Alternative 1	Alternative 2	Alternative 3	Alternative 4	Alternative 5	Alternative 6
Slow	0.52	0.64	1.00	0.68	0.72	0.79	0.73
Moderate	0.35	0.36	1.00	0.37	0.40	0.43	0.41
Fast	0.17	0.17	0.99	0.18	0.21	0.22	0.20
Very Fast	0.10	0.11	1.00	0.11	0.16	0.13	0.12

5. Results and Discussion

The current situation (baseline) and the selected intervention strategies were analysed and evaluated with respect to a number of key performance indicators (KPI) for a 30-year planning horizon. The KPIs used here for comparison of the different selected strategies are total water demand, percentage of water demand coverage (i.e., fraction of water demand delivered), consumption per capita, energy consumption, and costs (capital and O&M), for each growth scenario.

Figure 6 shows the results of the Puerto Ayora case study. These figures portray results of year 30, since based on the population and tourism growth scenarios, it has been considered as the most critical year. Obviously, the most severe growth scenario is the very fast growth scenario, which is driven by the governmental objective to optimise tourist revenues for the country.

Based on Figure 6a, it can be inferred that the current infrastructure would not suffice for any of the population growth scenarios, since a 70% coverage of demand with supply could hardly be reached even with the slow growth scenario, and in the very fast scenario hardly 20% coverage could be achieved. This also shows that the current situation is not as perceived, since based on the volume of water supplied, the current coverage is calculated as 91%. However, the local community considers the coverage to be less [42]. Figure 6b shows that only Strategy 2, which includes desalination, will fully cover the water demand by the end of the planning horizon, for all growth scenarios. This suggests that current growth trends are exorbitant and will generate a significant local population and tourism demand of water. Nevertheless, Strategy 5 (a combination of all alternatives, except desalination) will be sufficient, but only in the slow growth case scenario, which has been the one preferred by all NGOs and conservation authorities, but highly unlikely [6].

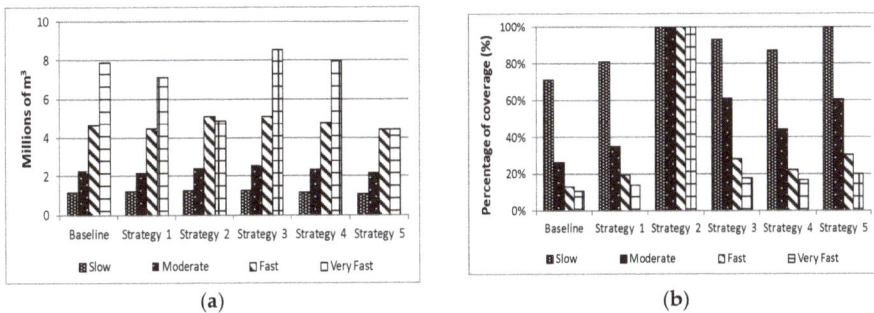

(a)　　　　　　　　　　　　　　　(b)

Figure 6. (a) Total water demand by 2030 for various growth scenarios and (b) Percentage of demand coverage by supply in 2030 for various intervention strategies in Puerto Ayora.

Strategy 2 increases water availability by installation of a new desalination plant, complying with water demand over the planning horizon. Therefore, even though the other strategies have been suggested to avoid such an investment and greater potential environmental impacts on the island, they cannot meet the demand in 30 years. The best strategy to save on total water demand, reducing pressure on the supply system and infrastructure, is Strategy 5, for all growth scenarios. This is because Strategy 5 is the only one that considers and contributes to proper demand management, reducing current specific demand by 40 L/cap/day at the end of the planning horizon. However, is still insufficient for 100% coverage of demand in year 2044 and therefore cannot cope with the future proposed growths; except in the slow growth scenario.

Regarding costs and energy use shown in Figure 7 as expected, Strategy 2 seems best, being significantly higher than the other intervention strategies. These costs refer to the total unit cost and the total water demand for the year 2044. The costs vary for each growth scenario, making the fast and very fast more unsuitable due to the enormous financial burden. This makes reference also to a higher investment (depending on the plant size calculated per growth scenario), as well as operation and management costs, implicated with a desalination plant and a much higher energy use for desalination treatment and process, compared to the other strategies, regarding this particular case study. Since this archipelago is located approximately 1000 km from the mainland, fuel for producing electricity needs to be imported from the mainland, adding extra costs to this option. Furthermore, GWR also has high costs due to pumping costs and investment per installed unit.It is assumed that all the pumps modelled here are operated and maintained by the municipality only; nevertheless, it is more environmentally friendly since it reduces wastewater and water demand. The most economical strategies are 1 and 4, which include leakage reductions, water meter installation, as well as rainwater and GWR. Even though Strategies 3 and 4 are pretty similar because they both have RWH and GWR, the influence of water meter installation is more positive and therefore the costs reduce. On the other hand, Strategy 5 has the highest costs after Strategy 3, because it includes all of the alternatives,; therefore, the costs for leakage reduction alternative (replacement of pipes and an automated leakage control system) as well as GWR are pretty significant. Regarding the energy consumption in year 30, Strategy 2 is three to four times higher than the other strategies (4 KwH/m^3). Regarding energy use, the best option is Strategy 5, because it decreases water demand. The other strategies do not seem to vary much in the energetic consumption compared to the current situation. The aggregated annual energy consumption for other strategies has minor change compared to the baseline due to low volume of water supplied by RWH or GWR compared to the total water supplied by the mains.

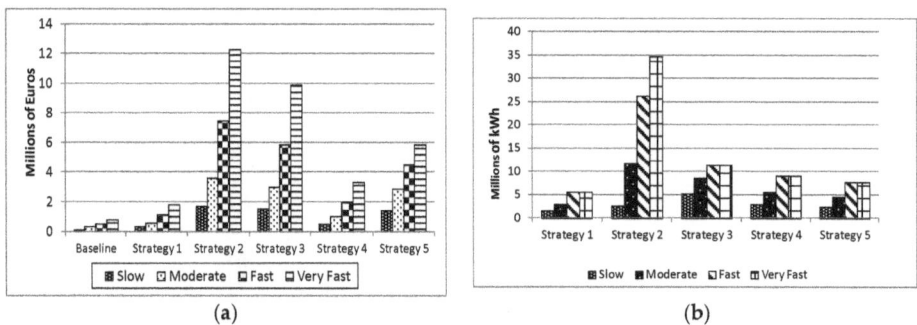

(a) (b)

Figure 7. (a) Total costs for year 30 and (b) Total energy use for year 30 for various intervention strategies and growth scenarios in Puerto Ayora.

Furthermore, Figure 8 shows the variations of water demand delivered to consumers over the planning horizon for different strategies and for the scenarios of slow and very fast growth only. As can be observed from this figure, the first peaks on both diagrams occur in year 3, when the alternatives are implemented. The rest of the peaks can be explained by the influence of meteorological data, making some years better for rainwater harvesting (and GWR) than the others. Therefore, based on historical precipitation rates, the methodology adopted predicts similar variation for the future, directly affecting rainwater collected by every individual household, making some years better than the others. Also, greywater is influenced, since in peak years, the amount of rainwater contributing to the water demand delivered is higher. Strategy 1's percentages of coverage have a decreasing tendency over the years, which is more abrupt in the very fast scenario. On the other hand, Strategy 2's percentages of coverage remain constant due to the increase in the transmission capacity and availability of water. In addition, Strategies 3–5 are influenced by rainfall levels, since rainwater is considered as an alternative in these three. The variations between these last three strategies are due to the combination of different alternatives, regarding water meter installation, leakage reduction and water demand reduction.

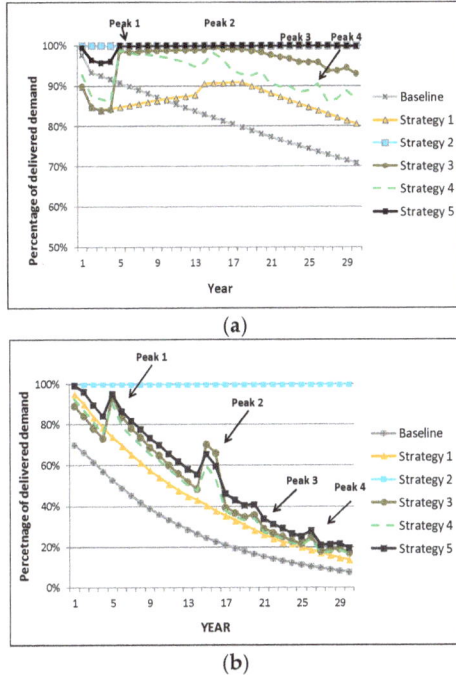

Figure 8. Coverage of demand with supply over the planning horizon for the (**a**) slow growth and (**b**) very fast growth scenarios in Puerto Ayora.

Figure 9 shows the impact and evolution of different intervention strategies regarding the calculated per capita demand. This was calculated based on the prognoses of population growth for every year and for each scenario. In the slow growth scenario, the specific demands for all strategies have more or less the same tendency to decrease at the end of the planning horizon, but not necessarily because of the reduction of consumption per capita, but due to the amount of total water divided by more people every year. The highest per capita consumption is observed for Strategies 2 and 3, which, as stated before, are the strategies that involve the increase of water availability. Moreover, Strategy 5 has a significant impact regarding the reduction of households' consumption and use, especially toward the end of planning horizon, where the per capita figure tends to decrease. In the case of very fast growth, the per capita consumption trends vary between strategies, reflecting each alternative selected and the type of population growth scenario. Unexpectedly, none of the strategies reduce per capita water consumption from the baseline scenario, but all of them will increase these figures over the years. Strategy 2 seems to increase per capita consumption 2–3 times more when compared to other strategies, but this means that this strategy allows the customers to be completely satisfied without the need to reduce it. Nevertheless, these values reduce at the end of the planning horizon and stay within reasonable margins because of the large projected population for the latest years.

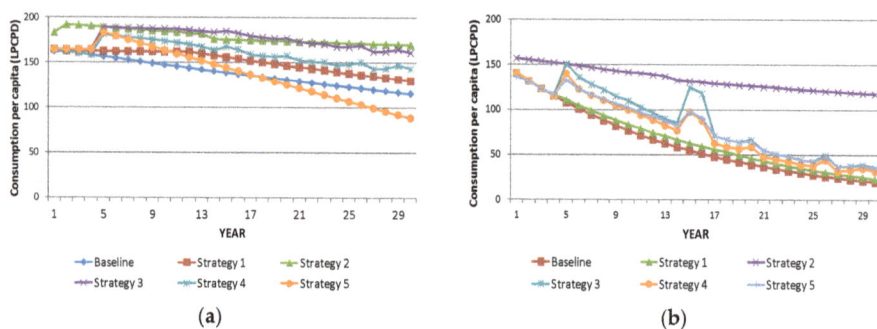

Figure 9. Consumption per capita over the planning horizon for the (**a**) slow growth and (**b**) very fast growth scenarios.

Also, current climate change impacts should be taken into account. This means that more extreme events of less or more precipitation may affect several strategies, such as the rainwater harvesting. Also, these possible events may affect even more the unbalance of the basal aquifer for the strategies that consider brackish water (strategies with leakage reduction, grey water recycling, water meter installation and per capita demand reduction). This may suggest that the quality of the water could reduce significantly, limiting its uses even more than in the current situation. Therefore, these strategies could be complemented by other individual alternatives, making them more complex. Furthermore, this suggests once again that the desalination option needs to be further considered due to the many factors already presented.

6. Conclusions

This paper addresses the issue of long-term water supply/demand balance for the main town in the Galápagos Island of Santa Cruz. To address this, a WaterMet^2 model capable of simulating the water balance over a 30-year period was built and calibrated. Five possible intervention strategies were defined by first considering six individual alternatives, aiming at either increasing supply by using alternative water sources (e.g., desalination water plant construction, rainwater harvesting and GWR) or reducing future water demand (leakage reduction, water meter installation and water demand management). The impact of these strategies on the water system performance was evaluated by using the aforementioned WaterMet^2 model with suitable KPIs and under four scenarios of population growth. The KPIs appear to be very sensitive to the population growth, actual water demand (domestic and tourist) and leakage levels, which have been estimated, based on other studies, but would need to be verified by further research.

Clearly, the current infrastructure would not be sufficient for any growth scenario, suggesting that fast and very fast scenarios are unsustainable and unaffordable. The results obtained show that the most viable strategy with respect to water demand coverage (i.e., fraction of future water demand cover by supply) in all population growth scenarios is to install a desalination plant. However, this would increase the energy consumption drastically; exerting extra pressure on the current thermal plant and implying additional fuel importation from the mainland, increasing costs and negative environmental impacts. Moreover, the disposal of brine is likely to be a potential problem and the costs of this option implicate a high investment that most likely cannot be afforded by the local municipality. As stated by the study by Dhakal et al. [43], desalination options are still the most energy intensive technology to produce drinking water and, being implemented most of the time as a last resort where conventional freshwater resources have been stretched to the limit. Furthermore, it produces considerable amounts of Green House Gas emissions if fossil energy sources are used.

None of the suggested strategies would suffice for the moderate, fast and the very fast scenarios, expect for Strategy 2 (a combination of desalination, water meter installation and leakage reduction). Because of high potential environmental impacts associated with this strategy, a more sustainable option is to apply Strategy 5 (a combination of all alternatives, except desalination. However, if the annual population and tourist growth continue as the governmental objectives suggest, this strategy would not meet the expected future water demand. Furthermore, the uncontrolled water abstraction from crevices resulting from Strategy 5 may lead to infiltration of seawater, compromising the water quality in the aquifer and making it unusable. As already identified by Pryet et al. [44], the current aquifer where La Camiseta crevice is located, has high infiltration potential, weak rainfall and probably negligible recharge. Finally, the water demand management program suggested in Strategy 5 would impact positively on the specific water consumption and hence would alleviate the need for additional supply.

A number of other analyses could and must be done in the future, which are out of the scope of this paper. A combination of other alternatives could be further investigated, such as desalination for potable water (drinking and cooking) and the use of brackish water for all other requirements, in combination with rainwater harvesting and GWR. Nevertheless, climate change could be a reason to reconsider rainwater harvesting; if the supply system becomes highly dependent on this source, it would consequently be dependent on the weather. This also implies a further calibration of the meteorological data before a decision involving rainwater can be taken with confidence. Finally, as observed in Strategy 5, the reduction of per capita water demand would have a considerable influence on the future supply/demand balance. Therefore, other water demand management strategies (e.g., using water-efficient appliances and fitting, especially inefficient toilets which account for 49% of total household demands) should be further investigated in the future researches.

The WaterMet^2 methodology has been shown to be a useful and practical software for data limited and small case studies. The current study also demonstrated that WaterMet^2 can provide a holistic approach in modelling urban water systems under current and various future scenarios.

Acknowledgments: We would like to thank Secretaría Nacional de Educación Superior, Ciencia y Tecnología (SENESCYT) for the fellowship granted. Also, we would like to thank the Department of Potable Water in Puerto Ayora-Galápagos, and special thanks to Delio Sarango for his input, time and availability for all data of the case study.

Author Contributions: Maria Fernanda Reyes collected the data, modelled it and wrote the paper. Kourosh Bezhadian contributed with the modelling, fixing of the model and writing of the paper. Nemanja Trifunovic, Saroj Sharma, Maria D Kennedy and Zoran Kapelan contributed to the writing of the paper, analysis and interpretation of results.

Conflicts of Interest: The authors declare no conflict of interest. The founding sponsors had no role in the design of the study; in the collection, analyses, or interpretation of data; in the writing of the manuscript, and in the decision to publish the results.

References

1. Epler, B. *Tourism, the Economy, Population Growth, and Conservation in Galapagos*; Report for Charles Darwin Foundation; Charles Darwin Foundation: Puerto Ayora, Ecuador, 2007.
2. Independent National Electoral (INEC). Censo de población y vivienda del Ecuador 2010. Ecuador, Instituto Nacional de Estadísticas y Censos, 2010. Available online: http://www.ecuadorencifras.gob.ec (accessed on 1 September 2015).
3. DPNG Dirección del Parque Nacional Galápagos. 2016. Available online: http://www.galapagospark.org/ (accessed on 5 February 2016).
4. González, J.A.; Montes, C.; Rodríguez, J.; Tapia, W. Rethinking the Galapagos Islands as a complex social-ecological system: Implications for conservation and management. *Ecol. Environ.* **2008**, *13*, 13. [CrossRef]
5. Pizzitutti, F.; Mena, C.F.; Walsh, S.J. Modelling tourism in the Galapagos Islands: An agent-based model approach. *J. Artif. Soc. Soc. Simul.* **2014**, *17*, 14. [CrossRef]

6. Mena, C. *Determination of Social, Environmental and Economic Relations which Allow the Development Based on Different Processes of Modelling, Potential Scenarios of Sustainability of the Socio-Ecological System of the Galápagos Islands with Emphasis on the Dynamic of the Flux of Tourist Visitors*; Report for the Ministry of Environment; the Ministry of Environment: Puerto Ayora, Ecuador, 2014; unpublished.

7. Reyes, M.; Trifunovic, N.; Sharma, S.; Kennedy, M. Data Assessment for Water Demand and Supply Balance in the Island of Santa Cruz (Galápagos Island). *Desalin. Water Treat J.* **2015**. [CrossRef]

8. Watkins, G.; Cruz, F. *Galapagos at Risk: A Socioeconomic Analysis*; Report for Charles Darwin Foundation; Charles Darwin Foundation: Puerto Ayora, Ecuador, 2007.

9. Proctor and Redfern CA. *Estudio De Provisión De Agua y Tratamiento De Aguas Residuales De Santa Cruz*; Proctor and Redfern International Limited: Santa Cruz-Galápagos, Ecuador, 2003.

10. Reyes, M.; Trifunovic, N.; Sharma, S.; Kennedy, M. Implications of Water Tariff Structure on Water Demand in Santa Cruz Island (Galapagos Archipelago). In Proceedings of the XVth World Water Congress, Edinburgh, Scotland, UK, 25–29 May 2015.

11. D'Ozouville, N.; Deffontaines, B.; Benveniste, J.; Wegmüller, U.; Violette, S.; De Marsily, G. DEM generation using ASAR (ENVISAT) for addressing the lack of freshwater ecosystems management, Santa Cruz Island, Galápagos. *Remote Sens. Environ.* **2008**, *112*, 4131–4147. [CrossRef]

12. Reyes, M.; Trifunovic, N.; d'Ozouville, N.; Sharma, S.; Kennedy, M. Quantification of urban water demand in the Island of Santa Cruz (Galápagos Archipelago). *Desalin. Water Treat.* **2017**, *64*, 1–11.

13. Behzadian, K.; Kapelan, Z.; Venkatesh, G.; Brattebø, H.; Sægrov, S.; Rozos, E.; Makropoulos, C. Quantitative UWS Performance Model: WaterMet^2, 2014, TRUST Report. Available online: https://ore.exeter.ac.uk/repository/bitstream/handle/10871/17062/d332-Final.pdf?sequence=1&isAllowed=y (accessed on 25 May 2016).

14. Mitchell, V.G.; Diaper, C. UVQ User Manual urban water balance and contaminant balance analysis tool. *CSIRO Version1* **2010**, *2*, 2005–2282.

15. Dos Santos, C.C.; Pereira Filho, A.J. Water Demand Forecasting Model for the Metropolitan Area of São Paulo, Brazil. *Water Resour. Manag.* **2014**, *28*, 4401–4414. [CrossRef]

16. Billings, R.B.; Jones, C.V. *Forecasting Urban Water Demand*; American Water Works Association: Denver, CO, USA, 2008.

17. Ajbar, A.; Ali, E. Prediction of municipal water production in touristic Mecca City in Saudi Arabia using neural networks. *J. King Saud Univ. Eng. Sci.* **2015**, *27*, 83–91. [CrossRef]

18. Soyer, R.; Roberson, J.A. *Urban Water Demand Forecasting: A Review of Methods and Models*; American Society of Civil Engineers: Reston, VA, USA, 2010.

19. Mitchell, V.; Duncan, H.; Inma, R.M.; Stewart, J.; Vieritz, A.; Holt, P.; Grant, A.; Fletcher, T.; Coleman, J.; Maheepala, S. State of the art review of integrated urban water models. *Novatech Lyon France* **2007**, *1*, 507–5014.

20. Makropoulos, C.K.; Natsis, K.; Liu, S.; Mittas, K.; Butler, D. Decision support for sustainable option selection in integrated urban water management. *Environ. Model. Softw.* **2008**, *23*, 1448–1460. [CrossRef]

21. Last, E. City Water Balance: A New Scoping Tool for Integrated Water Management Options. Ph.D. Thesis, Univeristy of Birmingham, Birmingham, UK, 2010.

22. Behzadian, K.; Kapelan, Z. Modelling metabolism based performance of an urban water system using WaterMet^2. *Resour. Conserv. Recycl.* **2015**, *99*, 84–99. [CrossRef]

23. Behzadian, K.; Kapelan, Z. Advantages of integrated and sustainability based assessment for metabolism based strategic planning of urban water systems. *Sci. Total Environ.* **2015**, *527*, 220–231. [CrossRef] [PubMed]

24. Sharma, S. *Urban Water Supply and Demand Management*; Lecture Notes; UNESCO-IHE (Institute for Water Education): Delft, The Netherlands, 2014.

25. Trifunovic, N. *Introduction to Urban Water Distribution: Unesco-IHE Lecture Note Series*; CRC Press: Boca Raton, FL, USA, 2006.

26. Reyes, M.; Trifunovic, N.; Sharma, S.; Kennedy, M. Assessment of Domestic Consumption in Puerto Ayora Intermittent Supply System (Santa Cruz Island-Galápagos). *Manuscr. Rev. J. Water Supply Res. Technol.* **2017**, under review.

27. D'Ozouville, N. Water Resource Management: The Pelican Bay Watershed; Galápagos Report 2007–2008, 2009. Available online: https://www.galapagos.org/wp-content/uploads/2012/04/biodiv11-water-resource-mgmt.pdf (accessed on 12 August 2017).

28. Domínguez, C.; Pryet, A.; Vera, G.M.; Gonzalez, A.; Chaumont, C.; Tournebize, J.; Villacis, M.; d'Ozouville, N.; Violette, S. Comparison of deep percolation rates below contrasting land covers with a joint canopy and soil model. *J. Hydrol.* **2016**, *532*, 65–79. [CrossRef]

29. Gobierno Autonomo Descentralizado Municipio de Santa Cruz. *Atlás Geográfico del Cantón Santa Cruz*; Secretaria Técnica de Planificación y Desarrollo Sustentable del Gobierno Autonomo Municipal Descentralizado de Santa Cruz: Santa Cruz- Galápagos, Ecuador, 2012; p. 50. (In Spanish)

30. Verrecht, B.; Maere, T.; Benedetti, L.; Nopens, I.; Judd, S. Model-based energy optimisation of a small-scale decentralised membrane bioreactor for urban reuse. *Water Res* **2010**, *44*, 4047–4056. [CrossRef] [PubMed]

31. Ghaffour, N.; Missimer, T.; Amy, G. Technical review and evaluation of the economics of water desalination: Current and future challenges for better water supply sustainability. *Desalination* **2013**, *309*, 197–207. [CrossRef]

32. Al-Karaghouli, A.; Kazmerski, L. Energy consumption and water production cost of conventional and renewable-energy-powered desalination processes. *Renew. Sustain. Energy Rev.* **2013**, *24*, 343–356. [CrossRef]

33. Watereuse Association Desalination Committee. Seawater Desalination Costs: White Paper, 2011. Available online: https://watereuse.org/wp-content/uploads/2015/10/WateReuse_Desal_Cost_White_Paper.pdf (accessed on 12 August 2017).

34. Lattemann, S.; Kennedy, M.D.; Schippers, J.C.; Amy, G. Global desalination situation. *Sustain. Sci. Eng.* **2010**, *2*, 7–39.

35. Tam, V.W.Y.; Tam, L.; Zeng, S.X. Cost effectiveness and tradeoff on the use of rainwater tank: An empirical study in Australian residential decision-making. *Resour. Conserv. Recycl.* **2010**, *54*, 178–186. [CrossRef]

36. Retamal, M.; Turner, A.; White, S. Energy implications of household rainwater systems. *Aust. Water Assoc.* **2009**, *38*, 70–75.

37. Hauber-Davidson, G.; Shortt, J. Energy Consumption of Domestic Rainwater Tanks. *Water J.* **2011**, *3*, 1–5.

38. Fletcher, H.; Mackley, T.; Judd, S. The cost of a package plant membrane bioreactor. *Water Res.* **2007**, *41*, 2627–2635. [CrossRef] [PubMed]

39. Boehler, M.; Joss, A.; Buetzer, S.; Holzapfel, M.; Mooser, H.; Siegrist, H. Treatment of toilet wastewater for reuse in a membrane bioreactor. *Water Sci. Technol.* **2007**, *56*, 63–70.

40. Gnirss, R.; Luedicke, C.; Vocks, M.; Lesjean, B. Design criteria for semi-central sanitation with low pressure network and membrane bioreactor—The ENREM project. *Water Sci. Technol.* **2008**, *57*, 403–410. [CrossRef] [PubMed]

41. Fountoulakis, M.; Markakis, N.; Petousi, I.; Manios, T. Single house on-site grey water treatment using a submerged membrane bioreactor for toilet flushing. *Sci. Total Environ.* **2016**, *551*, 706–711. [CrossRef] [PubMed]

42. Guyot-Tephany, J.; Grenier, C.; Orellana, D. *Uses, Perceptions and Management of Water in Galápagos*; Galapagos Report 2011-2012; GNPS, GCRG, CDF and GC: Puerto Ayora, Galápagos, Ecuador, 2013.

43. Dhakal, N.; Salinas Rodriguez, J.C.; Schippers, J.C.; Kennedy, M.D. Perspectives and challenges for desalination in developing countries. *IDA J. Desalin. Water Reuse* **2014**, *6*, 10–14. [CrossRef]

44. Pryet, A.; Dominguez, C.; Tomai, P.F.; Chaumont, C.; d'Ozouville, N.; Villacís, M.; Violette, S. Quantification of cloud water interception along the windward slope of Santa Cruz Island, Galapagos (Ecuador). *Agric. For. Meteorol.* **2012**, *161*, 94–106. [CrossRef]

MDPI

Article

Integrated Hydrological Model-Based Assessment of Stormwater Management Scenarios in Copenhagen's First Climate Resilient Neighbourhood Using the Three Point Approach

Sara Maria Lerer [1,*], Francesco Righetti [1,2] ⓘ, Thomas Rozario [1,3] ⓘ and Peter Steen Mikkelsen [1] ⓘ

1   Department of Environmental Engineering (DTU Environment), Technical University of Denmark, Bygningstorvet, Building 115, 2800 Kongens Lyngby, Denmark; psmi@env.dtu.dk
2   COWI A/S, Karvesvingen 2, 0579 Oslo, Norway; frrg@cowi.com
3   COWI A/S, Parallelvej 2, 2800 Kongens Lyngby, Denmark; troz@cowi.com
*   Correspondence: smrl@env.dtu.dk; Tel.: +45-5048-5947

Received: 14 September 2017; Accepted: 8 November 2017; Published: 12 November 2017

Abstract: The city of Copenhagen currently pursues a very ambitious plan to make the city 'cloudburst proof' within the next 30 years. The cloudburst management plan has the potential to support the city's aim to become more green, liveable, and sustainable. In this study, we assessed stormwater system designs using the Three Point Approach (3PA) as a framework, where an indicator value for each domain was calculated using state-of-the-art modelling techniques. We demonstrated the methodology on scenarios representing sequential enhancements of the cloudburst management plan for a district that has been appointed to become the first climate resilient neighbourhood in Copenhagen. The results show that if the cloudburst system is exploited to discharge runoff from selected areas that are disconnected from the combined sewer system, then the plan leads to multiple benefits. These include improved flood protection under a 100-years storm (i.e., compliance with the new demands in domain C of the 3PA), reduced surcharge to terrain under a 10-years storm (i.e., compliance with the service goal in domain B of the 3PA) and an improved yearly water balance (i.e., better performance in domain A of the 3PA).

Keywords: climate adaptation; combined sewer; cloudburst; flooding; hydraulic modelling; Three Point Approach (3PA); urban drainage

---

1. Introduction

The municipality of Copenhagen passed its first Climate Change Adaptation Plan (CCAP) in 2011 [1]. In terms of stormwater management, the plan focused on how to maintain the current level of service, i.e., assuring that the combined sewer system does not surcharge more than once every 10 years, despite a predicted 30% increase in the intensity of a 10-year storm 100 years into the future [2]. The main solution recommended by the CCAP was a disconnection of 30% of the impermeable surfaces from the combined sewer system, redirecting these surfaces to local stormwater control measures (SCMs) (also known as local diversion of stormwater or LAR in Danish [3]). The planned SCMs were primarily based on detention and infiltration, as in rain gardens and swales, thus also contributing to a "greening" of the city, but many details related to implementation have yet to be clarified.

Soon after passing the CCAP, in July 2011, a very rare and intense rainfall event caused unprecedented widespread flooding in the city, with total insurance claims exceeding 800 million Euros [4]. This increased the city's awareness of the importance of extreme events, and resulted in a new and complementary plan to the CCAP, the Cloudburst Management Plan (CMP) [5,6]. This plan set a completely new goal for the municipality: to ensure that a 100-years storm, 100 years into the future, will not cause more than 10 cm of water on the street surface anywhere in the city. Another amendment was a shift of focus from the detention of extreme events, i.e., directing the stormwater to temporary storage areas, towards solutions that convey stormwater all the way out of the city, i.e., to the surrounding harbour. The plan advocates for establishing transport solutions on the surface wherever possible, yet allowing for underground tunnels where terrain and existing infrastructure prevents surface solutions.

The Three Point Approach (3PA, see Figure 1) is a systems thinking concept that has proven useful when organizing and communicating climate adaptation plans, particularly when involving multifunctional solutions [7,8]. Using the 3PA terminology, the CMP reflects an expansion of Copenhagen's perception of the challenge of managing stormwater: there is a shift from focusing only on domain B (the domain of traditional urban drainage engineering, targeted at technical optimisation and standard regulations) to the inclusion of domain C (the domain of extreme events, where traditional drainage systems are designed to fail and city planners together with rescue forces are charged with protecting assets and lives from flooding). Both the CMP and the CCAP also reflect the city's ambition to give the new stormwater management systems added values (mainly stemming from their "greening" potential), which can be regarded as an inclusion of domain A (the everyday domain, where the focus is on seeing rainwater as a resource that can be used to enhance sustainability, liveability, etc.).

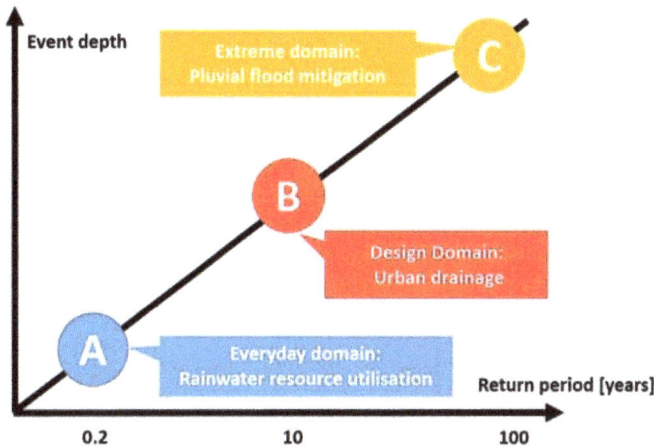

Figure 1. The Three Point Approach. Figure adapted from [8].

Copenhagen's climate adaptation plans are unique in their ambitiousness, as expressed by the many international prizes won, including i.a. the INDEX:Award in 2013 [9] and the C40 Cities Award in 2016 [10]. Our study aimed to assess the impact of the plans in a suitably ambitious manner, evaluating their performance in all three domains of the 3PA. To this end, we used advanced modelling techniques to calculate an indicator for each domain. Furthermore, we wished to investigate how modifications to the official CMP could improve the overall system performance. To this end, we developed four scenarios with gradually increasing investments. We demonstrate the methodology by application to a section of Copenhagen where some of the first climate adaptation projects have been built, and where combinations of aboveground green elements and underground tunnel systems are planned.

2. Materials and Methods

2.1. Case Study—The Ydre Østerbro Cloudburst Branch, Copenhagen

Copenhagen is the capital city of Denmark, overlooking the strait of Øresund, which connects the North Sea and the Baltic Sea. At 55 degrees north, the climate is oceanic, with an annual precipitation of about 600 mm rather evenly distributed throughout the year. The topography varies between 0 and 50 m above sea level, and has been substantially affected by human activity such as land reclamation and military defence construction. Hence, the urban hydrology is highly disturbed compared to natural conditions, and the original surface watercourses are no longer visible.

The CMP divides the city into seven overall surface runoff catchments, each containing multiple sub-catchments termed "cloudburst branches". These branches delineate the future planned flow paths of rainfall-runoff on the city surface in case of flooding due to cloudburst. The city is also served by an old combined sewer system that controls the underground flow of stormwater runoff and wastewater (the catchments formed by this system differ to varying degrees from the respective cloudburst branch delineations). The latest version of the CMP [11] uses a typology of five different measures to control the storage and flow of stormwater under cloudburst conditions. These include two measures based on conveyance (cloudburst roads/boulevards and cloudburst pipes/tunnels) and three measures based on different combinations of detention (i.e., temporary storage) and retention (i.e., hydrological losses in the form of evapotranspiration or infiltration). This CMP is still rather rough, indicating only the expected overall structure of the cloudburst management system and not detailing how each measure will be implemented.

The cloudburst branch "Ydre Østerbro", located in the north-eastern part of Copenhagen, was in 2011 declared "the first climate resilient neighbourhood" in the City of Copenhagen [12]. The ambition was to combine a pilot implementation of the CCAP with an ongoing regeneration project of the Sankt Kjelds district, which focussed on improving social well-being through improvements in housing and outdoor recreation options [13]. Some public space projects conceived in this phase have already been constructed, e.g., the Taasinge Square [14]. The latest version of the CMP for this branch is shown in Figure 2. It features a substantial number of cloudburst pipes, culminating in a large cloudburst tunnel that under extreme weather conditions conveys stormwater runoff directly to the harbour. There are also several cloudburst roads and retention roads, as well as a single retention space.

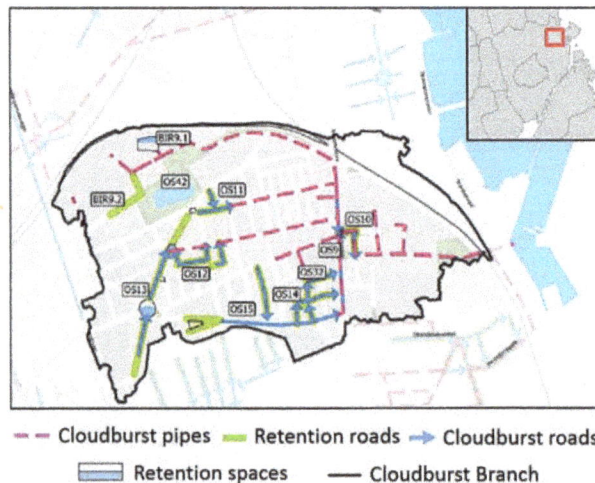

Figure 2. The cloudburst management solution for the cloudburst branch "Ydre Østerbro", adapted from Reference [11].

The cloudburst branch of Ydre Østerbro consists of approximately 118 ha of primarily residential areas, of which about 73 ha are impermeable (~62%). The branch slopes from west to east (towards the harbour), with the highest point at 12.5 m above sea level and the lowest point at 1 m above sea level. The combined sewage (i.e., stormwater runoff and wastewater) from the case study area and surrounding areas to the north and west is pumped to Lynetten wastewater treatment plant (WWTP) via a pumping station near the harbour. There are no overflow structures from the combined sewer system within the Ydre Østerbro branch, but several downstream from the outlet point of the branch.

2.2. Modelling Approach

A comprehensive 1D semi-distributed urban drainage network model representing the drainage system of the entire catchment of Lynetten WWTP, created using MIKE URBAN software (MU, from DHI, Hørsholm, Denmark [15]), served as a point of departure. It contained a total of 7454 catchments, 5618 nodes, and 5935 links for a total network length of 492 km. From this model we extracted a sub-model containing only the combined sewer system within the study area, i.e., the Ydre Østerbro cloudburst branch. The land use description in the extracted model had a high level of detail, containing 2942 catchments with a mean area of 436 m^2. The catchments were classified as roads, sidewalks, roofs facing inner yards, roofs facing the roads, green areas, railways, or "diverse paved". The final sewer network model included 413 manholes and a total pipe length of 24 km. We applied upstream boundary conditions to account for water discharged into the network from areas upstream from the study area, and downstream boundary conditions to account for flow limitations in the outlet points from the study area. We obtained the boundary conditions by running the entire original model and extracting time series of flow and water level at the points of interest.

Historical rainfall data from the rain gauge network of the Water Pollution Committee of The Society of Danish Engineers (SVK, in Danish) [16] were used as input for long-term continuous simulations. For single event analysis, we used synthetic rainfalls of 4-h duration modelled as symmetrical design storms based on the "Chicago Design Storm" (CDS) concept [17], derived from Danish regional intensity-duration-frequency curves [18] using a spreadsheet provided by the Water Pollution Committee of the Society of Danish Engineers [19]. This spreadsheet allows incorporating two types of safety factors in the CDS: a factor accounting for model uncertainty (due to e.g., lack of model calibration), which was set to 1.2, and a factor accounting for future climatic changes (climate factor), which was set to 1.3, in accordance with national guidelines [20]. The resulting storms had 10-min peak intensities of 32.58 μm/s and 55.41 μm/s for return periods of 10 years and 100 years, respectively.

For analysing flooding extent, we further developed the MU model of the study area into a MIKE FLOOD (MF) model by coupling the one-dimensional (1D) network model with a two-dimensional (2D) surface model [21]. We used the Danish national digital terrain model of 2015, which we modified to include buildings by lifting the surfaces falling within building polygons (using the national building registry). We also reduced the spatial resolution of the resulting raster from 1.6 m × 1.6 m to 3.2 m × 3.2 m in order to speed up computation time.

Table 1 describes each of the five types of cloudburst management elements in the CMP and how they were represented in our MU/MF model.

2.3. Scenario Development

Figure 3 provides an overview of the scenarios. In addition to the baseline scenario (BL) which represents the current sewer system, we developed four new scenarios (S1–S4), representing different enhancements of the cloudburst management system described in the latest CMP for the Ydre Østerbro branch. The main concept was to increase the fraction of surfaces disconnected from the combined sewer system and reconnected to the planned cloudburst system in a stepwise manner, thereby exploiting the latter also in non-extreme events. The modifications made in each scenario are explained below.

Table 1. Typology of measures used in the cloudburst management plan (CMP) [11] and how they were modelled in this study.

Solution Typology	Description	Modelling Approach
Cloudburst road	Main road that is re-profiled by making changes to terrain or raising the kerb in order to allow conveying stormwater on the surface during cloudburst events. Also referred to as cloudburst boulevard.	In the MIKE FLOOD model this was represented in the one-dimensional (1D) model by adding an open channel with the same dimensions as the road; water surcharging from the combined sewer system was directed to this channel, and if the channel reached its limit, the water was directed to the two-dimensional (2D) model.
Retention road	Road that retains and stores stormwater using elements such as bioretention units.	The sub-catchments representing these road surfaces were modified from 100% impermeable to include permeable stretches 1 m wide with an initial loss of 20 cm.
Retention space	Public or private space that is re-designed to provide some volume that can retain and store stormwater.	Same as retention roads.
Green road	Smaller road where there is space to retain all stormwater locally.	Not part of the CMP for Ydre Østerbro.
Cloudburst pipe	Underground solution for conveying and discharging stormwater mainly during cloudbursts, usually of larger dimension than a typical drainage pipe (thus also referred to as cloudburst tunnel), and applied only where existing city topography and infrastructure does not allow for conveying the necessary flows on the surface.	Normal MIKE URBAN (MU) pipe link element.

In scenario 1 (Figure 3-S1), we added a cloudburst management system to the existing combined sewer system, based on the latest CMP. Furthermore, we redirected runoff from the most obvious surfaces to the nearest cloudburst pipes, thus effectively disconnecting these areas from the combined sewer system. By most obvious surfaces, we mean the road surfaces designated as cloudburst roads or retention roads, and the road surfaces that are directly above a cloudburst pipe (the latter only if the road is classified as a low intensity traffic road, i.e., having an annual average daily traffic load of less than 5000 cars).

In this part of Copenhagen, runoff from the roofs that face the street is most often led through small gutters across the sidewalk directly to the street gutters. Therefore, the sub-catchments classified as "roof facing road" next to the abovementioned roads were redirected to the cloudburst system together with the roads. Ideally, this mixed roof and road runoff would first be treated in e.g., bioretention units before being discharged through the cloudburst pipes to the harbour. However, current guidelines in Copenhagen do not allow infiltration of road runoff to groundwater due to the risk of contamination of the groundwater with de-icing salts. Hence, bioretention units for road runoff would need to be sealed at the bottom, and their application would not affect the overall water balance (runoff would only be temporarily detained in them). Furthermore, current guidelines in Copenhagen allow for discharging runoff from low intensity traffic roads to flow into the harbour without treatment. Therefore, this scenario does not include any SCMs. The sum of areas redirected to the cloudburst system in this scenario is equivalent to 18% of the total impervious area in the cloudburst branch.

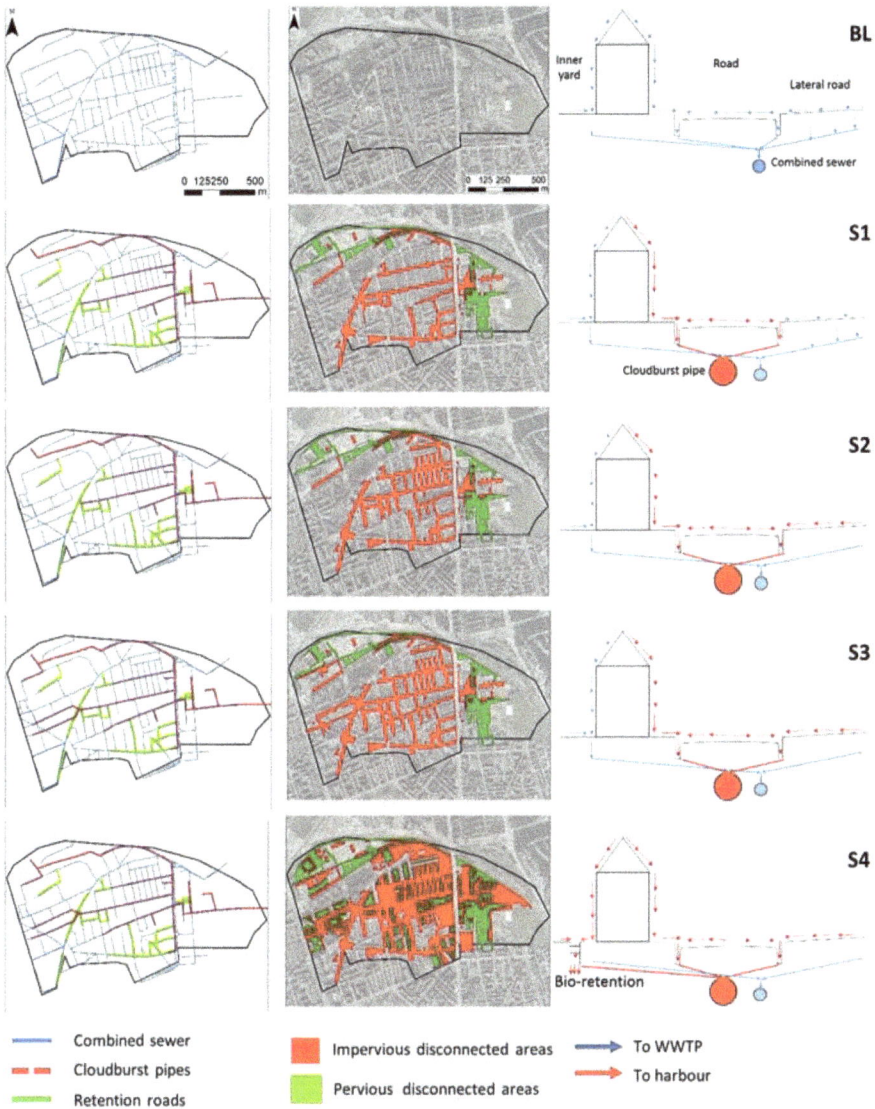

Figure 3. Overview of the different scenarios: (**left panel**) conveyance measures implemented in each scenario; (**centre panel**) areas disconnected from the combined sewer system in each scenario; (**right panel**) a typical street profile illustrating which surfaces are directed to the combined sewer system and to the cloudburst system, respectively.

Scenario 2 (Figure 3-S2) extends the fraction of surfaces redirected to the cloudburst system by adding all low intensity traffic lateral roads that slope towards any of the roads redirected in scenario 1 (including the roofs facing the street along these roads). Our rationale for adding this as an obvious next step is that this can be implemented using simple surface solutions (e.g., swales or gutters) at relatively low costs, especially if done in connection with ordinary road maintenance works (accepting a long implementation horizon). The sum of areas redirected to the cloudburst system in this scenario is equivalent to 25% of the total impervious area in the cloudburst branch.

In scenario 3 (Figure 3-S3), we extended the system of cloudburst pipes in order to address sewer surcharge problems that persisted in some parts of the cloudburst branch (according to results from scenario 2). These parts were too far away from the cloudburst pipes in the CMP. The extension of cloudburst pipes (and accompanying redirection of road- and roof-surfaces) was done gradually until a satisfactory solution was reached at an extension level corresponding to 15% of the originally planned length of pipes. The sum of areas redirected to the cloudburst system in this scenario is equivalent to 30% of the total impervious area in the cloudburst branch.

In scenario 4 (Figure 3-S4), we significantly increased the fraction of surfaces disconnected from the combined sewer system and redirected to the cloudburst system by adding all the inner yards, and roofs facing inner yards, in the buildings that are next to roads that were redirected in the previous scenarios. Runoff from the inner-yard-roofs was led through bioretention units with infiltration to groundwater, designed to manage 90% of the annual runoff volume, and modelled using the new Low Impact Development Screening module in MU [22]. The sum of areas redirected to the cloudburst system in this scenario is equivalent to 49% of the total impervious area in the cloudburst branch.

2.4. Scenario Assessment

To evaluate the performance of the different system designs with respect to domain A of the 3PA (the everyday domain), we wanted to examine the fate of all rainwater flowing through the systems. As shown earlier [8], domain A is mainly composed of rain events smaller than the typical design storm. Therefore, we ran the model in a continuous simulation mode throughout a whole year and summarised how much rainwater ended up at the pumping station (which primarily sends it onwards to the WWTP), how much was directed to the harbour via the cloudburst conveyance system, and how much evaporated or infiltrated to groundwater. We chose the year 2003 from a 37-year-long rain gauge series (1979–2015, SVK station number 5740) after ensuring this year had similar statistical properties to the entire series (in terms of mean annual precipitation and occurrence of extreme events).

To evaluate the performance of the different system designs with respect to domain B of the 3PA (the design domain) we wanted to examine how well the systems complied with the explicit service goal of limiting surcharge from the combined sewers to terrain to once every 10 years. Therefore, we ran the model with a 10-years CDS and calculated the number of manholes that surcharged during the simulation.

The baseline situation (BL) and the final system design (S4) were evaluated with respect to domain C of the 3PA, the extreme domain, by examining the extent of surface flooding. For this purpose, we ran the MF model with a 100-years CDS.

3. Results

3.1. System Performance in Domain A

Figure 4 and Table 2 illustrate the total water balance for each of the five scenarios during the selected year (year 2003). In the baseline scenario, approximately half of the yearly rainfall volume is evapotranspired or infiltrated to the groundwater, while the other half ends up at the WWTP. In reality, a small fraction of the latter overflows to the harbour before reaching the WWTP, but our model setup focusing entirely on the Ydre Østerbro cloudburst branch could not distinguish this fraction because all overflow structures are located outside the study area. This half-half division on a yearly basis is as expected, given that about 62% of the area is impermeable and the model includes initial losses (i.e., depression storage and evapotranspiration/infiltration) from the impermeable surfaces.

For each of the suggested system designs (S1–S4), an increasing fraction of the yearly rainfall ends up as direct outflow to the harbour via the cloudburst system, coupled with reduced flow to the WWTP. This is an expected result of the progressive disconnection of surface areas from the combined sewer system and their reconnection to the cloudburst system. In the most comprehensive redesign, S4, the discharge to the WWTP is reduced from 50% to 29% of the annual rainfall over the model area, i.e., a relative reduction of 42%. Such a reduction, if achieved throughout the entire catchment of the WWTP, is expected to have significant environmental and economic benefits at the WWTP due to reduced bypass, reduced pumping costs, etc. The slight increase in evapotranspiration and infiltration in scenario 4 is attributable to the implementation of bioretention units in back yards.

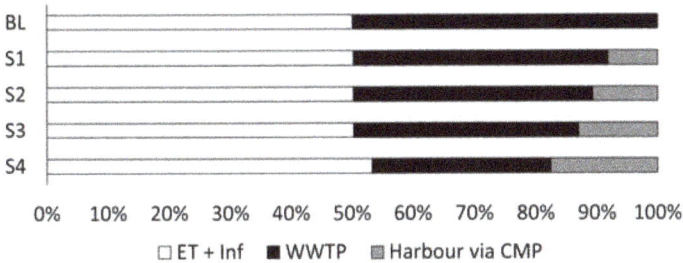

Figure 4. Total water balance for the different scenarios for one year of continuous simulation (2003); ET stands for evapotranspiration, Inf for infiltration, WWTP for wastewater treatment plant, and CMP for cloudburst management plan.

Table 2. System performance summarised for each scenario and each domain of the Three Point Approach (3PA).

	A Everyday Domain			B Design Domain			C Extreme Domain	
	ET and Inf * (%)	Harbour via CMP (%)	WWTP (%)	No. of Manholes within the Project Area	No. of Surcharging Manholes	Surcharging Manholes (%)	Water on Surface (1000 m^3)	Flooding Extent (m^2)
BL	50	0	50	413	190	46	43.1	143,349
S1	50	8	42	456	84	18		
S2	50	11	39	456	39	9		
S3	50	13	37	478	20	4		
S4	53	17	29	478	3	1	7.4	15,322

Note: * ET for evapotranspiration and Inf for infiltration.

3.2. System Performance in Domain B

In the baseline model, when forced with a 10-years CDS without climate factor, 58 of the manholes in the study area surcharge (approx. 15%, result not included in Table 2). This indicates that already today the system may not comply with the municipality's service goal. When forced with the required factors to account for model uncertainty (1.2) and future climate (1.3), 190 of the manholes (46%) surcharge (as shown in Table 2), distributed around the entire study area, with a high density along the north-south arterial road in the eastern part of the area, see Figure 5 (top, left panel). This indicates, as expected, that the capacity of the entire system is insufficient to deal with the expected increase in rainfall intensities.

The first suggested system redesign, S1 (involving the reconnection of surfaces directly affected by the construction of the officially proposed cloudburst system cf. Figure 3), achieves a significant reduction of surcharged manholes to around 18%. Further redesign reduces the proportion of surcharged manholes to 9%, 4%, and 1% for scenarios S2, S3, and S4, respectively. The few manholes that surcharge in S4, see Figure 5, S4 (top, right panel), are located in two different upstream locations in the network where the adjacent surfaces cannot easily be reconnected to the cloudburst system, indicating a need for local solutions outside the scope of this study.

Note that the upstream boundary conditions were the same for all scenarios. In reality, we also expect some retrofitting in the combined sewer system that is upstream of this area, causing a reduced flow into the study area in scenarios 1–4 compared with BL. This implies that some of the manholes might stop surcharging at a lower degree of surface reconnection than indicated by our results. Note also that the impermeable area disconnected from the combined sewer system in S3 corresponds to 30% of the impermeable area, the same degree of disconnection that was advised in the CCAP. In our simulations, this level still leaves 4% of the manholes surcharging, which is why we also included a fourth, more radical scenario (disconnecting 49% of the impermeable area). However, considering the expected effect of changing upstream boundary conditions and the need for some local solutions outside the scope of this study, our results support the recommendation of 30% disconnection level to mitigate climate change impacts.

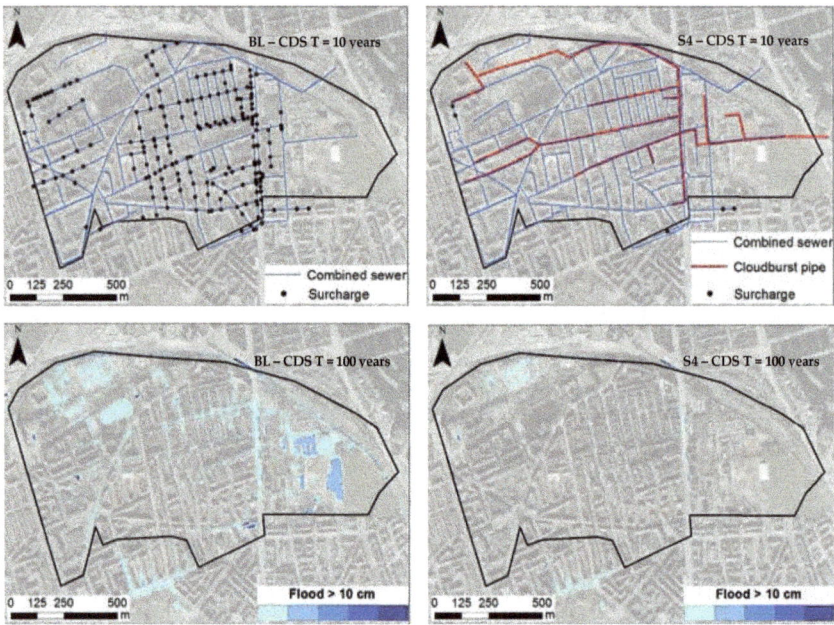

Figure 5. Extent of manhole surcharge under a 10-years Chicago Design Storm (CDS) (**upper panels**) and maximum flood levels above 10 cm under a 100-years CDS (**lower panels**) for the baseline scenario (BL, **left side**) and for the last design scenario (S4, **right side**).

3.3. System Performance in Domain C

The flooding simulation results for the baseline model (BL, current system configuration in future climate) and for the final system redesign (S4) are illustrated in Figure 5 (bottom panels) in terms of maximum depth of water on terrain in each model cell during a future 100-years CDS (showing only water depths above 10 cm). Vast inundations appear all over the study area and especially in the low-lying eastern part of the area with the given system design (BL), where more than 140,000 m^2 are covered by more than 10 cm of water (~12% of the area). In the final system design (S4), the inundated area is limited mainly to a few stretches along the north-south arterial road in the east and a bus terminal area in the north-west, totalling approx. 15,000 m^2 (~1% of the area). The persistent inundation on the arterial road reflects two main limitations of the approach taken in constructing the scenarios. First, as discussed for the domain B results, with better boundary conditions the surcharging under a 10-years storm can probably be mitigated at a reconnection level of around 30% (as in scenario S3).

At this level of reconnection, the cloudburst system is not expected to overflow. Secondly, if some reconnection of back yards still turns out to be necessary, it would be preferable (and more realistic) to make it stepwise rather than all-or-none, as was done here. The all-or-none approach serves to demonstrate the impact of an extreme scenario rather than pointing at an optimum.

4. Discussion of Modelling Approach

The modelling approaches used in this study are not new in themselves; it is the combination of several modelling approaches in a single study and how we chose to present the results, inspired by the 3PA, that is new. We find that the 3PA is a useful thinking system when dealing with the adaptation of stormwater systems to climate change, as proposed initially by Fratini et al. [7] and later used as a quantitative analytical frame by Sørup et al. [8] for simple hydrologic cases without the use of sophisticated modelling, as well as by Brudler et al. [23] for the lifecycle assessment of alternative SCMs. The 3PA inspired us to make a comprehensive assessment of the CMP that addresses all three domains, showing very few results for each domain. Thus, the 3PA helped to structure and present a very complex hydrologic and hydraulic analysis in a rather simple way, which does not reduce the complexity of the problem but supports effective communication about the solutions.

An important issue not covered by this approach is water quality, e.g., the impact of discharging untreated runoff from roofs and roads to the harbour. At first this impact may seem as a purely negative consequence, yet it must be weighed against the positive impact of reduced flow to the WWTP, which is expected to reduce combined sewer overflows and by-passes. A way to better address water quality within the framework could be to further refine the water balance in domain A, e.g., use a model that allows dividing the flow to the WWTP into how much overflows en route to the WWTP, how much is by-passed at the WWTP, and how much is effectively treated. These results could then be coupled with quantitative knowledge about the environmental impact of each fraction.

Due to the extensive modelling done in this study and the limited space in a journal paper, it was not possible to elaborate on all modelling choices we had to make. The most important choice in terms of how much uncertainty it adds to the results is probably the use of constant boundary conditions. The Ydre Østerbro branch is a separate catchment with regard to planned conveyance of cloudburst water, but the combined sewer system within the branch interacts with the combined sewer system outside the branch both upstream and downstream. Thus, climate adaptation measures implemented outside the branch will have an effect on the performance of both the combined sewer network and the cloudburst system within the branch, as discussed earlier.

Also, the compromise we chose for modelling retention roads and retention spaces may have some impact on the results. We planned to use the swale option in the new Low Impact Development (LID) module of MU 2016 for this purpose, but unfortunately this was not yet functioning. Turning a stretch of road into a pervious channel as we did misses the effect that a swale has on the runoff from the rest of the road. Thus, if an alternative modelling approach was used, it may have predicted less runoff from the retention roads ending up in the cloudburst system, and thus the desired performance in domains B and C may have been attained earlier (i.e., with less modifications to the original CMP).

Using a CDS of only 4-h duration is debatable in our models, especially for the 100-years CDS. However, given that the cloudburst system in the study area is mainly based on conveyance and not on detention or retention, the impact of larger rainfall volumes (entailed by longer duration CDS) is not considered significant.

5. Conclusions

In this study, we developed and applied an innovative and comprehensive modelling approach to assess the performance of increasingly ambitious urban stormwater system retrofits with respect to all three domains of the Three Point Approach (3PA), and we chose simple indicators to illustrate the results.

- Using a one-year long continuous simulation, we calculated the water balance and used this to indicate the system performance in domain A (the everyday domain).
- Using a simulation with a 10-years CDS, we calculated the percentage of surcharging manholes and used this to indicate the system performance in domain B (the design domain).
- Using a coupled 1D-2D simulation with a 100-years CDS, we calculated the extent of flooding and used this to indicate the system performance in domain C (the extreme domain).

Our study evaluates realistic combinations of "green" and "grey" SCMs that are partly visible on the surface and partly hidden underground, and which in combination lead to multiple benefits according to the 3PA. It furthermore shows how the 3PA, which was essentially developed for communicating and negotiating complex issues related to climate adaptation, can be combined with comprehensive state-of-the-art quantitative hydrologic engineering assessments.

The officially suggested cloudburst system for the case study area (Ydre Østerbro, Copenhagen) was, in scenario S1, enhanced by reconnecting obvious surfaces from the combined sewer system to the cloudburst system. This improved the system performance with respect to domain A (reducing the percentage of yearly rainfall on the case study area that is lead to the WWTP from 50% to 42%) and with respect to domain B (reducing the percentage of surcharging manholes from 46% to 18%). A reduction of the percentage of surcharging manholes to 1% (considered an acceptable compliance with demands in domain B) was predicted in scenario S4, which extended the cloudburst pipes with approximately 15% of the original pipe length and reconnected 49% of the study area to the cloudburst system. This final scenario also achieved a reduction of flow to the WWTP to 29% of the yearly rainfall (thus making a significant improvement in domain A), and a mere 1% of the area inundated in a 100-years storm (considered an acceptable compliance with demands in domain C). This scenario is substantially more ambitious than the current plan for the case study area, but if reconnections are implemented concurrently with other maintenance works over the next decades, it may not need to be substantially more expensive.

The extensive modelling approach suggested in this study allows for evaluating the different system designs in a structured, comprehensive manner that supports clear communication among stakeholders and decision-makers. The modelling results could be improved by including in the models the upstream areas that drain through the study area, implementing CMP-induced retrofits in the upstream areas at the same rate as in the study area. Nevertheless, however comprehensive the model, it is still limited to the hydrologic and hydraulic aspects of the system; the decision of which system design to choose will in reality be affected also by other aspects such as economics, aesthetics, biodiversity, demands for co-creation, politics, etc.

Acknowledgments: The authors would like to thank Greater Copenhagen Utility (HOFOR) for delivering data to the project. Special thanks go to Francisco Laurito Torres from DHI for his invaluable technical support. The rainfall dataset used is a product of The Water Pollution Committee of The Society of Danish Engineers made freely available for this research.

Author Contributions: Sara Maria Lerer designed the concept and methodology developed in this article and selected the case study, together with the other co-authors. Francesco Righetti and Thomas Rozario adapted the model boundaries to the defined purpose, elaborated the scenarios in detail and implemented them in the model, and ran the simulations and analysed and interpreted the results, under supervision of Sara Maria Lerer and Peter Steen Mikkelsen. Thomas Rozario and Francisco Righetti prepared the initial draft manuscript, Sara Maria Lerer wrote the final draft, with final illustrations made by Francisco Righetti and final editing by Peter Steen Mikkelsen. All authors read and approved the manuscript prior to submission.

Conflicts of Interest: The authors declare no conflict of interest.

References

1. City of Copenhagen. *Copenhagen Climate Adaptation Plan*; The Technical and Environmental Administration: Copenhagen, Denmark, 2011.

2. Gregersen, I.B.; Sunyer, M.; Madsen, H.; Funder, S.; Luchner, J.; Rosbjerg, D.; Arnbjerg-Nielsen, K. *Past, Present and Future Variations of Extreme Precipitation in Denmark: Technical Report*; DTU Environment: Kgs. Lyngby, Denmark, 2014.
3. Fletcher, T.D.; Shuster, W.; Hunt, W.F.; Ashley, R.; Butler, D.; Arthur, S.; Trowsdale, S.; Barraud, S.; Semadeni-Davies, A.; Bertrand-Krajewski, J.-L.; et al. SUDS, LID, BMPs, WSUD and more—The evolution and application of terminology surrounding urban drainage. *Urban Water J.* **2015**, *12*, 525–542. [CrossRef]
4. Arnbjerg-Nielsen, K.; Leonardsen, L.; Madsen, H. Evaluating adaptation options for urban flooding based on new high-end emission scenario regional climate model simulations. *Clim. Res.* **2015**, *64*, 73–84. [CrossRef]
5. City of Copenhagen. *Cloudburst Management Plan*; The Technical and Environmental Administration: Copenhangen, Denmark, 2012.
6. Ziersen, J.; Clauson-Kaas, J.; Rasmussen, J. The role of Greater Copenhagen Utility in implementing the city's Cloudburst Management Plan. *Water Pract. Technol.* **2017**, *12*, 338–343. [CrossRef]
7. Fratini, C.F.; Geldof, G.D.; Kluck, J.; Mikkelsen, P.S. Three Points Approach (3PA) for urban flood risk management: A tool to support climate change adaptation through transdisciplinarity and multifunctionality. *Urban Water J.* **2012**, *9*, 317–331. [CrossRef]
8. Sørup, H.J.D.; Lerer, S.M.; Arnbjerg-Nielsen, K.; Mikkelsen, P.S.; Rygaard, M. Efficiency of stormwater control measures for combined sewer retrofitting under varying rain conditions: Quantifying the Three Points Approach (3PA). *Environ. Sci. Policy* **2016**, *63*, 19–26. [CrossRef]
9. Index Danish Capital Adapts Successfully to Changing Climate. Available online: https://designtoimprovelife.dk/danish-capital-adapts-succesfully-to-changing-climate/ (accessed on 16 June 2017).
10. C40 Cities Awards 2016. Available online: http://www.c40.org/awards/2016-awards/profiles (accessed on 16 June 2017).
11. City of Copenhagen. *Climate Change Adaptation and Investment Statement*; The Technical and Environmental Administration: Copenhagen, Denmark, 2015.
12. Ydre Østerbro i København Får Klimatilpasset Kvarter. Available online: http://www.klimatilpasning.dk/aktuelt/nyheder/2012/juni2012/ydreoesterbro.aspx (accessed on 16 June 2017). (In Danish)
13. City of Copenhagen. *Områdefornyelse i Sankt Kjelds Kvarter—Et Kvarter i Bevægelse*; Områdefornyelsen Skt. Kjelds Kvarter: Copenhangen, Denmark, 2011. (In Danish)
14. City of Copenhagen. *Tåsinge Plads*; The Technical and Environmental Administration: Copenhangen, Denmark, 2015.
15. DHI. *Mike 1D—DHI Simulation Engine for 1D River and Urban Modelling—Reference Manual*; DHI: Hørsholm, Denmark, 2016.
16. Jørgensen, H.K.; Rosenørn, S.; Madsen, H.; Mikkelsen, P.S. Quality control of rain data used for urban runoff systems. *Water Sci. Technol.* **1998**, *37*, 113–120.
17. Keifer, C.J.; Chu, H.H. Synthetic storm pattern for drainage design. *J. Hydraul. Div. ASCE* **1957**, *83*, 1–25.
18. Madsen, H.; Gregersen, I.B.; Rosbjerg, D.; Arnbjerg-Nielsen, K. Regional frequency analysis of short duration rainfall extremes using gridded daily rainfall data as co-variate. *Water Sci. Technol.* **2017**, *75*, 1971–1981. [CrossRef] [PubMed]
19. Arnbjerg-Nielsen, K.; Harremoës, P.; Mikkelsen, P.S. Dissemination of regional rainfall analysis in design and analysis of urban drainage at un-gauged locations. *Water Sci. Technol.* **2002**, *45*, 69–74. [PubMed]
20. Bülow, I.G.; Madsen, H.; Linde, J.J.; Arnbjerg-Nielsen, K. *Skrift 30—Opdaterede Klimafaktorer og Dimensionsgivende Regnintensiteter*; Water Pollution Committee of the Society of Danish Engineers: Copenhagen, Denmark, 2014. (In Danish)
21. DHI. *MIKE FLOOD—1D-2D Modelling—User Manual*; DHI: Hørsholm, Denmark, 2016.
22. DHI. *MIKE URBAN Collection System*; DHI: Hørsholm, Denmark, 2016.
23. Brudler, S.; Arnbjerg-Nielsen, K.; Hauschild, M.Z.; Rygaard, M. Life cycle assessment of stormwater management in the context of climate change adaptation. *Water Res.* **2016**, *106*, 394–404. [CrossRef] [PubMed]

water

MDPI

Article

Pollution Removal Performance of Laboratory Simulations of Sydney's Street Stormwater Biofilters

James Macnamara and Chris Derry *

School of Science and Health, Western Sydney University, Penrith 2751, Australia;
16495845@student.westernsydney.edu.au
* Correspondence: c.derry@westernsydney.edu.au; Tel.: +61-2-449-631-226

Received: 13 September 2017; Accepted: 17 November 2017; Published: 22 November 2017

Abstract: The City of Sydney is constructing more than 21,000 square metres of street biofilter units (raingardens) in terms of their Decentralised Water Master Plan (DWMP), for improving the quality of stormwater runoff to Port Jackson, the Cooks River, and the historical Botany Bay. Recharge of the Botany Sand Beds aquifer, currently undergoing remediation by extraction of industrial chlorinated hydrocarbon pollutants, is also envisaged. To anticipate the pollution removal efficiency of field biofilter designs, laboratory soil-column simulations were developed by Western Sydney University partnered with the City. Synthetic stormwater containing stoichiometric amounts of high-solubility pollutant salts in deionised water was passed through 104 mm columns that were layered to simulate monophasic and biphasic field designs. Both designs met the City's improvement targets for total nitrogen (TN) and total phosphorus (TP), with >65% median removal efficiency. Prolonged release of total suspended solids (SS) on startup emphasised the need for specifications and testing of proprietary fills. Median removal efficiency for selected heavy metal ecotoxicants was >75%. The researchers suggested that Zinc be added to the targets as proxy for metals, polycyclic aromatic hydrocarbons (PAH) and oils/greases co-generated during road use. Simulation results suggested that field units will play an important role in meeting regional stormwater improvement targets.

Keywords: biofilter; raingarden; stormwater treatment; road runoff; WSUD; soil column

1. Introduction

Urban development effectively results in the short-circuiting of the hydrologic cycle, reducing stormwater retention in the terrestrial phase, and limiting natural purification by filtration, settlement, and biochemical stabilisation. Roads receive high-nutrient runoff from yards, parks, and pavements, while vehicles, road infrastructure, and adjacent roofs add metals, polycyclic aromatic hydrocarbons (PAHs), and oils/greases to the mix [1–3].

In Sydney, nutrients and metals in stormwater runoff exert a negative influence on receiving water ecosystems, and are thus the focus of ongoing research and intervention [4–6]. In 2012, the City of Sydney announced their Decentralised Water Master Plan (DWMP) as part of a water sensitive urban design (WSUD) strategy. To implement this, basic stormwater improvement targets, as shown in Table 1, were included in the City's general Development Control Plan (DCP) and Botany Bay Water Quality Improvement Plan (BBWQIP), as approved by the New South Wales (NSW) Office of Environment and Heritage [7,8]. Slight differences in the targets under these two plans relate to informed decision making regarding different environmental needs.

Table 1. Sydney's stormwater improvement targets.

Water Quality Parameter	Development Control Plan (DCP) Removal Target	Botany Bay Water Quality Improvement Plan (BBWQIP) Removal Target
Total Nitrogen (TN)	45%	45%
Total Phosphorus (TP)	65%	60%
Total Suspended Solids (SS)	85%	80%
Gross Pollutants	90% (>5 mm)	90% (>5 mm)

One initiative to implement the plans is the design and construction of 21,000 square meters of decentralised street-biofilter units to intercept and treat stormwater before discharge to receiving waters or the aquifer. A schematic of a basic monophasic biofilter design is shown in Figure 1, although the City use a number of design variations, including an unlined version where aquifer recharge is an aim. A more complex biphasic design involves the addition a 200 to 300 mm saturation sump to provide an anaerobic environment in which bacteria can carry out denitrification; the conversion of dissolved nutrient nitrogen to harmless atmospheric nitrogen in a carbon-rich environment [9]. This reduces algal and cyanobacterial (blue-green algal) nutrients in the stormwater discharged to receiving waters, with a reduction in eutrophication potential.

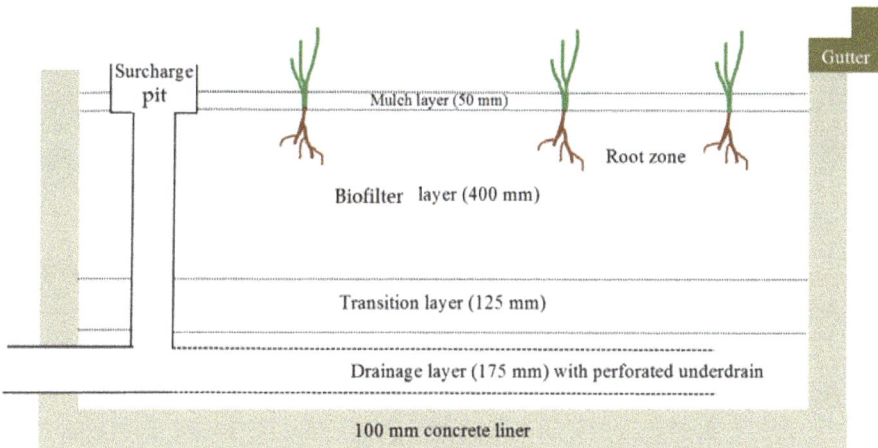

Figure 1. Schematic of a basic lined, monophasic field design showing filter layer depths.

Biofiltration units are often referred to as "raingardens" above the surface because of their ability to visually soften the streetscape with a bed of wetland plants, the active stem, and rootzone, of which maintain surface-layer permeability and provide attachment for nitrifying bacteria (Figure 2).

In 2013, The City of Sydney formed a research partnership with Western Sydney University (WSU) with the aim of conducting laboratory simulations to assess the potential pollutant removal efficiencies of the main field designs under controlled conditions, which is the topic of this paper. A further aim was to guide the development of a performance monitoring framework for the field units, which is a work in progress to be published later.

The use of 104 mm cylindrical columns facilitated a low-cost and simplified approach using off-the-shelf materials, offering potential for technology transfer to the City, and to urban administrations elsewhere. Given the growing interest in WSUD internationally, the system offered value in terms of the simplified testing of designs and of the available filtration material.

It was agreed that the metric that was applied in the simulation testing would be the regional removal targets, as shown in Table 1, with the addition of heavy metals zinc, copper, nickel, cadmium, lead, and chromium, because of their known association with road runoff and their potential ecotoxicity [10–12].

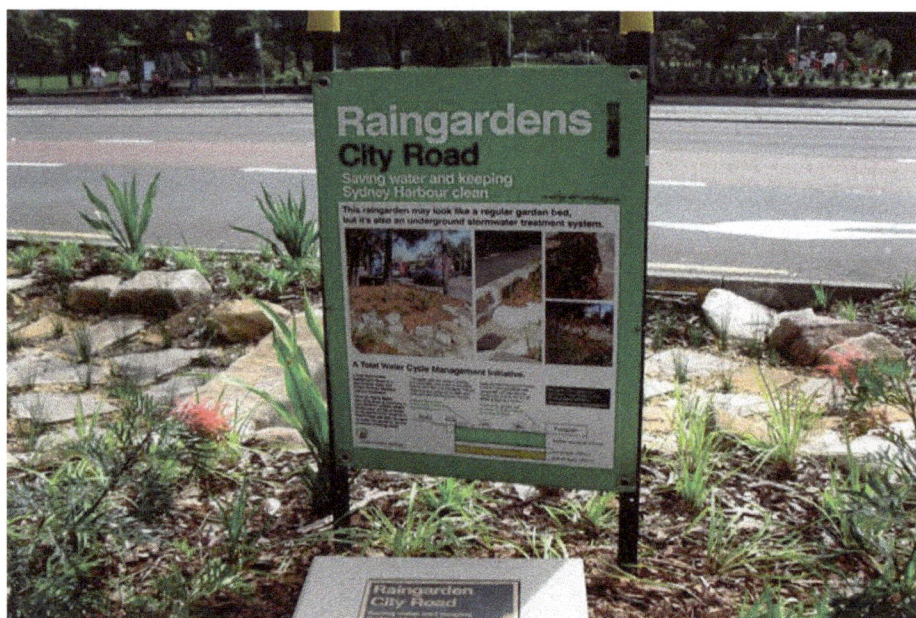

Figure 2. A City of Sydney raingarden.

Heavy metal indicators in road runoff have also been shown to exhibit strong correlation, with hydrocarbon indicators such as BTEXN (benzene, toluene, ethylbenzene, xylenes, and naphthalene), TRH (total recoverable hydrocarbons), and PAH [13,14]. This makes highly soluble heavy metals, such as zinc, good proxies for a range of co-generated pollutants in hydraulic column experiments where added non-polar hydrocarbons would coat filter particles, interfering with the removal of important polar pollutants, such as the metals and nutrients. Given these considerations, hydrocarbons were not included in the column simulation experiments, but given their variability and potential ecotoxicity, they will be included in the later field study of biofilter performance.

While total suspended solids (SS) could not be added directly to the synthetic stormwater that was fed to the soil columns for reasons discussed under Section 2.2, SS release on startup from two lots of proprietary raingarden fill material was assessed [15,16]. Measurement of gross pollutant (large flotsam) removal was impractical in terms of the size of the material in relationship to the narrow diameter of the columns.

2. Materials and Methods

2.1. Soil Columns

Soil columns that were identical in vertical section to the field biofilters were established using off-the-shelf 104 mm diameter PVC tubing, with the aim of producing a cost- and space-efficient experimental unit with a diameter that would limit potential edge effect and dispersivity in the columns.

Edge effect is the tendency for increased hydraulic flow to occur at the interface between a packed column and its retaining surface, resulting in the potential short-circuiting of the column [17,18].

This effect is likely to be amplified in narrower bore columns because of the increased retaining tube surface area to tube-diameter ratio. To minimize this effect, internal tube surfaces were lightly sanded to retard edge flow, and distribution of the synthetic stormwater took place over a mulch in the centre of the column [19].

In large diameter columns, the risk of dispersivity increases [20]. This involves the erosion of the filter medium itself through the establishment of meandering, lateral flow pathways, which later erode vertically, short-circuiting the filtration medium. Dispersivity is exacerbated by fine particles that are added to the fill to support plant growth, particularly in the presence of excess sodium ions.

All of the columns were established in tandem in terms of a common City of Sydney design where two units are linked by pipework under a central pedestrian access way leading to a safe road crossing point. Simulation of biphasic, lined field units was achieved by arranging the second column inlet 210 mm above the first column outlet with a 13 mm diameter link, simulating the inverted siphon arrangement used to maintain the saturation zone in field designs (Figure 3).

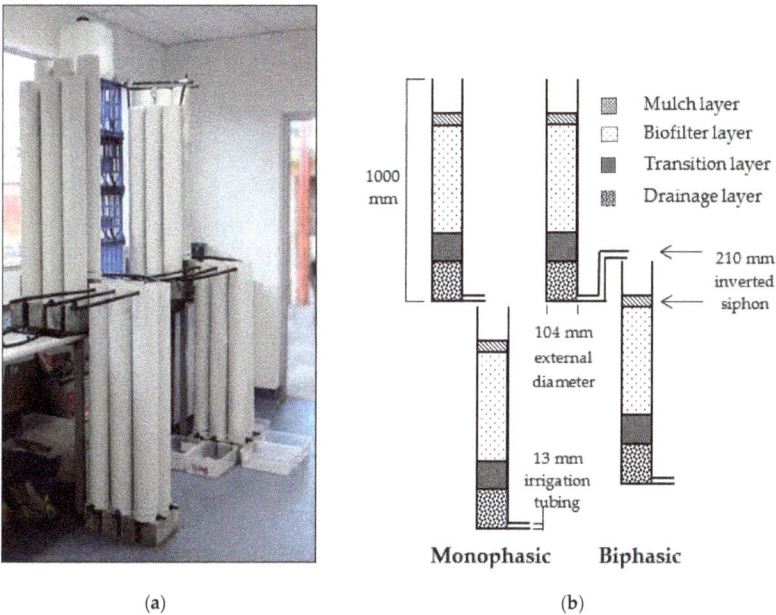

(a)   (b)

Figure 3. Laboratory columns: (**a**) view of general layout; (**b**) sectional schematic showing arrangement and layering.

While wetland plants and their related root zone are an integral part of field units, planting of the columns to represent the distribution and variety of plants found in raingardens was not feasible, given the small column diameter. This was, however, a scientifically conservative approach in that raingarden plants enhance pollution removal through biochemical activity in the rhizosphere of well-maintained field units [21]. In field settings, plant stem presence and root movement keeps surface channels open, and this was compensated for in the columns by using a distribution pebble-mulch and the avoidance of clogging material in the synthetic stormwater [22,23]. It was assumed that the necessary microorganisms for a range of biochemical processes would be present in the quarry fill media, which was commercially extracted from riverine sources [24].

During each monophasic experiment, synthetic stormwater was conducted from a hard-plastic header tank through an adjustable drip-feed system to the primary column, from which it drained to the secondary column, exiting at the column base for collection and analysis.

2.2. Synthetic Stormwater

The concentration of each pollutant in the synthetic stormwater was based on the median value that was reported for laboratory simulations elsewhere, averaged with the median value for six grab samples for first flush events at Sydney sites where field units were to be constructed [25–30]. It is important to note, that on this basis, synthetic stormwater formulation would vary for each City catchment and that some standardization is ultimately needed [31].

The synthetic stormwater was produced by dissolving stoichiometric amounts of high-solubility salts containing relevant pollutant ions in deionised water, followed by iterations of analysis and adjustment until a stable product complying with the required concentrations was achieved.

Suspended solids were not included in the synthetic stormwater because of the potential for adsorption of the positively charged ammonium and metal ions at an unknown rate onto these particles during storage, and the absence again of research standards relating to particulate size and ion-exchange properties. For this reason, the study followed the direction of a number of published studies in omitting SS from their versions of synthetic stormwater [25,27–30].

In terms of impact on results, the exclusion of SS represents a conservative scientific approach in that SS mobilized by stormwater in Sydney is known to contain a variable proportion of zeolite, which would be likely to improve removal efficiency for cations in the field units [32].

2.3. Column Performance Experimental Method

An experimental soil-column method was developed with the following phases:

1. Purging: To limit background column "noise", naturally occurring soluble ions were purged from the soil column using deionised-water flushing until the three-day median for electrical conductivity (EC), a proxy for ion content, fell to 5% of that of the synthetic stormwater. This typically took eight consecutive days for a fill that is compliant with City of Sydney biofilter material specifications, as discussed later. Post-purge draining for 16 h resulted in a median mass reduction of 70% reducing the risk of potential dilution of pollutants in the synthetic stormwater.

2. Dosing: Each primary column was dosed with 1.1 L of synthetic stormwater per hour based on the unsaturated hydraulic conductivity rate for the soil column, as determined by a Decagon Devices® mini disk portable tension infiltrometer. For each experimental run, dosing was maintained for three hours based on the dosing regime established elsewhere for effective cation capture [25,32], which complied with recorded median rainfall duration for the years 2010 to 2013 [33]. It must be noted that the dosing cycle was based principally on hydraulic and not climatological considerations given the endemically high rainfall variability on the Eastern seaboard relating to the El Nino Southern Oscillation, reducing the reliability of rainfall prediction since the year 2000 [34].

Following dosing, the columns were drained to an inert sampling container for a median interval of 16 h until all of the drainage ceased, yielding a median 2.3 L sample. Five days were then allowed to elapse, after which the dosing process was repeated, with a final repeat after five days in compliance with established hydrological principles [35–37]. This dosing process was repeated three times for both monophasic and biphasic designs, with the experiment run in tandem, giving a total of 18 results for each pollutant to provide sufficiently reliable data for analysis.

A conditioning period for the columns as applied by some researchers was considered unnecessary given that the filter material was commercially sourced from riverine quarries where hardy microbial species that were present in wetland ecosystems would be likely to be well represented. Given the organic component of the fill, it was hypothesized that there would be good retention with rapid regrowth [38].

3. Sample removal: A representative aliquot of the water drained from each column during each 19 h event was bottled for transport and analysis in compliance with Standard Methods [39].

In terms of quality control, no significant difference was noted in the level of individual pollutants taken at different stages of the sampling cycle (with the exception of copper), although a risk of dilution of "new" water with "old" water (such as purge) remaining in the biphasic columns has been reported in the literature [40]. Initial high copper release may, however, have been an artefact resulting from the leaching of nutrients and metals from organic material in the columns, offsetting the dilution of pollutants by residual purge water in initial column runs. The amount of water that was retained in the saturation sump wasshown to be <0.64 L by experimentation; offering potentially low dilution.

2.4. Laboratory Analysis, Data Capture and Statistical Analysis

Analysis of samples was carried out by Australian Laboratory Services (ALS) Environmental, in terms of a City of Sydney policy requirement that all of the environmental samples be sent to laboratories maintaining NATA accreditation. Analysis was performed on each sample for Total Nitrogen (TN) (APHA 4500 Norg/NO_3 method), Total Phosphorus (TP) (APHA 4500 P-F method), and metals cadmium (Cd), chromium (Cr), lead (Pb), nickel (Ni), copper (Cu), and zinc (Zn) (inductively-coupled plasma mass spectrometry (ICP-MS)).

The returned data were saved using Microsoft Excel® and were assessed for parametry for each pollutant by probability plotting using Minitab 18 Express with the application of Anderson-Darling testing for normality at the assumed significance limit ($p = 0.05$). After stripping outliers, data for all of the parameters except copper were found to be normally distributed enabling a two-tailed, heteroscedastic t-test to be applied to assess the difference in removal efficiency between monophasic and biphasic units. For visual examination of data spread, box-and-whisker plots were constructed using R® open source software environment for PC.

3. Results and Discussion

Median concentration removal efficiency for each of the synthetic stormwater pollutant parameters for two designs are shown in Table 2, followed by box-and-whisker plots (Figures 4–6) with a discussion of results. In the plots, the box limits containing the median (horizontal bar) represent the interquartile range Q1–Q3, and the whiskers the total range.

Table 2. Pollutant concentrations and removal efficiency by monophasic and biphasic simulations.

Pollutant Parameter	Source Compounds in Synthetic Stormwater	Pollutant Concentrationin Synthetic Stormwater (mg/L)	Pollutant Concentration (Median) after Biofiltration (mg/L)		Removal Efficiency (Median) for Biofiltration Simulation (%)	
			Mono-Phasic	Bi-Phasic	Mono-Phasic	Bi-Phasic
TN	Ammonium, nickel and lead nitrates	16.19	2.58	1.78	84.1	89.0
TP	Trisodium phosphate	10.00	2.22	3.15	77.8	68.5
Zn	Zinc chloride	0.690	0.098	0.103	85.8	85.1
Cu	Copper sulphate	0.140	0.044	0.042	68.6	70.0
Ni	Nickel nitrate	0.070	0.006	0.008	91.4	88.6
Cd	Cadmium chloride	0.013	0.0004	0.0003	96.9	97.7
Pb	Lead nitrate	0.300	0.024	0.025	92.0	91.8
Cr	Potassium chromate	0.050	0.006	0.004	88.0	92.0

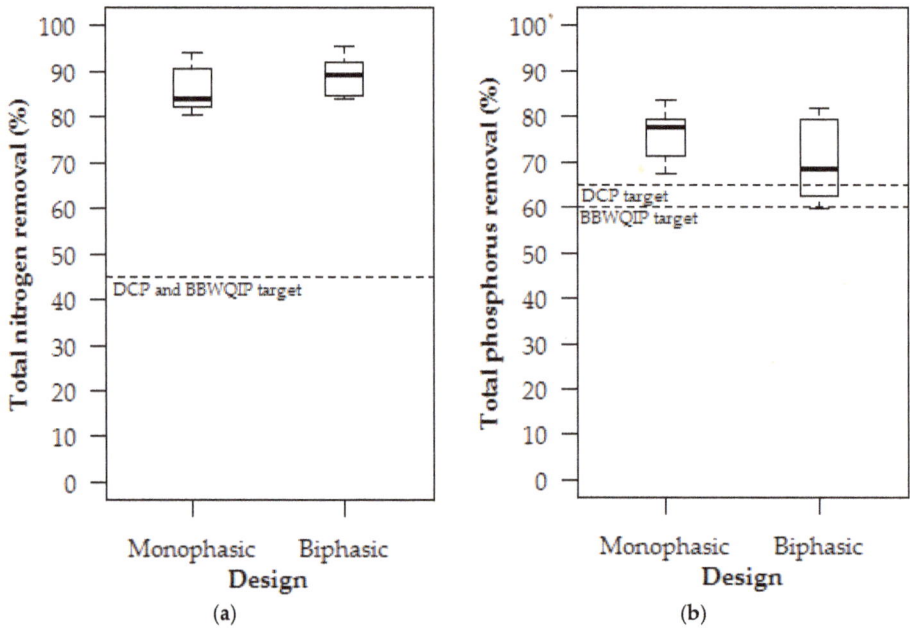

Figure 4. Median nutrient removal efficiency by design type: (**a**) total nitrogen (TN); (**b**) Total phosphorus (TP).

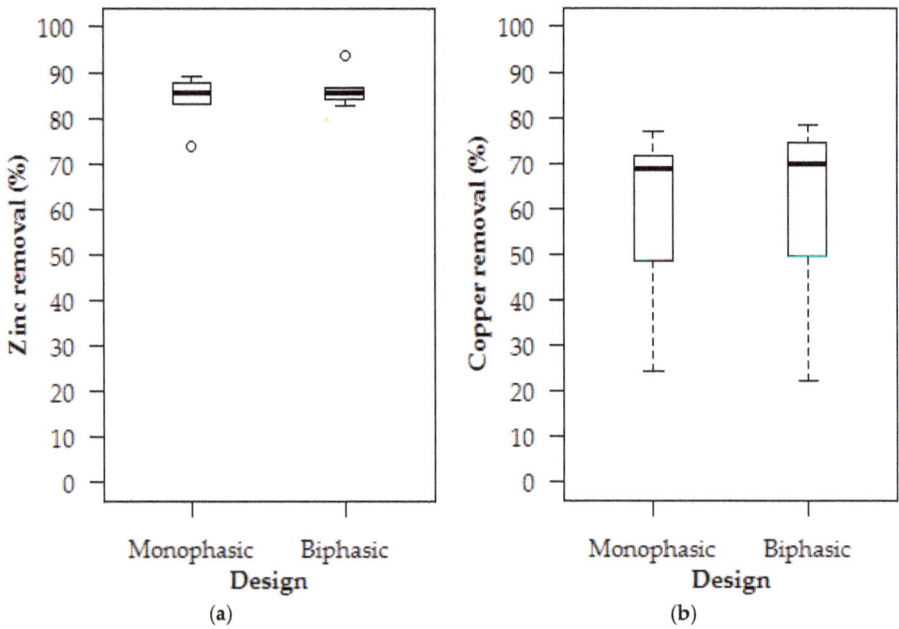

Figure 5. *Cont.*

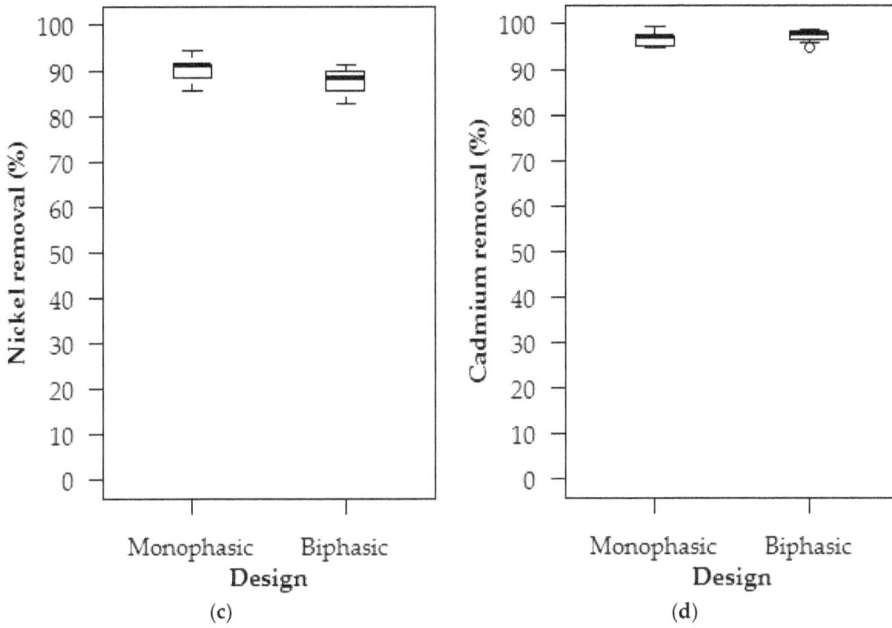

Figure 5. Dissolved-phase heavy metal removal efficiency by design type: (**a**) zinc (Zn); (**b**) copper (Cu); (**c**) nickel (Ni); and, (**d**) cadmium (Cd). "o" indicates outlier.

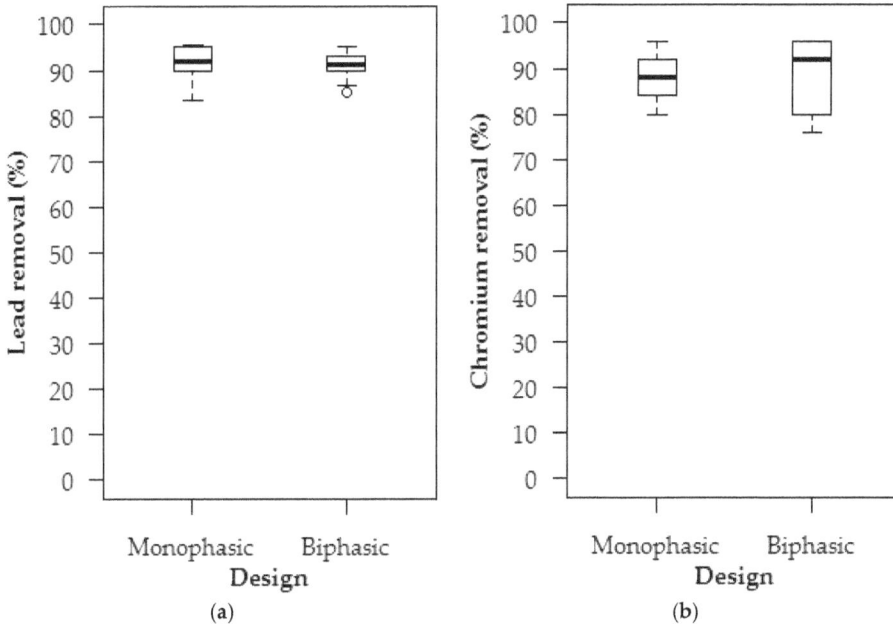

Figure 6. Suspended- or settled-phase heavy metal removal by design type: (**a**) lead; (**b**) chromium. "o" indicates outlier.

3.1. Nutrient Removal: Total Nitrogen (TN) and Total Phosphorus (TP)

The simulation columns achieved a median 84.1% and 89.0% TN removal efficiency for monophasic and biphasic designs, respectively, suggesting the achievement of the DCP and BBWQIP target of 45% to be well within reach for equivalent field units (Figure 4a). The higher median nitrogen removal efficiency for the biphasic unit suggested that nitrification-denitrification had been initiated, although this was not statistically significant in this short-term simulation. Nitrification-denitrification would be likely to be enhanced in biphasic field units with an established root zone and an elevated dissolved carbon input from road runoff [41,42].

A comprehensive biphasic study elsewhere using rectangular soil columns of a larger cross-sectional area (460 × 610 mm) achieved very similar TN removal efficiency (82%), suggesting the validity of using the economical, narrower-bore design [43].

For TP the median removal efficiencies for the two designs were 77.8% and 68.5%, respectively, again both complying with the City's DCP and BBWQIP targets (Figure 4b). While the full range of experimental results for the monophasic experiments complied with both targets (range = 67.6% to 83.8%), some of the biphasic results in the interquartile range lay outside of the DCP target. It is possible that this target is too stringent for onsite systems, being based on the performance of centralised water treatment works where chemical coagulation or orthophosphate uptake by biomass ensures a high level of phosphate removal [44]. Further research into the full field-treatment train involving bilfiltration, solids removal through terrestrial settlement and mechanical street sweeping is, however, needed before changing the guidelines.

While biphasic removal efficiency was very similar to that achieved elsewhere by larger rectangular columns (66%), it was distinctly better than that achieved elsewhere for small-diameter (64 mm) columns (58% median, 47% minimum) [45]. It is possible that where the column diameter is reduced below a critical level, factors such as edge effect and biofilm presence exert undue influence on removal efficiency.

Of interest was the increased median removal efficiency of TN, but not of TP, in biphasic as compared to monophasic units. This supports the existence of nitrification-denitrification in the laboratory monophasic units, given that there is no biochemical equivalent for phosphate removal under anaerobic conditions [38,46].

3.2. Heavy Metal Removal

The ecotoxicity of metals in stormwater relates in part to their field partitioning in the dissolved, suspended, or settled phase [47]. Predominance in the dissolved phase makes them readily bioavailable and hence more ecotoxic than those in suspended (particulate) or settled (sediment) phases. It must be pointed out, however, that metals readily cross these boundaries depending on the redox or pH status of the aquatic environment [48]. In making up the synthetic stormwater, metal salts of high solubility were used to ensure availability in the columns.

3.2.1. Dissolved-Phase Heavy Metals

Zinc, copper, nickel, and cadmium are the most important dissolved ecotoxicants in stormwater, which, through their bioavailability and bioaccumulation in aquatic and marine food chains, may result in reduction of biodiversity and ultimately the contamination of human water and food supplies [49,50].

Zinc and copper have been shown to exhibit the highest intrinsic ecotoxicity in biomonitoring experiments using representative macroinvertebrates, fish and algae, as based on a number of runoff stages during rainfall events [51]. In terms of intrinsic toxicity (LD_{50}), zinc is not as poisonous as the other metals in the dissolved-phase, but occurs in much higher concentrations in road runoff than other heavy metals, making it the most important ecohazard [52]. It is likely to be released by a wide range of road infrastructural sources, including galvanized crash barriers (undergoing frequent gouging impact), fences, traffic light gantries, and street lighting poles, as well as vehicular sources, including car bodies, automotive components, and recycled oil. It is an important vulcanising agent in tyres, with deposits being left on the road in rubber globules causing prolonged release, and is leached by soft rainwater from roofs, piping, and other infrastructure in urban areas [2,51].

Given these factors, zinc deserves a high profile in research relating to urban stormwater contamination, whereas copper might be a preferred indicator for atmospheric-mediated stormwater pollution, particularly where smelting of associated heavy metals occurs. Although chromium occurs at relatively low levels within this grouping, the reputation of hexavalent forms as human carcinogens makes it highly relevant in certain industrial settings [52].

In the simulations, median monophasic and biphasic removal efficiencies for metals were impressive, as shown in Figure 5 as read with Table 2. There was, however, no statistically significant difference between median monophasic and biphasic removals. The large interquartile and total ranges for copper may have resulted from copper leaching from biofiltration media in the early part of the experiment based on its known redox-dependent attachment with organic material [53].

The median monophasic removal efficiency results for zinc and copper were extremely similar to results for the field-scale biofiltration simulations that were carried out elsewhere in Australia (84% and 67%, respectively); although results are not available in the publication for nickel and cadmium [54].

The current simulation study suggests that the addition of zinc to stormwater improvement targets, as a heavy metal indicator with particular relevance to road runoff, is necessary because of its ecotoxicity and ubiquity. It is not only an animal ecotoxin, but it is highly toxic to plants, including the wetland species that are relied upon to secure hydraulic infiltration and pollutant removal in the raingardens themselves [55]. In this regard, a joint WSU-City of Sydney report implicated the accumulation of zinc in the City's oldest raingardens as being the likely cause of observed leaf stress and dieback of plant cover after about 13 years of street stormwater treatment [8]. Research for identifying the locus of zinc capture in field unit filters would inform strategies for improving the service life, such as the skimming and removal of the surface layer or root zone, or the addition of zeolite or recycled blast furnace slag (BFS) to surface layers [56].

3.2.2. Suspended- or Settled-Phase Heavy Metals

Lead and chromium salts on road surfaces are mainly particle-bound, so SS removal by raingardens may be an important strategy in avoiding accumulation of these substances in receiving water sediments [57]. Tetraethyl lead was liberally used (up to 0.15 g/L) as an anti-knock ingredient in Australian petrol from 1921 to 2002, and chromium release still occurs from wear and tear to automotive steel, chromium plating, paints, corrosion coatings, brake linings, and catalytic converters. Certain industrial plating processes act as a potential source of carcinogenic hexavalent chromium compounds, and these could also be environmentally generated through the conversion of the non-carcinogenic trivalent form [52].

In the column simulations impressive median monophasic and biphasic removal efficiencies were again obtained, as shown in Figure 6 as read with Table 2.This represented a reduction of hexavalent chromium in the synthetic stormwater to <10 μg L^{-1}, which, in a field setting, would increase the percentage of aquatic species protected from 80% to 90%, were treatment of 100% of receiving water to be achieved [58].

Given the potential for the sudden release of metals from sediments that are subjected to storm scouring or pH change, and the potential threat to biodiversity in effected rivers and dams, field research relating to suspended- or settled-phase metals is indicated.

3.3. Total Suspended Solids (SS) Release

While the addition of suspended particulate matter to the simulated stormwater was not considered to be viable for the reasons discussed, the problem of sometimes prolonged and variable SS release by raingarden filter media on start-up required investigation [59,60].

Two samples of proprietary filter media (fills) intended for City raingardens were put through purging in 104 mm monophasic and biphasic columns and the contemporary SS data for each type was pooled. Fill 1 was purchased prior to the implementation of Council's physical-sizing specifications, as shown in Table 3, while Fill 2 was purchased after after implementation.

Table 3. City of Sydney specifications for biofilter material (fill).

Biofilter Layer and Liner	Specification (Based on FAWB Guidelines [61])
Mulch	Washed aggregate, 10 mm to 20 mm grade
Filtration	Sandy loam mix, with saturate hydraulic conductivity > 100 mm/h to 300 mm/h. Total clay and silt content < 3%. Organic content < 5%
Transitional	Washed, recycled glass-sand, or coarse washed river sand with little or no fines
Drainage	No-fines drainage gravel, 2 mm to 5 mm grade
Outer liner	Concrete

For both of the fills, the final SS level that was achieved was significantly different from the original level ($p < 0.01$), but nevertheless remained above the ecological trigger value of <50 NTU (equivalent here to an SS of 78.3) within a week of startup (Figure 7) [58]. Of particular concern was the persistence and variation in SS release for Fill 1 as compared to Fill 2, suggesting that even if low absolute SS levels were achieved at startup, the matrix might remain unstable. A basic soil examination of the fills suggested that instability might have resulted from low particulate traction based on the high smooth-to-sharp fines ratio in the mix, as a result of the inclusion of river-dredged sand.

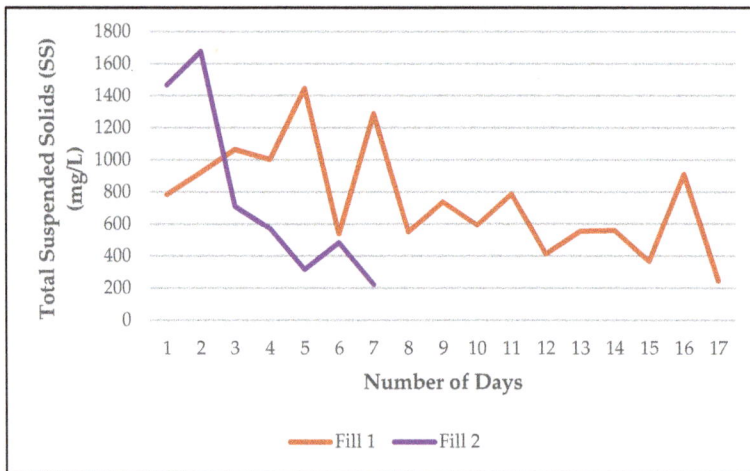

Figure 7. Mean total suspended solids (SS) release from two fills during daily purging.

While the table provides a good starting point for fill material selection based largely on physical considerations, there are also chemical properties to be taken into account, zeolite content enhancing cation retention, sodicity aggravating clogging, and compost introducing unknown chemical factors [62]. Given the difficulty of establishing a metric for chemical content, the authors suggest that it would be expedient to carry out empirical performance testing on all proprietary fills that are intended for large-scale projects, using the cost-effective methods that are described in this paper.

4. Conclusions

The laboratory simulation provided a low-cost, feasible approach for assessing the City of Sydney's biofiltration designs and intended fill material in terms of a limited but locally-relevant set of monitoring parameters. This approach is currently being extended to an additional six local authorities in the Sydney Metropolitan region where further field biofilters are being constructed in terms of the Cook's River Alliance initiative.

Removal of TN by the simulation units was found to be highly efficient, with the results of all test runs showing compliance with the City's targets. The removal of TP by the units was less efficient with a few biphasic test runs failing to comply, although the median results for both monophasic and biphasic designs showed compliance. It is possible, however, that the target for onsite TP removal has been set too high, but this will only be confirmed following research of field units and the treatment train of which they are an integral part.

More than 75% of heavy metals were removed with the exception of copper, which may have been leached from organic matter in the fill material, as recorded elsewhere. Of particular relevance was the removal of more than 80% of zinc as an important metallic ecotoxicant in terms of the large quantity generated and its intrinsic ecotoxicity. Given its potential use as proxy for heavy metals, PAHs and oils/greases that are co-generated during road use, the authors have suggested that zinc be added to existing improvement targets, which in NSW typically identify nutrients but not toxicants.

Removal efficiencies for nutrients and metals by the simulations are likely to have been conservative in terms of that which might be obtained in the field, using fully functional planted units within a treatment train that includes mechanical street sweeping and passive stormwater retention structures.

The research supported concerns that excessive and variable SS release from biofilters occurs on startup, but suggested that this can be limited in duration through the rigorous application of basic physical specifications for fills, backed up with low-cost simulation testing, possibly as an in-house activity.

The use of street biofilters remains an important strategy in achieving stormwater quality improvement, with simplified laboratory simulations providing a means for anticipating the potential performance of field designs and quality of intended fill material.

Acknowledgments: The authors wish to acknowledge the funding of the project in terms of a Western Sydney University/City of Sydney research partnership grant.

Author Contributions: Both authors designed the study. James Macnamara designed, built and operated the simulation columns, compounded and sampled the synthetic stormwater, and analysed and presented the data; Chris Derry supervised the research, advised on interpretation of the data and wrote the paper.

Conflicts of Interest: The authors declare no conflict of interest.

References

1. Todeschini, S. Hydrologic and environmental impacts of imperviousness in an industrial catchment of northern Italy. *J. Hydrol. Eng.* **2016**, *21*, 05016013. [CrossRef]
2. Yuan, Q.; Guerra, H.B.; Kim, Y. An investigation of the relationships between rainfall conditions and pollutant wash-off from the paved road. *Water* **2017**, *9*, 232. [CrossRef]
3. Al Ali, S.; Debade, X.; Chebbo, G.; Béchet, B.; Bonhomme, C. Contribution of atmospheric dry deposition to stormwater loads for PAHs and trace metals in a small and highly trafficked urban road catchment. *Environ. Sci. Pollut. Res.* **2017**, 1–16. [CrossRef] [PubMed]
4. Spooner, D.R.; Maher, W.A.; Otway, N. Trace metal concentrations in sediments and oysters of Botany Bay, NSW, Australia. *Arch. Environ. Contam. Toxicol.* **2003**, *45*, 92–101. [CrossRef] [PubMed]
5. Birch, G.F.; Cruickshank, B.; Davis, B. Modelling nutrient loads to Sydney estuary (Australia). *Environ. Monit. Assess.* **2010**, *167*, 333–348. [CrossRef] [PubMed]

6.   Nath, B.; Chaudhuri, P.; Birch, G. Assessment of biotic response to heavy metal contamination in Avicennia marina mangrove ecosystems in Sydney Estuary, Australia. *Ecotoxicol. Environ. Saf.* **2014**, *107*, 284–290. [CrossRef] [PubMed]

7.   City of Sydney Council. *Decentralised Water Master Plan 2012–2030*; City of Sydney Council: Sydney, Australia, 2012.

8.   Derry, C.; Macnamara, J.W.; Davis, K. *A Performance Monitoring Framework for Raingardens as Decentralised Stormwater Treatment Systems in the City of Sydney*; University of Western Sydney: Richmond, Australia, 2013.

9.   Zinger, Y.; Blecken, G.-T.; Fletcher, T.D.; Viklander, M.; Deletić, A. Optimising nitrogen removal in existing stormwater biofilters: Benefits and tradeoffs of a retrofitted saturated zone. *Ecol. Eng.* **2013**, *51*, 75–82. [CrossRef]

10.   Maltby, L.; Forrow, D.M.; Boxall, A.; Calow, P.; Betton, C.I. The effects of motorway runoff on freshwater ecosystems: 1. Field study. *Environ. Toxicol. Chem.* **1995**, *14*, 1079–1092. [CrossRef]

11.   Gillis, P.L. Cumulative impacts of urban runoff and municipal wastewater effluents on wild freshwater mussels (lasmigona costata). *Sci. Total Environ.* **2012**, *431*, 348–356. [CrossRef] [PubMed]

12.   Zhang, J.; Hua, P.; Krebs, P. Influences of land use and antecedent dry-weather period on pollution level and ecological risk of heavy metals in road-deposited sediment. *Environ. Pollut.* **2017**, *228*, 158–168. [CrossRef] [PubMed]

13.   Lau, S.-L.; Han, Y.; Kang, J.-H.; Kayhanian, M.; Stenstrom, M.K. Characteristics of highway stormwater runoff in los angeles: Metals and polycyclic aromatic hydrocarbons. *Water Environ. Res.* **2009**, *81*, 308–318. [CrossRef] [PubMed]

14.   Markiewicz, A.; Björklund, K.; Eriksson, E.; Kalmykova, Y.; Strömvall, A.-M.; Siopi, A. Emissions of organic pollutants from traffic and roads: Priority pollutants selection and substance flow analysis. *Sci. Total Environ.* **2017**, *580*, 1162–1174. [CrossRef] [PubMed]

15.   Gunawardana, C.; Goonetilleke, A.; Egodawatta, P.; Dawes, L.; Kokot, S. Role of solids in heavy metals buildup on urban road surfaces. *J. Environ. Eng.* **2012**, *138*, 490–498. [CrossRef]

16.   Liu, X.M.; Li, H.; Li, R.; Tian, R.; Hou, J. A new model for cation exchange equilibrium considering the electrostatic field of charged particles. *J. Soils Sediments* **2012**, *12*, 1019–1029. [CrossRef]

17.   Cameron, K.C.; Harrison, D.F.; Smith, N.P.; McLay, D.A. A method to prevent edge-flow in undisturbed soil cores and lysimeters. *Aust. J. Soil Res.* **1990**, *28*, 879–886. [CrossRef]

18.   Brown, R.A.; Hunt, W.F. Impacts of media depth on effluent water quality and hydrologic performance of undersized bioretention cells. *J. Irrig. Drain. Eng.* **2011**, *137*, 132–143. [CrossRef]

19.   Corwin, D.L. Evaluation of a simple lysimeter-design modification to minimize sidewall flow. *J. Contam. Hydrol.* **2000**, *42*, 35–49. [CrossRef]

20.   Lewis, J.; Sjöstrom, J. Optimizing the experimental design of soil columns in saturated and unsaturated transport experiments. *J. Contam. Hydrol.* **2010**, *115*, 1–13. [CrossRef] [PubMed]

21.   Gude, V.G.; Truax, D.D.; Magbanua, B.S. Natural treatment and onsite processes. *Water Environ. Res.* **2013**, *85*, 1232–1261. [CrossRef]

22.   Read, J.; Fletcher, T.D.; Wevill, T.; Deletic, A. Plant traits that enhance pollutant removal from stormwater in biofiltration systems. *Int. J. Phytoremediat.* **2010**, *12*, 34–53. [CrossRef]

23.   Virahsawmy, H.K.; Stewardson, M.J.; Vietz, G.; Fletcher, T.D. Factors that affect the hydraulic performance of raingardens: Implications for design and maintenance. *Water Sci. Technol.* **2014**, *69*, 982–988. [CrossRef] [PubMed]

24.   Hurst, C.J.; Crawford, R.L.; Garland, J.L.; Lipson, D.A. *Manual of Environmental Microbiology*, 3rd ed.; American Society for Microbiology Press: Washington, DC, USA, 2007.

25.   Yang, H.; McCoy, E.L.; Grewal, P.S.; Dick, W.A. Dissolved nutrients and atrazine removal by column-scale monophasic and biphasic rain garden model systems. *Chemosphere* **2010**, *80*, 929–934. [CrossRef] [PubMed]

26.   Read, J.; Wevill, T.; Fletcher, T.; Deletic, A. Variation among plant species in pollutant removal from stormwater in biofiltration systems. *Water Res.* **2008**, *42*, 893–902. [CrossRef] [PubMed]

27.   Wan, Z.; Li, T.; Shi, Z. A layered bioretention system for inhibiting nitrate and organic matters leaching. *Ecol. Eng.* **2017**, *107*, 233–238. [CrossRef]

28.   Davis, A.P.; Shokouhian, M.; Sharma, H.; Minami, C. Water quality improvement through bioretention media: Nitrogen and phosphorus removal. *Water Environ. Res.* **2006**, *78*, 284–293. [CrossRef] [PubMed]

29. Sun, X.; Davis, P. Heavy metal fates in laboratory bioretention systems. *Chemosphere* **2007**, *66*, 1601–1609. [CrossRef] [PubMed]

30. Henderson, C.; Greenway, M.; Phillips, I. Removal of dissolved nitrogen, phosphorus and carbon from stormwater by biofiltration mesocosms. *Water Sci. Technol.* **2007**, *55*, 183–191. [CrossRef] [PubMed]

31. Davis, A.P.; Shokouhian, M.; Minami, C.; Winogradoff, D. Water quality improvement through bioretention: Lead, copper, and zinc removal. *Water Environ. Res.* **2003**, *75*, 73–82. [CrossRef] [PubMed]

32. Birch, G.F.; Fazeli, M.S.; Matthai, C. Efficiency of an infiltration basin in removing contaminants from urban stormwater. *Environ. Monit. Assess.* **2005**, *101*, 23–38. [PubMed]

33. Bureau of Meteorology. *Daily Rainfall Data for Observatory Hill, Sydney Town, Period 3 February 2010 to 29 January 2013*; Bureau of Meteorology: Melbourne, Australia, 2015.

34. Attwater, R.; Anderson, L.; Derry, C. Agricultural risk management of a peri-urban water recycling scheme to meet mixed land-use needs. *Agric. Water Manag.* **2016**, *176*, 266–269. [CrossRef]

35. Rusciano, G.M.; Obropta, C.C. Bioretention Column Study: Fecal Coliform and Total Suspended Solids Reductions. *Am. Soc. Agric. Biol. Eng.* **2007**, *50*, 1261–1269. [CrossRef]

36. Davis, A.P.; Shokouhian, M.; Sharma, H.; Minami, C. Laboratory study of biological retention for urban stormwater management. *Water Environ. Res.* **2001**, *73*, 5–14. [CrossRef] [PubMed]

37. Vanderlinden, K.; Giráldez, J.V. Field Water Capacity. In *Encyclo of Agrophys*; Gliński, J., Horabik, J., Lipiec, J., Eds.; Springer: Dordrecht, The Netherlands, 2011.

38. Pelissari, C.; Guivernau, M.; Viñas, M.; de Souza, S.S.; Sezerino, P.H.; Ávila, C.; García, J. Unraveling the active microbial populations involved in nitrogen utilization in a vertical subsurface flow constructed wetland treating urban wastewater. *Sci. Total Environ.* **2017**, *584–585*, 642–650. [CrossRef] [PubMed]

39. American Public Health Association; American Water Works Association; Water Environment Federation. *Standard Methods for the Examination of Water and Wastewater*; American Public Health Association: Washington, DC, USA, 2005; Volume 2.

40. Payne, E.G.I.; Pham, T.; Cook, P.L.M.; Fletcher, T.D.; Hatt, B.E.; Deletic, A. Biofilter design for effective nitrogen removal from stormwater—Influence of plant species, inflow hydrology and use of a saturated zone. *Water Sci. Technol.* **2014**, *69*, 1312–1319. [CrossRef] [PubMed]

41. Björklund, K.; Li, L. Removal of organic contaminants in bioretention medium amended with activated carbon from sewage sludge. *Environ. Sci. Pollut. Res.* **2017**, *24*, 19167–19180. [CrossRef] [PubMed]

42. Stouthamer, A.H.; de Boer, A.P.N.; van der Oost, J.; van Spanning, R.J.M. Emerging principles of inorganic nitrogen metabolism in paracoccus denitrificans and related bacteria. *Anton. Leeuwenhoek* **1997**, *71*, 33–41. [CrossRef]

43. Ergas, S.; Sengupta, S.; Siegel, R.; Pandot, A.; Yao, Y.; Yuan, X. Performance of nitrogen-removing bioretention systems for control of agricultural runoff. *J. Environ. Eng.* **2010**, *136*, 1105–1112. [CrossRef]

44. Riffat, R. *Fundamentals of Wastewater Treatment and Engineering*; International Water Association CRC Press: London, UK, 2012.

45. Hsieh, C.-H.; Davis, A.P.; Needelman, B.A. Bioretention column studies of phosphorus removal from urban stormwater runoff. *Water Environ. Res.* **2007**, *79*, 177–184. [CrossRef] [PubMed]

46. Le Fevre, G.H.; Paus, K.H.; Natarajan, P.; Gulliver, J.S.; Novak, P.J.; Hozalski, R.M. Review of dissolved pollutants in urban storm water and their removal and fate in bioretention cells. *J. Environ. Eng.* **2015**, *141*, 04014050. [CrossRef]

47. Huber, M.; Welker, A.; Helmreich, B. Critical review of heavy metal pollution of traffic area runoff: Occurrence, influencing factors, and partitioning. *Sci. Total Environ.* **2016**, *541*, 895–919. [CrossRef] [PubMed]

48. Westerlund, C.; Viklander, M. Particles and associated metals in road runoff during snowmelt and rainfall. *Sci. Total Environ.* **2006**, *362*, 143–156. [CrossRef] [PubMed]

49. Bennasir, H.; Sridhar, S. Health hazards due to heavy metal poisoning and other factors in sea foods. *Int. J. Pharm. Sci. Rev. Res.* **2013**, *18*, 33–37.

50. Hrubá, F.; Strömberg, U.; Černá, M.; Chen, C.; Harari, F.; Harari, R.; Horvat, M.; Koppová, K.; Kos, A.; Krsková, A.; et al. Blood cadmium, mercury, and lead in children: An international comparison of cities in six european countries, and china, ecuador, and morocco. *Environ. Int.* **2012**, *41*, 29–34. [CrossRef] [PubMed]

51. Kayhanian, M.; Stransky, C.; Bay, S.; Lau, S.L.; Stenstrom, M.K. Toxicity of urban highway runoff with respect to storm duration. *Sci. Total Environ.* **2008**, *389*, 386–406. [CrossRef] [PubMed]

52. Klaassen, C.D. *Casarett and Doull's Toxicology: The Basic Science of Poisons*; McGraw-Hill: New York, NY, USA, 2013; Volume 1236.

53. Li, H.; Davis, A.P. Water quality improvement through reductions of pollutant loads using bioretention. *J. Environ. Eng.* **2009**, *135*, 567–576. [CrossRef]

54. Hatt, B.E.; Fletcher, T.D.; Deletic, A. Hydrologic and pollutant removal performance of stormwater biofiltration systems at the field scale. *J. Hydrol.* **2009**, *365*, 310–321. [CrossRef]

55. Ross, S.M. *Toxic Metals in Soil-Plant Systems*; John Wiley and Sons Ltd.: Hoboken, NJ, USA, 1994.

56. Hatt, B.E.; Fletcher, T.D.; Deletic, A. Hydraulic and pollutant removal performance of fine media stormwater filtration systems. *Environ. Sci. Technol.* **2008**, *42*, 2535–2541. [CrossRef] [PubMed]

57. Gunawardena, J.; Ziyath, A.M.; Egodawatta, P.; Ayoko, G.A.; Goonetilleke, A. Sources and transport pathways of common heavy metals to urban road surfaces. *Ecol. Eng.* **2015**, *77*, 98–102. [CrossRef]

58. Australian and New Zealand Environment and Conservation Council; Agriculture and Resource Management Council of Australia and New Zealand. *Australian and New Zealand Guidelines for Fresh and Marine Water Quality, Canberra*; Australian and New Zealand Environment and Conservation Council: Canberra, Australia, 2000.

59. Davis, A.P. Field performance of bioretention: Water quality. *Environ. Eng. Sci.* **2007**, *24*, 1048–1064. [CrossRef]

60. Subramaniam, D.N.; Egodawatta, P.; Mather, P.; Rajapakse, J. Stabilization of stormwater biofilters: Impacts of wetting and drying phases and the addition of organic matter to filter media. *Environ. Manag.* **2015**, *56*, 630–642. [CrossRef] [PubMed]

61. Facility for Advancing Water Biofiltration. *Guidelines for Filter Media in Biofiltration Systems (Version 3.01)*; Facility for Advancing Water Biofiltration: Melbourne, Australia, 2009; p. 8.

62. Duong, T.T.T.; Penfold, C.; Marschner, P. Differential effects of composts on properties of soils with different textures. *Biol. Fertil. Soils* **2012**, *48*, 699–707. [CrossRef]

water

MDPI

Article

Urban Floods and Climate Change Adaptation: The Potential of Public Space Design When Accommodating Natural Processes

Maria Matos Silva * [ID] and João Pedro Costa

CIAUD, Centro de Investigação em Arquitetura, Urbanismo e Design, Faculdade de Arquitetura, Universidade de Lisboa, Rua Sá Nogueira, Pólo Universitário do Alto da Ajuda, 1349-063 Lisboa, Portugal; jpc@fa.ulisboa.pt
* Correspondence: m.matosilva@fa.ulisboa.pt; Tel.: +351-21-361-5884

Received: 28 December 2017; Accepted: 7 February 2018; Published: 9 February 2018

Abstract: Urban public space is extraordinarily adaptable under a pattern of relatively stable changes. However, when facing unprecedented and potentially extreme climatic changes, public spaces may not have the same adaptation capacity. In this context, planned adaptation gains strength against "business as usual". While public spaces are among the most vulnerable areas to climatic hazards, they entail relevant characteristics for adaptation efforts. As such, public space design can lead to effective adaptation undertakings, explicitly influencing urban design practices as we know them. Amongst its different intrinsic roles and benefits, such as being a civic common gathering place of social and economic exchanges, public space may have found an enhanced protagonism under the climate change adaptation perspective. In light of the conducted empirical analysis, which gathered existing examples of public spaces with flood adaptation purposes, specific public space potentialities for the application of flood adaptation measures are here identified and characterized. Overall, this research questions the specific social potentiality of public space adaptation in the processes of vulnerability tackling, namely considering the need of alternatives in current flood management practices. Through literature review and case study analysis, it is here argued that: people and communities can be perceived as more than susceptible targets and rather be professed as active agents in the process of managing urban vulnerability; that climate change literacy, through the design of a public space, may endorse an increased common need for action and the pursuit of suitable solutions; and that local know-how and locally-driven design can be considered as a service with added value for adaptation endeavors.

Keywords: urban floods; climate change; adaptation; interdisciplinarity; public space

---

1. Introduction

The idea of public space may be apprehended by a set of two meanings: (1) a conceptual meaning commonly used in political and social science, in which public space, "brings together all the processes that configure the opinion and collective will" as characterized by Innerarity, ([1] p. 10) (author's translation) and (2) a physical meaning, commonly used in urban planning and design, regarding where and how previous actions are developed [2,3]. More specifically, a space that enables and promotes community life, such as streets, sidewalks, plazas, coffee shops, parks or museums, and that potentially offers services with wide-ranging benefits with tangible and intangible values. Other authors have additionally highlighted pubic space as multidimensional space, with a central social, political and cultural significance [4]. In regards to its physical characteristics, it has been particularly argued on the long-lasting permanence of public space as a structuring urban space [5,6] of interdisciplinary nature [7,8]. Overall, public space may be defined by Hanna Arendt's communal table: it "gathers us together and yet prevents our falling over each other, so to speak." ([9], p. 52).

In the present article, all these previously mentioned facets of public space are embraced and its specific role in urban adaptation processes is furthermore discussed, namely by arguing that through the application of effective measures in public spaces, communities are facilitated to comprehend, learn, engage and mobilize for climate action.

As highlighted in previous publications [10,11], the distinctiveness of urban territories as major centers of communication, commerce, culture and innovation may empower successful processes and outcomes in the climate change adaptation agenda, due to "interchange processes of products, services and ideas that are processed and expressed in their public spaces" ([2], p. 120, author's translation). As Jane Jacobs pointed out, in a pioneering critical criticism on modern urbanism orthodoxy that quickly became mainstream faith, "lively, diverse, intense cities contain the seeds of their own regeneration, with energy enough to carry over for problems and needs, outside themselves" ([12], p. 448). What is regularly overlooked in large scale planning and policy—often guided by questionable interests—is a community's inherent resilience. Not only do people want to be the main actors in the urban space, but also want to be at the center of space design concerns. The currently ongoing project "PSSS—Public Space Service System" [13] acknowledges this recognition of public space potential roles, and is working on new critical concepts for public space assessment, with new criteria within a systemic logic focusing on service and value creation, which is useful for actions calling for urban adaptability.

Concerning the recurrent phenomenon of urban flooding, climate change research has been warning to the fact that traditional flood management practices must be reassessed, namely if projected impacts are to be managed, such as the likely increased frequency and greater intensity of storms (precipitation and storm surges) together with a rise in sea level. This is an increasing threat that affects all the people in a community, particularly the most vulnerable (elderly, children, poor, among others). Indeed, it is possible to verify an emerging change from the conventional focus on reducing the probability to experience floods to the aim to reduce society's vulnerabilities. This former notion inevitably promoted the emergence of new flood management approaches that started to integrate risk and uncertainty in its practice, notably by fully acknowledging and welcoming the processes of the natural water cycle. Among others, Best Management Practices (BMPs), Green Infrastructure (GI), Integrated Urban Water Management (IUWM), Low Impact Development (LID) or Sustainable Urban Drainage Systems (SUDS) [14] are well representative of these new approaches. All of which implies changes in the relationship between the city and (its) water.

Considering public space as a communal space system that develops spatial services [15], as a collective entity of shared concerns, a new claim for climate change adaptation is presented: the claim that public space may additionally serve as a social beacon for change. In light with Pelling's findings, people and communities are not only targets but also active agents in the management of vulnerability [16]. Ulrich Beck also highlighted that "what was made by people can also be changed by people" ([17], p. 157). Correspondingly, not only it is in the public space where hazards become tangible to a community, but it may also be where adaptation initiatives may strive. It therefore comes as no surprise that a new variety of insurgent citizenship is arising within public spaces as the urgent matter of climate change adaptation is recognized among our societies. Regardless, bearing in mind the potential severity of the projected impacts that are expected to become increasingly more unavoidable, many authors agree that our society is still not responding accordingly [18].

Some societies have shown to be reluctant of the need to face impending threats of climate change due to an absence of common understanding of what is the "common good". Hesitant communities may be driven by the fact that climate change is still a much-politicized issue or by the fact that adaptation is still a fairly recent strategy of response. Regardless of the causes in climate change suspicions, some cases provide evidence of an inclination to prioritize other values. Even within developed countries that have suffered direct consequences of severe climate impacts, some communities, such as the case of New Orleans local society [19], have rejected initiatives towards a more adapted urban environment. Other communities, in other situations, also did not initially welcome adaptation actions. This is namely the case of the first attempt to implement the currently internationally recognized concept of the Water Plaza, which can be briefly characterized by being a low-lying square that is submerged only during storm events. Despite a promising start—with an idea that had won the first prize of the 2005 Rotterdam Biennale competition—the first pilot project failed. Risks, such as of children drowning, triggered strong emotional reactions from local citizens who started naming the idea as the "drowning plaza" ([20], pp. 121–122). Having wisely learned from the encountered barriers that prevented the implementation of the first project, the second pilot project had not only a new location, but, more importantly, a new approach towards technical criteria and social participation and is currently considered an exemplary case of concrete climate change adaptation in a highly urbanized area.

Cases like these evidence that social, cultural and emotional factors can be more valued and respected than the need of physical safety or ecological services of public spaces. This fact is one that strengthens the importance of continual community evolvement alongside additional and distinct methods for the dissemination of scientific knowledge with within the agenda of climatic adaptation. According to Van Der Linden, persuasive communication about climate change is only successful when based on an integrated acknowledgement of the psychological processes that control pro-environmental behavior [21]. However, public spaces seem to offer what Van Der Linden considered as fundamental for climate change adaptation engagement: through public spaces and public space design, local aspects of climate change can be made visible and thus meaningful for citizens and their livelihoods.

Furthermore, public spaces provide a source of knowledge and information (besides the mainstreamed sources of science and media) that may be apprehended as an autonomous and independent process. A process with direct learning experiences, based on deep-rooted traditional experience and know-how, in a public domain that is naturally subject to social control. In other words, public spaces may provide extended opportunities for experiential learning that are influenced by specific contexts and social pressures. Through a medium that is closer to people, "climate change literacy" may more likely endorse a common need to search for solutions. As highlighted by CABE (Commission for Architecture and the Built Environment) adaptation of cities to climate-driven threats is strongly dependent on "well-designed, flexible public spaces"([22], p. 2).

When integrating local expertise as well as scientific and technical knowledge in a flexible and clear-cut design, public spaces are not only able to promote adaptation action and reduce risk of disaster, but also improve awareness on climate change. The physical and the social components combined make public spaces favored interfaces for adaptation action. In public spaces, people may "be" as well as "become" both producers and managers of adaptation action. People may "be" producers and managers of adaptation through autonomous, individual or collective involvements—from art manifestations to community-based projects. In addition, people may also "become" both producers and managers of adaptation when awareness is raised through direct consequence of the formerly mentioned processes or through institutional endeavors, by the message of public art and other participative or deliberative actions evolved in public space design. In this line of reasoning, the design of public space sees itself enhanced in the face of impending weather events, being here considered as a determinant for the adaptation of urban territories when facing climate change.

Underlying this assessment lies the argument that public spaces support the new emerging tendency on urban flood management [23], where the precedent goal to effectively and rapidly avoid or convey stormflows is being gradually replaced by the goal to incorporate storm water within the city and through the enhancement of the whole natural water cycle. In other words, that public space helps promote a change of paradigm for more flood-adapted cities that aim to reduce vulnerabilities while integrating environmental, social and economic concerns.

2. Methodology

Climate change adaptation initiatives are still faced with numerous challenges. Most common barriers to adaptation can be associated with "short term thinking of politicians and long term impacts of climate change", "little finance reserved/available for implementation", "conflicting interests between involved actors", "more urgent policy issues need short term attention" or "unclear social costs and benefits of adaptation measures" ([20], p. 139). However, every day successful adaptation examples grow in number, and, as argued by Howe and Mitchell, it is increasingly important to see more empirical studies of adaptation examples rather than just dwell on the barriers to change [24]. New and innovative adaptation projects can be exploited as a creative laboratory, which can serve to propose, assess and monitor solutions through an ongoing learning process that may serve to inform future decisions and reduce generalized hindering constraints.

This research advances from an empirical analysis, which gathered existing examples of public spaces with flood adaptation purposes, and is targeted at identifying and characterizing specific public space potentialities for the application of flood adaptation measures. The presented analysis is based on comprehensive case studies highlighted in research projects, bibliographical reviews, interviews with specialists, networking or in site visits. Besides including main or secondary functions related to flood vulnerability reduction, the chosen range of examples also aimed to select "good quality" cases among a comprehensive group of public space typologies. This research approach is named here "Portfolio Screening".

For Jan Jacob Trip, public space may be the element of urban development that is most difficult to plan and design as it relates to so many intangible qualities inherent to the quality of place itself [25]. Difficulties in this purpose arise for there is no consensual formula that would define the quality of public space design. Although it is commonly accepted that good design must develop from a sensible understanding of its situation and all its encompassing contexts (environmental, cultural, social, economic and political) [26], and therefore is likely to result in a place that is valued "in general", a rational may be required in the assessment of adaptation practices.

Based on an exhaustive evaluation of thousands of public spaces worldwide, the non-profit organization Project for Public Spaces (PPS) provides evidence that "great places" generally share four principal attributes, namely: sociability, uses and activities, access and linkages and comfort and image. One may anyhow presume that public spaces that entail flood adaptation measures are likely to include at least one of PPS's intangible qualities of being "vital", "useful", "sustainable" or "safe".

The Portfolio Screening, i.e., the range of empirically collected examples, is therefore based on the abovementioned references and emphasized attributes. It further encompasses a comprehensive range of public space typologies. For this purpose, the typology of public spaces identified by Brandão was used, namely the differentiation regarding "Layout spaces" (plazas, streets, avenues), "Landscape spaces" (gardens, parks, belvederes, viewpoints), "Itinerating spaces" (stations, interfaces, train-lines, highways, parking lots, silos), "Memory spaces" (cemeteries, memory and monumental spaces), "Commercial spaces" (markets, shopping malls, arcades, temporary markers, kiosks, canopies) and "Generated spaces" (churchyard, passage, gallery, patio, cultural, sports, religious, children's, lighting, furniture, communication, art) ([27], p. 35). The range of presented examples in the Portfolio Screening therefore covers all the aforementioned types of public spaces.

The gathered range of cases aimed to further provide a geographically representative scope, yet, inevitably, projects with greater dissemination and improved access to information were privileged. The examples presented in the Portfolio Screening specifically involve 19 countries and 72 cities. Together, they are not meant to offer an exhaustive collection but rather a significant sample of designed solutions that endorse further reliable research and decision-making.

3. Public Space Potentialities for the Application of Flood Adaptation Measures

Table 1 advances the conducted empirical analysis, which gathered existing examples of public spaces with flood adaptation purposes, together with the identification of public space potentialities for the application of flood adaptation measures for each presented example.

In several of the examples that enabled this analysis, adaptation measures were unrecognized as such. The existing functional qualities of some cases were rather associated to other, more prevailing, conceptual approaches such as sustainability or flood protection. However, in light of the previous findings, namely in regard to the concept of adaptation [28], all presented examples are considered as adaptation measures. Not only do all examples entail the transposition of uncertainty, and its apparent impediments, into public spaces of multifunctional qualities, but also all examples serve as solid grounds for the assessment of adaptation action.

In light of the examples presented in the Portfolio Screening, it was possible to identify six public space potentialities specifically directed towards the application of flood adaptation measures, namely: (1) the favoring of interdisciplinary design; (2) the possibility to embrace multiple purposes; (3) the promotion of community awareness and engagement and interaction; (4) the comprehension within an extensive physical structure; (5) the possibility to expose and share value and (6) the opportunity to diversify and monitor flood risk. Each of these features will be analyzed in the following pages and further reinforced by the association particular cases from the Portfolio Screening.

Table 1. Portfolio Screening: Public space potentialities for the application of flood adaptation measures.

Project Name	Location	Construction	Public Space Typologies [1]	Interdisciplinary Design	Multiple Purposes	Community Engagement and Interaction	Extensive Physical Structure	Expose and Share Value	Diversify and Monitor Risk
1 Caixa Forum plaza	Madrid	2006	L, I, M, G	X	X	X			
2 Westblaak' car park silo	Rotterdam	2010	I		X	X			
3 Woolworths Shopping playgr.	Walkerville	2013–2014	C, G	X	X	X			
4 North Road	Preston	2009	C, G	X	X				
5 Expo Boulevard	Shanghai	2010	L, G		X	X			
6 Jawaharlal Planetarium Park	Karnataka	2013	Ld, G		X	X			
7 'Water Table/Water Glass'	Washington	2001	L, G	X	X	X		X	
8 Whole Flow'	California	2009	L, I	X	X	X		X	
9 Dakpark	Rotterdam	2009–2014	I, C, G	X	X	X			
10 Promenade Plantée	Paris	1993	Ld, I, M, G	X	X	X	X		
11 European Patent Office	Rijswijk	2001	Ld	X	X	X		X	X
12 Womans University campus	Seoul	2008	L, Ld, G	X	X	X	X		
13 High Line Park	New York	2006–2009	L, Ld, I, M		X	X		X	X
14 Waltebos Complex	Apeldoorn	2000–2007	Ld, C	X	X	X		X	X
15 Stephen Epler Hall	Portland	2001–2003	Ld, I, G	X	X	X		X	X
16 Parc de Diagonal Mar	Barcelona	2002	L, Ld, I, M, G	X	X	X		X	X
17 Parc del Poblenou	Barcelona	1992	L, G	X	X	X		X	X
18 Benthemplein square	Rotterdam	2012–2013	L, Ld, G	X	X	X		X	X
19 Tanner Springs Park	Portland	2005	L, Ld, G	X	X	X		X	X
20 Parc de Joan Miró	Barcelona	2003	L, Ld, I, C	X	X	X			
21 Escola Industrial	Barcelona	1999	Ld, G	X	X	X			
22 Potsdamer Platz	Berlin	1994–1998	L, I, C, G	X	X	X	X		
23 Museumpark car park	Rotterdam	2011	Ld, I, G	X	X	X			
24 Place Flagey	Brussels	2005–2009	L, I	X	X	X			
25 Stata Center	Massachusetts	2004	L, Ld, I, C	X	X	X			
26 The Circle	Illinois	2010	L, Ld, I, G	X	X	X			
27 Georgia Street	Indianapolis	2010–2012	L, I, C, G	X	X	X	X	X	X
28 Parque Oeste	Lisbon	2005–2007	Ld	X	X	X		X	X
29 Qunli park	Haerbin	2009–2010	Ld		X	X		X	X
30 Emerald Necklace	Boston	1860s	Ld, M, C, G		X	X	X	X	
31 Quinta da Granja	Lisbon	2011	L, Ld, C		X	X		X	X
32 Parque da Cidade	Porto	1993	L, Ld, C, G		X	X		X	X
33 Trabrennbahn Farmsen	Hamburg	1995–2000	L, Ld, I, M, C, G	X	X	X	X	X	X
34 Elmhurst parking lot	New York	2010	L, I	X	X	X		X	
35 Ecocity Augustenborg	Malmö	1997–2002	L, Ld, G	X	X	X	X	X	X
36 Museum of Science	Portland	1990–1992	L, G		X	X		X	
37 High Point 30th Ave	Seattle	2001–2010	I	X	X	X		X	
38 Moor Park	Blackpool	2008	Ld, I, G	X	X	X	X	X	X
39 Ribblesdale Road	Nottingham	2013	L, Ld, I, C, G	X	X	X	X	X	X
40 South Australian Museum	Adelaide	2005	L, Ld, I, G	X	X	X	X	X	X
41 Columbus Square	Philadelphia	2010	L, Ld, I	X	X	X	X	X	X

Table 1. *Cont.*

	Project Name	Location	Construction	Public Space Typologies [1]	Interdisciplinary Design	Multiple Purposes	Community Engagement and Interaction	Extensive Physical Structure	Expose and Share Value	Diversify and Monitor Risk
42	Derbyshire Street	London	2014	L, Ld, I, C, G		X	X		X	X
43	Onondaga County	New York	2010	L, Ld, I		X	X	X	X	X
44	Edinburgh Gardens	Melbourne	2011–2012	Ld, I, G		X	X		X	
45	Taasinge Square	Copenhagen	2014	L, Ld, G	X	X	X		X	X
46	Australia Road	London	2013–2015	L, Ld, G		X	X		X	
47	East Liberty Town Square	Pittsburgh	2013–2014	L, M, G	X	X	X			
48	Can Caralleu	Barcelona	2006	L, Ld, I, C, G		X				
49	Zollhallen Plaza	Freiburg	2011	L, I, C		X	X	X	X	X
50	Green park of Mondego	Coimbra	2000–2004	Ld, C		X	X			X
51	Bakery Square 2.0	Pittsburgh	2015	L, Ld, I		X	X	X	X	X
52	Praça do Comércio	Lisbon	2010	L, Ld, I, M, C, G		X				
53	Percy Street	Philadelphia	2011	L	X	X				
54	Greenfield Elementary	Philadelphia	2009–2010	G	X	X	X		X	
55	Etna Butler Street	Pittsburgh	2014	L, Ld, I, C, G	X	X	X	X	X	X
56	Community College	Philadelphia	2005	Ld, G		X	X	X	X	X
57	Elmer Avenue Neighbourhood	Los Angeles	2010	L, Ld	X	X	X	X	X	
58	Green gutter	Philadelphia	2016	L, Ld, I		X	X	X	X	
59	Ribeira das Jardas	Sintra	2001–2008	L, Ld	X	X	X	X	X	X
60	Ahna	Kassel	2003–2004	L, Ld		X	X		X	
61	River Volme	Hagen	2006	L		X	X		X	
62	Promenada	Velenje	2014	L	X	X	X		X	X
63	Catharina Amalia Park	Apeldoorn	2013	L, Ld, C		X	X	X	X	X
64	Kallang River	Bishan Park	2009–2012	Ld, I	X	X	X	X	X	X
65	Alb	Karlsruhe	1989–2004	Ld, I		X	X	X	X	X
66	Westersingel	Rotterdam	2012	L, Ld, I		X	X		X	
67	Thornton Creek	Seattle	2003–2009	Ld, G	X	X	X	X	X	X
68	Cheonggyecheon River	Seoul	2003–2005	L, Ld, I, M, G	X	X	X	X	X	X
69	Soestbach	Soest	1992–2004	L, Ld		X	X		X	
70	Banyoles	Girona	1998–2008	L, I, M		X	X		X	
71	Freiburg Bächle	Freiburg	13th century	L, I, M, C, G		X	X	X	X	
72	Roombeek	Enschede	2003–2005	L, Ld		X	X		X	
73	Solar City streets	Linz	2004–2006	L		X	X	X	X	X
74	Pier Head	Liverpool	2009	I, M	X	X				
75	Olympic park	London	2012	L, Ld, I	X	X				
76	Kronsberg	Hannover	1998–2000	L, Ld, I		X	X	X	X	X
77	Renaissance Park	Tennessee	2006	Ld, M		X	X			
78	21st Street	Paso Robles	2010–2011	L	X	X		X	X	X
79	West India Quay	London	1996	L, Ld, I, G		X				
80	Ravelijn Bridge	Bergen op Zom	2013–2014	L, G	X	X	X		X	
81	Yongning River Park	Taizhou	2002–2004	Ld		X	X			X
82	Landungsbrücken pier	Hamburg	1980?	L, I, C		X	X			X
83	Spree Bathing Ship	Berlin	2004	G	X	X	X			X

Table 1. *Cont.*

	Project Name	Location	Construction	Public Space Typologies [1]	Interdisciplinary Design	Multiple Purposes	Community Engagement and Interaction	Extensive Physical Structure	Expose and Share Value	Diversify and Monitor Risk
84	Leine Suite	Hannover	2009	G	X	X	X			
85	Rhone River Banks	Lyon	2004–2007	Ld	X	X	X		X	X
86	Parque fluvial del Gallego	Zuera	2000–2001	Ld,I	X	X	X	X	X	X
87	Río Besòs River Park	Barcelona	1996–1999	Ld,I	X	X	X	X	X	X
88	Buffalo Bayou Park	Houston	2006	Ld	X	X	X	X	X	
89	Parc de la Seille	Metz	1999	L,Ld,G	X	X	X		X	
90	Park Van Luna	Heerhugowaard	1997–2003	Ld		X				
91	Passeio Atlântico	Porto	2001–2002	L,Ld,G		X	X			X
92	Quai des Gondoles	Choisy-le-Roi	2009	I,G		X	X	X		X
93	Elster Millraces	Leipzig	1996	L,Ld,I,M	X	X	X	X	X	
94	Terreiro do Rato	Covilhã	2003–2004	L,Ld,G		X	X		X	
95	Waterfront promenade	Bilbao	?	L		X				X
96	Tagus Linear Park	Póv. de Sta. Iria	2013	Ld		X	X		X	X
97	Elbe promenade	Hamburg	2006–2012	L,Ld,I,M	X	X	X		X	X
98	Dike of 'Boompjes'	Rotterdam	2000–2001	L,Ld	X	X				X
99	Zona de Banys del Fòrum	Barcelona	2004	L,Ld,I,M,G	X	X	X		X	X
100	Molhe da Barra do Douro	Porto	2004–2007	Ld,M		X	X		X	
101	Jack Evans Harbour	Tweed Heads	2011	Ld	X	X	X		X	
102	Schevenigen	The Hauge	2006–2009	L,Ld,I,M,C	X	X				X
103	Sea organ	Zadar	2005	L,Ld,M	X	X	X			X
104	Main riverside	Miltenberg	2009	L,Ld,I,C	X	X	X			X
105	Blackpool Seafront	Blackpool	2002–2008	L,Ld,M	X	X				X
106	Westhoven	Cologne	2006	Ld	X	X	X		X	X
107	Waalkade promenade	Zaltbommel	1998	L,Ld	X	X				
108	Kampen waterfront	Kampen	2001–2003	L,G	X	X				X
109	Landungsbrücken building	Hamburg	2009?	M,G	X	X	X			
110	Corktown Common	Toronto	2006–2014	Ld,I	X	X				
111	Westzeedijk	Rotterdam	12th century	L,Ld,I,M		X				
112	Anfiteatro Colina de Camões	Coimbra	2008	Ld,M,G	X	X			X	X

[1] Public Space Typologies according to [29]: L—Layout spaces; Ld—Landscape spaces; I—Itinerating spaces; M—Memory spaces; C—Commercial spaces; G—Generated spaces; Highlighted in bold is the most evident applicable public space typology.

3.1. Interdisciplinary Design of Public Spaces

Acknowledging that public space ethics concept may be interpreted as "it is of everyone", its design therefore is "not a matter of one sole profession, entity or interest group" ([26], p. 19, author's translation). Likewise, Madanipour argues that public spaces should be created by different professionals from different disciplines of the built, natural and social environments or by any professional with multi-disciplinary concerns and awareness [7]. As Lefébvre acutely states, "ultimate illusion: to consider the architects, urbanists or urban planners as experts in space, the greatest judges of spatiality..." ([30], p. 30, author's translation).

Recalling Horacio Capel's argument, since the nineteenth century, the subject of urbanism has been excessively controlled by a fierce competition between engineers on the one side and architects on the other. While the first would define and design major infrastructures, the latter would define and design interventions in streets, buildings or green areas. However, as the author highlights, "all this should be at the service of social needs" ([31], p. 92, author's translation).

Back in the 1870s, Frederick Law Olmsted designed Boston's Emerald Necklace (#30) (Henceforth, the mentioned examples will be additionally identified with their corresponding number presented in Table 1 to facilitate the access to further information. Emerald Necklace, for example, is number 30 and so it is identified within the text by #30.) with the goal to resolve engineering problems of drainage and flood control together with the fulfilment of the increasing social needs for leisure and recreation opportunities in a growing population. Simply put, Olmsted demonstrated that it was possible to integrate complex connections between natural and technical processes together with and improvement on the quality of life of the surrounding populations. For Cynthia Zaitzevsky, "Olmsted foresaw that such a comprehensive approach embraced planning, engineering and architecture and that, to bring the disciplines together to create the best solution, needed the unifying instincts of the new profession of landscape architecture" ([32], p. 43). Today, Emerald Necklace parks include land and water features, engineering structures, public buildings and ecological designs that are merged together in a rational and balanced design.

Further examples of interdisciplinary public space designs can be namely seen in the city of Barcelona. It was likely due to Barcelona's urban regeneration grounding ideals from the 1980s that, in the beginning of the twenty first century, the city decided to integrate the infrastructural construction of underground reservoirs underneath different types of public spaces (#20, #21). An interdisciplinary approach that required for multiple professional areas to share their expertise throughout all procedural planning stages. By contrast, other municipalities have chosen to solely focus on one technical discipline. As a result, similar infrastructures were designed as isolated monofunctional facilities fenced from its surroundings [33]. Through Barcelona's integrated approach, it was further possible to enclose parallel advantages from a grand urban intervention, namely the creation of more public spaces. Putting it simply, Barcelona turned the constraint of a required great drainage improvement into the opportunity to build more public spaces for its citizens and all its potential succeeding side-benefits.

Other successful public spaces, particularly known for its interdisciplinary design that further entailed multiple purposes, is Postdamer Platz in Berlin (#22). Situated in an important area of the city, near the Berliner Philharmonie and the Berlin State Library, Postdamer Platz has an approximate area of 1.2 hectares. Its design, composed of a series of urban pools, reveals an integrated approach between ecological, aesthetical and civil engineering functions. The large water features are fed uniquely by rainwater. In summer, water surfaces lower the ambient temperature and improve microclimates. Roofs from the surrounding buildings capture rainwater and store it in underground cisterns. The collected water is then used for topping up the pools, flushing toilets and for irrigating green areas [34].

One can further provide evidence of a growing tendency for interdisciplinary design, specifically when interventions consider the need for climate change adaptation. One of the most recent examples of urban realm to have been created in light of the disseminated climate change projections is the Olympic Park (#75), or more precisely "Queen Elizabeth Olympic Park" in London (Figure 1). In brief, the Park's landscape design priorities included "Great amenity; Improved micro-climate; Biodiversity; Integrated water management; Energy generation; Resource management; Waste management and minimization; Local food production; ... " ([35], p. 1), among others. Priorities that involved the inclusion of additional and uncommon disciplines to be actively involved in the design process.

Figure 1. Olympic Park, London. Image credits: Maria Matos Silva, 2 March 2017.

Among other challenges, the adaptation to new environments created by climate change requires a new integrated and interdisciplinary approach [36]. Solutions arising from interdisciplinary designs are very diverse and combine the use of a wide range of approaches such as technical, social, economic, ecological, among others. With regards to adaptation, there is much we can learn from our civilizational past, which has surpassed other great turbulences. We must also humbly accept that the impending future will require new outsets and new paradigms. New ideas that will most likely arise from a common effort of multiple, shared and applied expertise. Public spaces, as spaces that particularly favor interdisciplinary convergence, may serve to promote and explore technological reinventions or innovations. A continuing learning process that, in face of climate change, searches for new design solutions that increase adaptability and reduce vulnerability.

3.2. Multiple Purpose Public Space

All the examples highlighted in the presented table entail multiple purposes for the basic fact that all were gathered following the basic premise of being a public space with flood adaptation capacities, and as such already encompass both the purpose of being a common space for community encounters as well as an infrastructural space for the management of flood waters.

Indeed, the resulting combination of an interdisciplinary design that integrates flood adaptation functions with public space design offers side purposes among other sectorial needs such as recreation, microclimatic melioration or energy use and efficiency. The more interdisciplinary design is, the more adjacent functions the resulting public space will comprise.

Traditional drainage infrastructure, for instance, such as large-scale underground retention chambers disconnected from public space, is only useful occasionally during the year, namely during heavy rainfall. In contrast, other source control measures, such as green walls, bioretention basins or rain gardens, when applied within public spaces, may not only serve its prime infrastructural function, but may also serve to improve local environment and quality of life as well as vulnerability reduction and local awareness [37]. The side benefits that result from reconfiguring drainage infrastructure within public space design thus generally gathers recurring advantages all year long.

"The Circle", in a Roundabout at Uptown Normal, Illinois (#26) serves to provide evidence of this argument. It is a green water square in a roundabout that collects, stores and purifies storm water runoff from the nearby streets. Besides the aesthetical and leisure characteristics, the water feature masks surrounding traffic noise while purified water is used to spray nearby streets and thus lessen heat stress [34]. The square is further used as a meeting place situated near a multimodal transportation centre and a children's museum.

The previously mentioned 'Queen Elizabeth Olympic Park' in London (#75) is one other example of an interdisciplinary design that consequently embraced multiple purposes. That is namely the case of the included treatment process that turn Londoner's wastewater from an outfall sewer into water suitable for irrigation, flushing toilets and as a coolant in the Park's energy centre [38].

In addition, some breakwaters or wave-breakers, such as the one existing at the Zona de Banys del Fòrum in Barcelona (#99), are here understood as a multi-purpose public spaces, as they not only entail the infrastructural function to ease the power of waves but also include the possibility to be used as a sightseeing route—two encompassing functions that are additionally combined in a sculptural design that is aesthetically appealing (Figure 2).

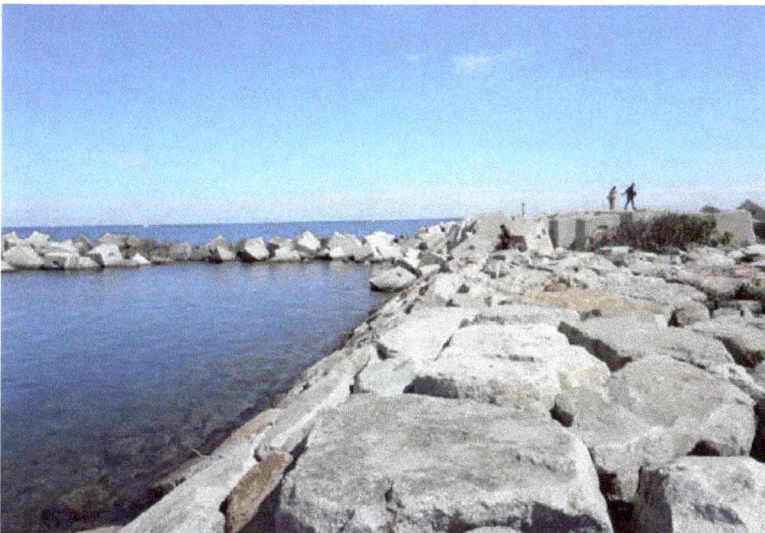

Figure 2. Wave-breakers designed by Beth Gali, Architect, at the Zona de Banys del Fòrum in Barcelona. Image credits: Maria Matos Silva, 26 April 2014.

3.3. Public Space for Actor's Awareness and Engagement

Social and political engagements are a particularly important factor in the success of adaptation endeavors, namely when acknowledging that adaptation is a learning process of continued assessment. Indeed, benefits may be gained from a multifaceted approach that includes social enquiry and stakeholder engagement [39]. Moreover, community engagement may be reached through community involvement, emotional connection and a design that makes visible the invisible. When adaptation actions are applied within a public space, where design can make visible certain intangibles, endeavors are no longer an abstract phenomenon for people and communities. Community engagement practices in public space design and management gain, therefore, a new dimension.

Considering flood events, which are expected to increase in light of future climatic extremes, the roughly intangible water cycle can be made visible through design. Particularly through the design of public spaces that, due to their inherent values, provide the opportunity to approximate and connect people with water and thus potentially raise awareness and overall engagement. Indeed, as it is for Ashley et al., the challenge of appropriate drainage systems for a changing climate is as much sociological as it is technological [40].

The Benthemplein square (#18) (Figure 3) and Tanner Springs Park (#19) may serve as examples that corroborate the argument that public spaces are rarely "mute" and may serve to connect people with water. Both cases encompass the concept of a "water plaza" previously mentioned, which intentionally unveils part of the urban water cycle dynamics for the citizens that use that public space. As mentioned in the Rotterdam Climate Proof report "Water disarms and binds people. In adaptation projects in the city, citizens and different cultures come together. This can reinforce social ties and the sense of safety" ([41], p. 7).

Figure 3. Benthemplein square, Rotterdam. Image credits: Maria Matos Silva, 14 June 2014.

In regard to the case of Tanner Springs Park in Portland, its design comprised the restoration of a wetland into the setting of an urban context. Inspired by the area's original natural state, the park is composed of a pond at its lowest point, to which rainwater from the surroundings is conveyed. The design therefore combined several objectives among the fields of ecology, water management, art and participation. Some of its main characteristics include reintroduced groundwater, water features, appropriate vegetation and site-specific artwork that provides evidence of the biological beings from the former wetland.

To promote community engagement, it is further important to highlight the need to create places that people can value and connect emotionally too. Likewise, the success of community engagement processes is strongly related with the development and value of local identity. In this sense, the presence of water in urban design, and more specifically in the design of public spaces, has particular symbolic dimensions (emotional, aesthetic, and cultural) that should not be overlooked.

One of the oldest representations of water is Genesis' description of the Garden of Eden and its four structuring rivers that give life to this mythical space ([42], Gn 2:10–14). However, as we are all aware of, water is not only the source of life, but it is also a permanent threat. In addition, the fear of water is also tattooed onto our civilizations worldwide. Genesis's flood narrative in the Bible is one of many flood myths found in our cultures.

Water's symbolic dimensions should therefore be enhanced in a public space design that aims to connect people with water. This exercise is particularly evident in the works of Atelier Dreiseitl, here represented by the examples #19, #22, #49, #61, #64, #73 and #76. Atelier Dreiseitl is an office that recurrently uses water beyond its decorative features. Through their designs, water is rather integrated with other systems and other functions, always bearing in mind the final purpose of aesthetic appreciation and public perception of the value of water as a resource.

Another way to promote community engagement on the urging need to adapt our urban spaces in the face of climate change is through direct community involvement and interaction—specifically because adaptation is not one in a lifetime project. On the contrary, adaptation processes require ongoing collaborations and organization between and among government, institutions and its citizens.

Greenfield Elementary in Philadelphia (#54) is a good example of the fruitful results that may arise from collaborative design among stakeholders. More specifically, parents, teachers, students, school administrators, designers from Community Design Collaborative and the Philadelphia School District. More importantly, the all planning stages of the project until its end result worked as a living laboratory that teaches anyone who passes by about overall features of environmental processes. The plan aimed to convert the school yard, used previously as a parking lot, into a green space with sustainable concerns. The improvements included the installation of a flood management system with indigenous vegetation, the removal of impervious pavement, a permeable recycled play surface, an agriculture zone as well as solar shading. A storm water bioretention area with a rain garden was also installed.

Communal management also occurred in New Orleans after the destruction and desolation of Hurricane Katrina. More precisely, an "extraordinary new level of civic and community engagement" ([43], p. 61) helped the city towards recovery and rebuilding, through a process that retained a strong connection to the cities' history while also looking forward in addressing future challenges such as climate change.

In accordance with the report developed by the Institute for Sustainable Communities (ISC) in partnership with the Center for Clean Air Policy, the community embraced the idea that the best approach to endure future climatic extremes is to become a greener city that, consequently, promotes safety and enriches attractiveness for business and residents [43]. One of the implemented projects aimed to transform the constraint of having more than 60,000 vacant lost lots in the city, transforming some of them into a network of urban farms and public gardens.

Furthermore, it is important to provide evidence that, in New Orleans' recovery, governments' investments alone would have had a reduced impact. The city was able to recover, and it is able to take forward its strategic plans, because of a creative and energized community, because of public and private partnerships and because of a comprehensive cooperation among national and international experts [44].

3.4. Public Space as an Extensive Physical Structure and System

Reflecting upon the perception of public space as a structuring element of the urban form [5,6], additional reasons promptly lead to further conclude that these are particularly favorable places for the implementation of flood adaptation measures.

Public spaces have a fundamental role in city life as they enable formal and environmental continuity, accessibility and legibility, contributing to the reinforcement of social and economic centralities [45]. In the series of lectures "O Urbano e a Urbanística ou os tempos das formas" [46], Nuno Portas highlighted that, in the history of cities, public spaces are more durable than buildings; that buildings are stable elements, but not durable elements; and that, after public spaces, the most durable elements are the buildings that are transformed into monuments, i.e., transformed into public spaces. Indeed, public spaces are determinant elements in the form and identity of a city.

Public space, as a structuring layer of urban form, is organized in a systemic way, and can be seen as performing two complementary roles: as hardware, it provides a physical setting, making connections and furthering an infrastructural base for urban functions; as software, it incorporates relations and interactions that make urban life—a social-cultural dimension representing its society or community—as a space for expression and sharing. According to Borja, "The fact that public space is the determining element of the urban form is enough to attribute it the role of structurer of urbanism and, firstly, its urban fabric" ([2], p. 137, author's translation). One can thus claim that public spaces are not only the means of social, economic and cultural dynamics, but are also a physical structuring element of the urban fabric. A structuring network that is able to construct a "recognizable and lasting image of an individual unity, which arises from a system of complementary parts, as various and as unorganized as they may be" ([6], p. 17, author's translation).

By conforming a structural network based on the local scale, public space offers a decentralized and expansive means to tackle flood management. An approach that strongly contrasts with the traditional method that tends to be linear and centralized. This distinction, together with an assessment of exploratory nature regarding the adaptation measure's infrastructural efficiency, was particularly emphasized in the article "Urban Flood Adaptation through Public Space Retrofits: The Case of Lisbon (Portugal)" [37].

Moreover, public spaces offer a network that not only supports the urban fabric, but also connects its different urban spaces, from buildings and infrastructures to natural structures such as the ecological network. For Portas, this communicating network of public spaces "cannot be reduced to a simplistic addition of segments, unconnected streets, detached to the territories they cross, more or less urbanized" ([6], p. 17, author's translation). In other words, public spaces must not be understood by its individual elements, but rather as a "coherent structure that encompasses different territorial scales (from the neighborhood to the metropolitan city)" ([47], p. 1). The same can be said about hydrographic basins and other structures that support dispersed urban settlements. Indeed, one of the main causes of urban floods is related to the manipulation of natural watersheds through forced interruptions or divisions into smaller parts. These approaches do not consider the fact that their effective functioning is highly dependable on a system that is comprehensive by nature.

It is therefore equitable to conceive water systems equally converged within the network of public spaces. One can easily identify episodes were water systems' networks have met with public spaces. However, most of the time, it is an event that is neither planned nor wanted. Considering, for instance, drainage overflows resulting from heavy rainfall, in this situation, storm water generally flows along the next available spaces, generally "open" spaces and mostly public spaces. If this "encounter" could be looked upon through a different perspective, one that would capitalize from the inherent values of public space, the excess of water could be integrated within designs as an opportunity to potentiate a comprehensive adaptation in an extensive and decentralized network (Figure 4).

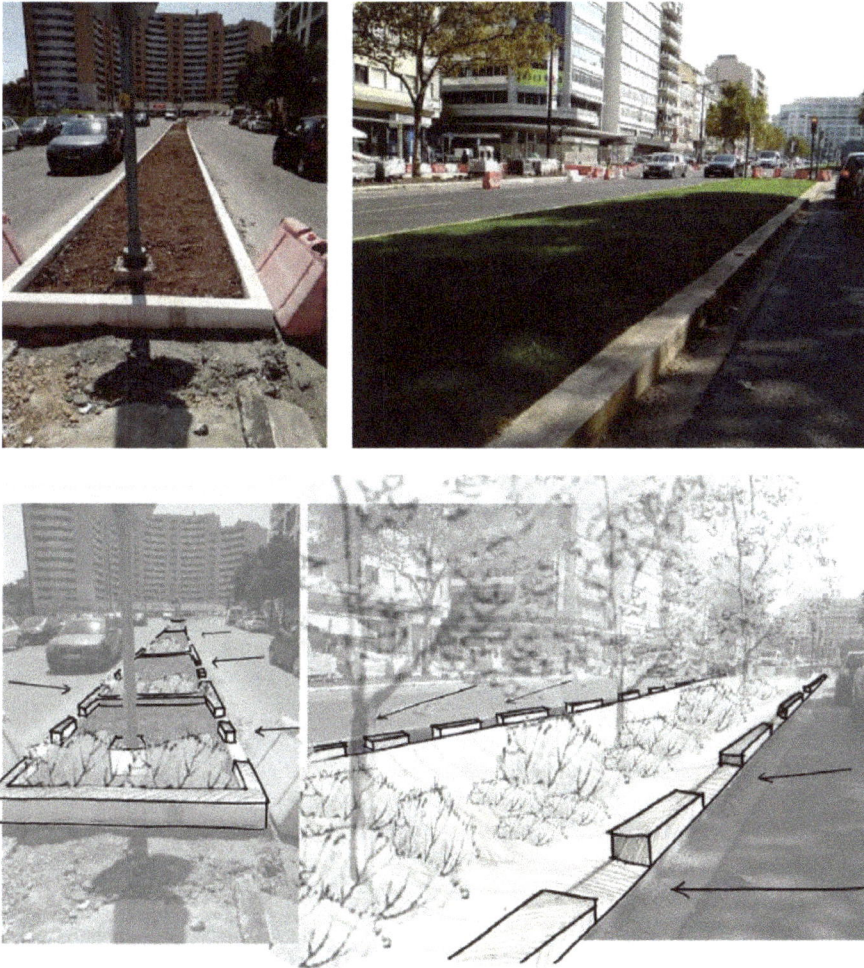

Figure 4. This example aims to highlight how significant benefits could have been gained, with little added investment, if municipal undertakings, such as the Lisbon's Municipality public space rehabilitation programmes of "Pavimentar Lisboa 2015–2020" or "Uma Praça em cada Bairro, would have considered including flood adaptation measures in their design, such as "check dams" (on the left) or "bioswales" (on the right). For more information regarding the implementation of each of these measures, please consult the preliminary design studies developed in [37]. Although they do not correspond to the same areas of intervention, their design encompasses many similarities, such as the potential depth and materiality of the soil and storage layers, the necessary inclusion of an outlet drainage tube or specifications regarding appropriate native vegetation. Image credits: Maria Matos Silva, 26 November 2017.

A representative undergoing example that takes advantage of the benefits offered by the extensive physical structure of public spaces is probably New York's Green Infrastructure plan launched in 2010. In brief, this plan aimed to offer a more sustainable alternative to the conventional "grey" infrastructure by proposing integrated structure that combined solutions such as: rooftop detention, green roofs, subsurface detention and infiltration, swales; street trees, permeable pavement, rain gardens, engineered wetlands, among others.

While New York's program is illustrated by the example of Elmhurst parking lot (#34), many other examples fit within its overall approach. More specifically, examples such as the bioretention planters on Ribblesdale Road in Nottingham, United Kingdom (#39), the open drainage system in Trabrennbahn Farmsen residential area in Hamburg (#33), Germany or the drainage systems of the Ecocity Augustenborg in Malmö, Sweden (#35).

The bioretention planters of Ribblesdale Road in Nottingham were a pilot retrofit project of sustainable urban drainage. They were therefore created for its design and construction to be documented and evaluated in order to assess its comprehensive application. A total of 148 m^2 of bioretention planters were implemented within an existing urban road setting. Among the main objectives of this intervention, it is worth mentioning the following: (1) maximize surface water interception, attenuation and infiltration; (2) encourage participation from local residents in the design and future management of the rain gardens; and (3) evaluate the effectiveness of the scheme as an engagement tool around the sources of urban diffuse pollution and flood risk [48].

Trabrennbahn Farmsen is an example of a newly built residential area that comprised the application of a particularly interesting open drainage system. Because its implementation area has little infiltration capacity, designers chose to implement an open water system that would retain and convey rainwater. In accordance, storm water is collected from surrounding streets as well as from the building's roofs. Overall, the system is composed of grassed swales, storm water channels and two retention ponds [49]. The greatest highlight of this example is the autonomy of this natural system to manage all storm water from the Trabrennbahn Farmsen residential area on-site, providing evidence of a reduced importance of underground sewers for rainwater.

3.5. Expose and Share Value through Public Space

By integrating infrastructure in the design of a public space, instead of camouflaging it underground or in an isolated impenetrable area, a public investment is exposed and shared with a community. A shared value that may instigate further opportunities such as amenity or environmental quality. For example, while in the common mainstream urban drainage approach investments are camouflaged underground, frequently encompassing a sole function and use for storm water alone; investments on urban drainage could be applied in infrastructure that is integrated within the public space itself. In the second option, value is not only exposed to all, but also shared among everyone using that space. Sustainable Urban Drainage systems (SUDs) clearly illustrate and make the case.

Through dispersed, yet extensive, small-scale investments within public space design, urban amenities may be further created while taking advantage of ecological and economic opportunities along the way. While buried culverts may be a missed opportunity for the enhancement of the quality of public space, obsolete and no longer necessary flood walls may likewise hide valuable water assets [50]. Indeed, there are many opportunities for infrastructure renovation and necessary landscape improvements throughout urban areas: from the need to provide alternatives to reduce the load of obsolete drainage infrastructure to vacant lots that can be used to store water. Bearing in mind the ever exceeding costs of traditional infrastructure repair alongside expected climate change extreme projections, a wide range of literature argues that established methods are no longer affordable nor sustainable, such as [51–53]. As such, new alternatives, supportive of an integrated water management, should be considered not only as a necessary immediate investment, but, more importantly, as an investment in our future.

Uncovering small scale storm water drainage systems, such as in Banyoles, Girona (#70) or in the 13th century Freiburg Bächle (#71), is one way of exposing and sharing the expressed value of water in an urban environment.

In brief, the regeneration of the old city center of Banyoles, designed by Miàs Arquitectes, envisaged two main purposes: (1) to repave the town center and define a new pedestrian area and (2) to reclaim the irrigation canals that used to feed medieval private gardens. These waterways used to run in open channels from the lake and throughout the town. With the loss of these gardens, the canals were progressively covered and water quality worsened. The resulting public space offers a new main square and adjoining streets composed by linear travertine paving stones that are "cut-off" by open channels, which undercover the presence of the water.

Projects that "bring into light" buried pre-existing water lines are another example of adaptation measures that aim to expose and share value. This is the case of Westersingel channel (#66) and the Soestbach River neat Soest (#69), besides the representative example of the Cheonggyecheon river (#68) previously analyzed.

Rotterdam's Westersingel channel, which had been formerly sunken, was redesigned by Dirk van Peijpe from the De Urbanisten office. The resulting promenade almost disguises its capacity to sustain and retain excesses of water when needed (Figure 5). Its banks are mostly made of brick as well as grass and trees. All materials, including urban equipment such as benches and lamps, are designed to endure occasional overloads of the canal. Currently, this public space is enriched with sculptures by well-known artists such as Rodin, Carel Visser, Joel Shapiro and Umberto Mastroianni.

Figure 5. Rotterdam's Westersingel channel. Image credits: Maria Matos Silva, 13 June 2014.

3.6. Public Spaces as a Means to Diversify and Monitor Flood Risk

According to the Intergovernmental Panel on Climate Change (IPCC), "The main challenge for local adaptation to climate extremes is to apply a balanced portfolio of approaches, as a one-size-fits-all strategy may prove limiting for some places and stakeholders" ([54], p. 291). In other words, in light of climate change, the sole investment in one isolated infrastructure is not recommended, built to fulfill only its particular purpose. If plan A fails, the risk will be great and generalized. However, if investments are diversified, risk is dissipated through the reliance on parallel plans.

In addition, when massive infrastructures are kept out of site, people do not remember their existence and thus will not expect their failure. This unpreparedness, led by a false sense of safety that is usually termed "levee-effect", may further exacerbate vulnerability and increment potential impacts. Contrastingly, if approaches are implemented within public spaces, some risks are more closely acknowledged and thus less unexpected.

Regardless, research aimed at analyzing the social construction of risk or social risk perception is rather complex. As made evident by Sergi Valera [55], social theories of risk namely suggest that the causes and consequences of risk are mediated by the subjective criteria of individual processes (or psychological), social (psychosocial) behaviors and culture. The same way the design of a public space may reduce or exacerbate a risk through "rational and scientific" processes, it may also reduce or exacerbate the perception of that risk through "subjective-social" processes. It is further important to note that the social construction of a risk may influence the degree of the risk itself, minimizing or maximizing it. Risk perception is thus a very important factor that must be taken into consideration, namely in the design process of a public space with flood adaptation purposes, so that the resulting outcome does not contradict the initial purpose.

As argued by Evers, Höllermann et al., one must focus on a pluralistic approach in order to incorporate Human–Water relations [56]. Through the diversification of risk, by investing in more than one great mono-functional strategy, the communal management among government, institutions, communities and private companies is also promoted; unlike traditional management that is essentially based on government's actions on behalf of communities. As a result of communal management, risk is further shared and communities are more likely involved in the management and monitoring of implemented infrastructures. In addition, local knowhow is explored, citizens are empowered to act before the need for safety and the identity of vulnerable places is reinforced.

One illustrative example is the Passeio Atlântico at Porto designed by Manuel de Solà-Morales and others. This submergible pathway, which develops between Montevideu Avenue and the Atlantic coast, encompassing both submergible boardwalks as well as submergible concrete pathways, demonstrates how a storm surge defense slope, may be integrated within the design of a multifunctional public space (Figure 6).

(a) (b)

Figure 6. Submergible pathway at the Passeio Atlântico in Porto, Portugal: (**a**) submergible concrete pathway; (**b**) submergible boardwalks. Image credits: Maria Matos Silva, 1 July 2007.

Through this project, people are more connected to the intense coastal water dynamics and thus more aware of its nature. By sharing the value of the infrastructure through its common use as a public space, not only is the awareness of the power of nature promoted, but also a certain sense of responsibility and appropriation is reinforced. While the first aspect may lead to the respect and willingness to adapt, the second aspect may lead to active management and monitoring of the infrastructure itself.

Comprised of three levels of parallel and undulating waterfront boulevard, this space offers more than its functional requirement to protect The Hauge from coastal floods. More specifically, it articulates other programs such as coastal life and recreation (bars and restaurants), public and private circulation (bicycles and cars) and connections with the urban fabric [57].

4. Discussion

While the most attractive adaptation strategies are usually those that offer development benefits in the short term and reductions of vulnerabilities in the long term, extensive literature has been highlighting that not all adaptation responses are benign. Among others, the IPCC in its report "Managing the Risks of Extreme Events and Disasters to Advance Climate Change Adaptation" states that "there are trade-offs, potentials for negative outcomes, competing interests, different types of knowledge, and winners and losers inherent in adaptation responses" ([58], p. 443). Selecting the optimal adaptation strategy or measure for a particular situation is therefore neither easy nor straightforward. This determining process is particularly complicated given the specific ramifications and secondary impacts related to adaptation processes. In some cases, results can be critical, namely when adaptation does not fulfil its designated objective and ultimately leads to increased vulnerability. This phenomenon is generally called Maladaptation, which can be described as "action taken ostensibly to avoid or reduce vulnerability to climate change that impacts adversely on, or increases the vulnerability of other systems, sectors or social groups" ([59], p. 211). For example, a bioswale with the right combination of phytopurification plants that are placed in the wrong location, may not only be an unnecessary expenditure, a lesser effect is that they may give rise to stagnant water that can be very dangerous to public health (namely through exposed untreated contaminants and mosquito breeding).

However, maladaptation cannot be considered as a hindering factor supporting "business as usual" as uncertainties can be minimized through the ongoing adjustments of continuous assessment. Facing an unprecedented area of action, concepts, paradigms or structures are expected to change overtime, as are the functions, appearance and complexities of public spaces with flood adaptation measures. The design of these spaces must therefore encompass an ongoing process that is fundamentally grounded on the need to learn, reflecting upon mistakes and generating experience while dealing with change [60]. In the words of Jordi Borja, today we must "Accept the challenges with the intent to provide answers and with the modesty of providing them with uncertainty, with the audacity to experiment and with the humility to admit mistakes" ([2], p. 140, author's translation).

It is furthermore essential to highlight that, while the analyzed initiatives have counterbalanced the inevitable uncertainties of global models and the generalized "top-down" policies, local action must be connected to the global scientific findings and its encompassing strategies. Otherwise, applied adaptation measures may get lost in scale, lose its value, and thus fail its purpose. While local scale action is presently acknowledged as a fundamental element for effective urban climate change adaptation, its greater challenge therefore relies on finding the balance and exploring the benefits from the arising synergies between local collective actions and national and international strategies. The same way local adaptation strategies must not be dissociated from global adaptation strategies, so too do the processes of public space design, which must follow objectives and strategies of regional and national levels, otherwise " ... actions will not contribute, in an effective way, to the achievement of community expectations in the safeguard public interests and collective resources" ([26], p. 19, author's translation). Through the inclusion of flood adaptation measures within public space design, new challenges arise before contemporary urbanism and urban design practices. Likewise, although this research is specifically focused upon adaptation measures applied in public spaces, there are several other areas of opportunity that may additionally provide significant contributions in the development of flood adapted cities. More specifically, disciplines such as building design, governance or landscape architecture have been extending their literature regarding this specific subject matter, suggesting further developments namely on floating buildings, transdisciplinary and transboundary consortiums or in blue and green corridors.

Sustained by an empirical analysis based on specific examples, this article emphasizes the particular advantages offered by public space itself as a means where flood adaptation measures can be implemented. However, it is important to note that the presented findings are unintended to serve as restrictive boundaries. It is not here advocated that flood adaptation endeavors can only be considered as such if comprising all the mentioned potential advantages offered by public space, nor that they are only successful if comprising all these mentioned advantages. What is argued is that the above-mentioned characteristics are only potential and are considered as an additional asset either alone or combined. Ultimately, it is reasoned that public space is an ideal interface for adaptation action. Consequently, it is further questioned whether the evaluation of adaptation initiatives should consider: (1) if the design of a public space comprises adaptation measures and, on a reverse perspective, (2) if the application of adaptation measures comprises the design of a public space.

Nuno Portas in [61] reflected about different phases of urban projects that have led to different ways on how public spaces were produced. In the described first phase, most interventions were held in heritage areas, entailing projects such as the pedestrianisation of historic centers or creation of public spaces as a replacement of old industrial uses. The second phase entailed urban projects that were induced by events such as the Olympic Games, Capital of Culture or International Exhibitions. These projects had in common the aim to generate new facilities suited for leisure, culture or sports as flag/brand attracting projects. Proposing a further prospective discussion, a final question arises: are we at the fringe of a third phase, in which, in a changing climate, urban projects of diverse territorial nature will also aim to produce public spaces that are prepared to adapt to future impending weather events?

5. Conclusions

Within the multi-scaled scope of adaptation action, local scale, from the bottom-up, adaptation is particularly relevant, not only because it very likely influences global climate, but also because it entails immediate repercussions on the reduction of society's vulnerability. Not only are hazards more acutely felt at the local level, but it is also within local communities that have the most know-how and experience to promptly deal with existing vulnerabilities. Competent and politically autonomous municipalities that are close with its citizens are therefore more likely to conduct effective adaptation action with wide ranging positive repercussions. In this line of reasoning, and bearing in mind the particular advantages of local scale adaptation, it is argued that the quality of our future cities will be influenced by the quality of future adaptation measures in public spaces.

Public space enables and promotes community life. It further potentially offers wide-ranging benefits such as place-making, sense of place or local identity. As a civic space, with communal and shared, a new variety of insurgent citizenship is arising within public spaces as the urgent matter of climate change adaptation is now broadly recognized. Furthermore, as emphasized, social, cultural and emotional factors can be more valued and respected within a community than the need of physical safety or ecological services.

Through public spaces and public space design, local climate change can be made visible and consequently meaningful for citizens and their livelihoods. Public spaces additionally provide a different source of knowledge and information, besides the mainstreamed sources of science and media, which may be apprehended an autonomous and independent process. Accordingly, public spaces provide extended opportunities for experiential learning inherent to adaptation processes.

In this line of reasoning, the design of public space sees itself enhanced in the face of impending weather events, being considered as a key factor in the adaptation of urban territories when facing climate change, and flood events in particular.

The examples here evidenced enabled the reasoning that, besides providing the means to include flood adaptation features, public spaces per se entail further specific connotations that are advantageous in adaptation endeavors. As evidenced by the analyzed examples, potential benefits may specifically arise from the characteristics of public space to:

- Favor interdisciplinary design—in places founded through interdisciplinary means, innovative thinking more easily emerges;
- Embrace multiple purposes—by combining flood adaptation measures with public space design, adjacent purposes arise among other sectorial needs such as water depuration, recreation or microclimatic melioration;
- Promote community awareness and engagement—by engaging the community in the design and use of a public space, not only awareness about climate change may be promoted but also the self-determining willingness for adaptation action is enhanced.
- Be supported by an extensive physical structure and system—by conforming a communicating structural network, public space offers the advantage of a decentralized and expansive means to tackle flood management;
- Expose and share value—by integrating flood management infrastructure in a public space design, instead of camouflaging it underground or in an isolated impenetrable area, a public investment is exposed and shared. A shared value that may instigate further opportunities such as amenity or environmental improvements.
- Promote risk diversification and communal monitoring—by investing in flood adaption measures applied in public space in addition to the conventional approaches, risk is dissipated and diversified and thus reduced. Moreover, through the diversification of risk, communal management among varied stakeholders is promoted. This way, communities are more involved, and the sense of responsibility and appropriation is stimulated, thus potentially leading to autonomous management and monitoring of shared implemented infrastructures.

People and communities can thus be perceived as more than susceptible targets and rather be professed as active agents in the process of managing urban vulnerability; climate change literacy, through the design of a public space, may endorse an increased common need for action and the pursuit of suitable solutions; and local know-how and locally-driven design can be considered as an added value for adaptation endeavors.

Acknowledgments: This research was supported by the Portuguese Foundation for Science and Technology funded by 'Quadro de Referência Estratégico Nacional—Programa Operacional Potencial Humano (QREN–POPH), Tipologia 4.1–Formação Avançada, comparticipado pelo Fundo Social Europeu e por fundos nacionais do Ministério da Ciência, Tecnologia e Ensino Superior (MCTES)', under the research project (CIAUD_UID/EAT/04008/2013) from the Research Centre for Architecture, Urbanism and Design (CIAUD), University of Lisbon, Portugal.

Author Contributions: Maria Matos Silva conceived and designed the research concept, research questions, data collection, data analysis, writing the manuscript and selecting the references. João Pedro Costa contributed to results analysis. The disclosed article introduces content that is developed in the doctoral thesis of Maria Matos Silva, with the title "Public space design and flooding: facing the challenges presented by climate change".

Conflicts of Interest: The authors declare no conflict of interest. The founding sponsors had no role in the design of the study; in the collection, analyses, or interpretation of data; in the writing of the manuscript, and in the decision to publish the results.

References

1. Innerarity, D. *El Nuevo Espacio Público*; Espasa Calpe: Madrid, Spain, 2006. (In Spanish)
2. Borja, J. *La Ciudad Conquistada*; Alianza Editorial: Madrid, Spain, 2003. (In Spanish)
3. Cowan, R. *The Dictionary of Urbanism*; Streetwise Press: Chicago, IL, USA, 2005.
4. Ricart, N.; Remesar, A. Reflexiones Sobre el Espacio Publico. 2013. Available online: http://repositorio.pucp.edu.pe/index/bitstream/handle/123456789/11961/reflexiones_espacio_Saenz.pdf?sequence=1&isAllowed=y (accessed on 8 February 2018). (In Spanish)

5. Martin, L. The grid as generator. In *Urban Design Reader*; Carmona, M., Tiesdell, S., Eds.; Architectural Press: Oxford, UK, 2007; pp. 70–82.
6. Portas, N. Espaço público e a cidade emergente-os novos desafios. In *Design de Espaço Público: Deslocação e Proximidade*; Brandão, P., Remesar, A., Eds.; Centro Português de Design: Lisboa, Portugal, 2003; pp. 16–19. (In Portuguese)
7. Madanipour, A. Ambiguities of urban design. *Town Plan. Rev.* **1997**, *68*, 363–383. [CrossRef]
8. Brandão, P. *Tica e Profissões, no Design Urbano. Convicção, Responsabilidade e Interdisciplinaridade. Traços da Identidade Profissional no Desenho da Cidade*; Universitat de Barcelona: Barcelona, Spain, 2004. Available online: http://hdl.handle.net/2445/35424 (accessed on 7 September 2010).
9. Arendt, H. *The Human Condition*, 2nd ed.; University of Chicago Press: Chicago, IL, USA, 1998.
10. Costa, J.P.; Sousa, J.F.D.; Matos Silva, M.; Nouri, A. Climate change adaptation and urbanism: A developing agenda for lisbon within the twenty-first century. *Urban Des. Int.* **2014**, *19*, 77–91. [CrossRef]
11. Matos Silva, M. Public Space Design for Flooding: Facing the Challenges Presented by Climate Change Adaptation. Ph.D. Thesis, Universitat de Barcelona, Barcelona, Spain, 2016.
12. Jacobs, J. *The Death and Life of Great American Cities*; Random House: New York, NY, USA, 1992; 458p.
13. Pedro Brandão (Coordinator). Research Project "Psss—public space service system" ptdc/ecm-urb/2162/2014. Funded by FEDER through the Operational Competitiveness Programme-COMPETE and by National Funds through FCT-Portuguese Foundation for Science and Technology: Universidade de Lisboa, Instituto Superior Técnico, CERIS, Investigação e Inovação em Engenharia Civil para a Sustentabilidade, 2016.
14. Fletcher, T.D.; Shuster, W.; Hunt, W.F.; Ashley, R.; Butler, D.; Arthur, S.; Trowsdale, S.; Barraud, S.; Semadeni-Davies, A.; Bertrand-Krajewski, J.L.; et al. Suds, lid, bmps, wsud and more—The evolution and application of terminology surrounding urban drainage. *Urban Water J.* **2015**, *12*, 525–542. [CrossRef]
15. Brandão, P.; Brandão, A.; Ferreira, A.; Travasso, N.; Remesar, A. What Is Public Space's Service Value? Some Relevant Research Questions. Available online: http://aesop2017.pt/images/Congresso/proceedings/BookofProceedings20170926.pdf (accessed on 7 December 2017).
16. Pelling, M. What determines vulnerability to floods; a case study in georgetown, guyana. *Environ. Urban.* **1997**, *9*, 203–226. [CrossRef]
17. Beck, U. *Risk Society: Towards a New Modernity*; SAGE Publications: Thousand Oaks, CA, USA, 1992.
18. Intergovernmental Panel on Climate Change (IPCC). Summary for Policymakers. In *Climate Change 2014: Impacts, Adaptation, and Vulnerability. Part A: Global and Sectoral Aspects*; Contribution of Working Group Ii to the Fifth Assessment Report of the Intergovernmental Panel on Climate Change; Cambridge University Press: Cambridge, UK; New York, NY, USA, 2014; 32p.
19. Couzin, J. Living in the danger zone. *Science* **2008**, *319*, 748–749. [CrossRef] [PubMed]
20. Biesbroek, G.R. *Challenging Barriers in the Governance of Climate Change Adaptation*; Wageningen University: Wageningen, The Netherlands, 2014.
21. Van Der Linden, S. Towards a new model for communicating climate change. In *Understanding and Governing Sustainable Tourism Mobility: Psychological and Behavioural Approaches*; Cohen, S.A., Higham, J.E.S., Stefan, G., Peeters, P., Eds.; Routledge, Taylor and Francis Group: Oxfordshire, UK, 2014; pp. 243–275.
22. Commission for Architecture and the Built Environment (CABE). *Public Space Lessons—Adapting Public Space to Climate Change*; CABE Space: London, UK, 2008; 8p.
23. Matos Silva, M.; Costa, J. Flood adaptation measures applicable in the design of urban public spaces: Proposal for a conceptual framework. *Water* **2016**, *8*, 284. [CrossRef]
24. Howe, C.; Mitchell, C. *Water Sensitive Cities*; IWA Publishing: London, UK, 2012.
25. Trip, J.J. *What Makes a City? Planning for 'Quality of Place'. The Case of High-Speed Train Station Area Redevelopment*; IOS Press: Amsterdam, The Netherlands; Delft University Press: Delft, The Netherlands, 2007.
26. Brandão, P.; Carrelo, M.; Águas, S. *O Chão da Cidade. Guia de Avaliação do Design de Espaço Publico*; Centro Português de Design: Lisboa, Portugal, 2002; 199p. (In Portuguese)
27. Brandão, P. *O Sentido da Cidade. Ensaios Sobre o Mito da Imagem Como Arquitectura*; Livros Horizonte: Lisboa, Portugal, 2011. (In Portuguese)
28. Nouri, A.S.; Matos Silva, M. Climate change adaptation and strategies: An overview. In *Green Design, Materials and Manufacturing Processes*; Bártolo, H., Bartolo, P.J.D., Alves, N.M.F., Mateus, A.J., Almeida, H.A., Lemos, A.C.S., Craveiro, F., Ramos, C., Reis, I., Durão, L., et al., Eds.; Taylor and Francis: Lisbon, Portugal, 2013; pp. 501–507.

29. Brandão, P. *La Imagen de la Ciudad: Estrategias de Identidad y Comunicación*; Publicacions i Edicions de la Universitat de Barcelona: Barcelona, Spain, 2011. (In Spanish)
30. Brandão, P. *Entrevista. ArqA—Arquitectura e Arte*; Novas Colectividades: Lisboa, Portugal, 2013; pp. 28–30.
31. Capel, H. *El Modelo Barcelona: Un Examen Crítico*; Ediciones del Serbal: Barcelona, Spain, 2005. (In Spanish)
32. Zaitzevsky, C. The "emerald necklace": An historic perspective. In *Emerald Necklace Parks: Master Plan*; Walmsley, T., Pressley, M., Eds.; Commonwealth of Massachusetts, Department of Environmental Management: Boston, MA, USA, 2001; pp. 27–42.
33. Matos Silva, M. El Modelo Barcelona de Espacio Público y Diseño Urbano: Public Space and Flood Management/Dipòsits D'aigües Pluvials. Universitat de Barcelona: Spain, 2011. Available online: http://hdl.handle.net/2445/17762 (accessed on 8 February 2018).
34. Pötz, H.; Bleuzé, P. *Urban Green-Blue Grids for Sustainable and Dynamic Cities*; Coop for Life: Delft, The Netherlands, 2012.
35. London Legacy Development Corporation (LLDC). Landscaping the Park. Available online: http://queenelizabetholympicpark.co.uk/ (accessed on 22 January 2016).
36. Malano, H.; Maheshwari, B.; Singh, V.P.; Purohit, R.; Amerasinghe, P. Challenges and opportunities for peri-urban futures. In *The Security of Water, Food, Energy and Liveability of Cities: Challenges and Opportunities for Peri-Urban Futures*; Maheshwari, B., Purohit, R., Malano, H., Singh, V.P., Amerasinghe, P., Eds.; Springer: Dordrecht, The Netherlands, 2014; pp. 3–10.
37. Matos Silva, M.; Costa, J.P. Urban flood adaptation through public space retrofits: The case of lisbon (portugal). *Sustainability* **2017**, *9*, 816. [CrossRef]
38. European Environment Agency (EEA). *Towards Efficient Use of Water Resources in Europe*; Report No. 1/2012; European Environment Agency: Copenhagen, Denmark, 2012; 70p.
39. Sharma, A.; Pezzaniti, D.; Myers, B.; Cook, S.; Tjandraatmadja, G.; Chacko, P.; Chavoshi, S.; Kemp, D.; Leonard, R.; Koth, B.; et al. Water sensitive urban design: An investigation of current systems, implementation drivers, community perceptions and potential to supplement urban water services. *Water* **2016**, *8*, 272. [CrossRef]
40. Ashley, R.M.; Faram, M.G.; Chatfield, P.R.; Gersonius, B.; Andoh, R.Y.G. Appropriate drainage systems for a changing climate in the water sensitive city. In *Low Impact Development 2010: Redefining Water in the City*; ASCE: Listeria, VA, USA, 2010.
41. Rotterdam Climate Initiative (RCI). *Rotterdam Climate Proof Adaptation Programme. The Rotterdam Challenge on Water and Climate Adaptation*; Rotterdam Climate Initiative: City of Rotterdam, The Netherlands, 2009; 22p.
42. *Bíblia Sagrada*, 5th ed.; Difusora Bíblica: Lisboa, Portugal, 1991. (in Portuguese)
43. Institute for Sustainable Communities (ISC). *Promising Practices in Adaptation & Resilience. A Resource Guide for Local Leaders*; Institute for Sustainable Communities, Produced in Partnership with Center for Clean Air Policy: Washington, DC, USA, 2010; 98p.
44. Dutch Dialogues. About Dutch 'Dialogues'. Available online: www.dutchdialogues.com (accessed on 2 April 2011).
45. Pinto, A.J. Coesão Urbana: O Papel das Redes de Espaço Público. Universitat de Barcelona: Spain, 2015. Available online: http://hdl.handle.net/2445/67852 (accessed on 5 April 2011). (In Portuguese)
46. Portas, N. O urbano e a Urbanística ou os Tempos das Formas. Available online: http://www.culturgest.pt/actual/01/01-nunoportas.html (accessed on 29 January 2012).
47. Pinto, A.J.; Remesar, A.; Brandão, P. Networks and Anchors: From Morphology to the Strategy of Urban Cohesion. In Proceedings of the 1st Conference of the Portuguese Network of Urban Morphology "Urban Morphology in Portugal: Approaches and Perspectives", Lisbon, Portugal, 8 June 2011; International Seminar on Urban Form, Ed.; pp. 1–3.
48. Susdrain. Susdrain Case Studies. Available online: http://www.susdrain.org/case-studies (accessed on 14 April 2016).
49. Howe, C.A.; Butterworth, J.; Smout, I.K.; Duffy, A.M.; Vairavamoorthy, K. *Sustainable Water Management in the City of the Future*; Findings from the SWITCH Project 2006–2011, Ed.; UNESCO-IHE: Delft, The Netherlands, 2011.
50. Papacharalambous, M.; Davis, M.S.; Marshall, W.; Weems, P.; Rothenberg, R. *Greater New Orleans Urban Water Plan: Implementation*; Waggonner & Ball Architects: New Orleans, LA, USA, 2013; 225p.
51. White, I.; Howe, J. The mismanagement of surface water. *Appl. Geogr.* **2004**, *24*, 261–280. [CrossRef]

52. Hartmann, T.; Driessen, P. The flood risk management plan: Towards spatial water governance. *J. Flood Risk Manag.* **2013**, 1–10. [CrossRef]
53. Lennon, M.; Scott, M.; O'Neill, E. Urban design and adapting to flood risk: The role of green infrastructure. *J. Urban Des.* **2014**, *19*, 745–758. [CrossRef]
54. Intergovernmental Panel on Climate Change (IPCC). *Managing the Risks of Extreme Events and Disasters to Advance Climate Change Adaptation*; Cambridge University Press: Cambridge, UK; New York, NY, USA, 2012; 582p.
55. Valera, S. La percepció del risc. In *Com "Sentim" el Risc*; Mir, N., Ed.; Beta Editorial: Barcelona, Spain, 2001; pp. 235–261.
56. Evers, M.; Höllermann, B.; Almoradie, A.; Garcia Santos, G.; Taft, L. The pluralistic water research concept: A new human-water system research approach. *Water* **2017**, *9*, 933. [CrossRef]
57. Solà-Morales, M.D. Scheveningen Den Haag, 2006–2012. Available online: http://manueldesola-morales.com (accessed on 26 January 2016).
58. Intergovernmental Panel on Climate Change (IPCC). *Managing the Risks of Extreme Events and Disasters to Advance Climate Change Adaptation—Summary for Policymakers*; IPCC: Cambridge, UK; New York, NY, USA, 2012; 19p.
59. Barnett, J.; O'Neill, S. Maladaptation. *Glob. Environ. Chang.* **2010**, *20*, 211–213. [CrossRef]
60. Berkes, F.; Colding, J.; Folke, C. *Navigating Social-Ecological Systems. Building Resilience for Complexity and Change*; Cambridge University Press: Cambridge, UK, 2003.
61. Brandão, P.; Remesar, A. *Design de Espaço Público: Deslocação e Proximidade*; Centro Português de Design: Lisboa, Portugal, 2003.

water

MDPI

Article

Urban Flood Simulation Using Synthetic Storm Drain Networks

Robert Bertsch *, Vassilis Glenis and Chris Kilsby [ORCID]

School of Engineering, Newcastle University, Newcastle upon Tyne NE1 7RU, UK;
vassilis.glenis@newcastle.ac.uk (V.G.); chris.kilsby@newcastle.ac.uk (C.K.)
* Correspondence: r.bertsch@newcastle.ac.uk; Tel.: +44-(0)-790-369-7834

Received: 28 September 2017; Accepted: 24 November 2017; Published: 28 November 2017

Abstract: Recent developments in urban drainage modelling allow for a more realistic coupling of the two-dimensional (2D) surface and one-dimensional (1D) sub-surface drainage domain exchanging water through storm drain inlets instead of a sub-catchment approach based on manholes. Experience has shown, however, that comprehensive records of storm drain inlet locations are often missing or incomplete, preventing users accessing the full benefit of these modelling capabilities. Therefore, this study developed a GIS routine to generate synthetic storm drain inlet locations for the purpose of urban flood modelling. Hydrodynamic model results for a synthetically generated and surveyed storm drain inlet network were obtained using the CityCAT 1D/2D system. On a catchment scale the flow field (surface and flow captured by inlets) simulated by the network of synthetic storm drainage inlets shows satisfactory results when compared with that simulated using the actual network. The results also highlight the sensitivity of the inflows to relatively small changes in terms of the location of storm drain inlets and the effectiveness of storm drain inlets in ponding areas.

Keywords: storm drain inlet; pluvial flooding; urban drainage 1D/2D modelling; GIS; CityCAT

1. Introduction

1.1. Background to Urban Drainage Models

The urban drainage system is often described by a dual drainage concept formed from the surface and sub-surface domains [1–4]. Under normal conditions (i.e., non pressurised sub-surface system), water from the surface drains into the sub-surface domain via storm drain inlets (sometimes referred to as gullies) (Figure 1). In case of a pressurised sub-surface system, a reverse flow [5] can lead to surcharging conditions at storm drain inlets and/or manholes. Manholes are located between pipe sections of different diameters, changes in direction or gradient, junctions of pipes, and at designed intervals that are required to provide access to the sub-surface domain for maintenance and inspection purposes [6].

Modelling the flow dynamics in between the two domains is complex, particularly when the flow is discontinuous. For example, modelling flows where shock waves are present requires the use of advanced numerical techniques and the appropriate system of conservation laws, see [7,8]. Additionally, modelling unsteady mixed (free surface—pressurised) flows in pipes is challenging because free surface flows and pressurised flows are described by different equations and the transition between these two flow states is difficult to capture, see [9].

Therefore, commonly applied software tools for modelling urban drainage and floods, such as Infoworks CS (Collection Systems) or ICM (Integrated Catchment Modelling) [10] adopt simplifications in two respects. Firstly, the numerical solutions are usually implemented by using the Preissmann slot with free-surface flow to simulate pressurised conditions in a pipe [11–13]. Secondly, in the way the drainage system is represented inside the model, where instead of using the storm drain inlets, a selection of nodes (usually manholes) are used to link the sub-surface system with pre-defined sub-catchments. These sub-catchments are usually delineated manually and are assigned a number of parameters to reflect the sub-catchment drainage characteristics. Based on these characteristics, rainfall-runoff hydrographs are established for each sub-catchment to transfer water volumes into the sub-surface domain [14]. This means that no two-dimensional (2D) surface flow routing is simulated at this stage, but surcharging nodes can be treated differently. For example, Infoworks CS adopts the concept of virtual reservoirs [15], which assign a pre-defined flood volume to each node. Surcharging water volumes from the nodes are stored in this virtual reservoir without conducting any 2D free surface flow simulations. More recently, Infoworks ICM allows for the coupling of nodes with a TIN (Triangular Irregular Network) mesh, which is derived from a terrain model [10]. This provides the opportunity to simulate 2D surface flow of the surcharged water volumes.

From a practical perspective, the simplifications outlined are justifiable as they significantly reduce model complexity, computational times, and data requirements. Also, recent versions of commercial modelling tools provide the option to incorporate the storm drain inlets and simulate 2D surface runoff directly from rainfall. However, there is a lack of fully coupling the one-dimensional (1D) sub-surface and the 2D surface domain by explicitly modelling all of the storm drain inlets within the entire urban catchment. This prevents an in-depth understanding of the flow dynamics between the drainage domains, which requires an implementation of all storm drain inlets. Particularly for detailed source-pathway-receptor analysis across the surface and sub-surface domain only fully coupled 1D/2D models provide the necessary results. Research presented by [1,16] introduced the multiple-linkage-element concept, which was a first attempt to address the coupling of the surface and sub-surface domain using grouped storm drain inlets. At the same time, the recent advances in cloud computing provide access to sufficient computational power to make city-wide large and complex models, such as CityCAT, practicable [17].

Figure 1. Examples of typical storm drain inlets.

1.2. CityCAT 1D/2D: Fully Coupled 1D/2D Modelling

CityCAT—City Catchment Analysis Tool [17]—is a newly developed modelling software for urban drainage and flood modelling purposes. The overland flow component of CityCAT is based on the shallow water equations, and the solution is obtained using high-resolution finite volume shock-capturing schemes [18]. The pipe flow component of CityCAT is based on the mathematical model for mixed flows in pipes (MFP) presented by [9]. The MFP uses the St Venant equations and a conservative system of equations for pressurised flows that are derived from the compressible Euler equations. It can model sub-atmospheric pressures and large pressure wave celerities (>1000 m/s). Additionally, the model can capture the transition between the free surface and pressurised flow.

The software does not depend on sub-catchments, but instead entire terrain-derived catchments are simulated along with buildings and green areas, allowing for a physical simulation of roof-drainage and infiltration. Building footprints are removed from the grid to reduce the cell numbers, and hence the run time that is required. Runoff from the roofs, however, is still kept in the system. By automatically generating the final computational grid CityCAT significantly simplifies the model setup process.

CityCAT can be deployed on servers or the cloud to access advanced computational resources, and therefore allow for large model domains to be simulated at fine grid resolutions.

By actively simulating all of the storm drain inlets and using high-resolution terrain data CityCAT 1D/2D offers new dimensions of accurately modelling the dynamic dual drainage system in urban areas. This is essential information for assessing the impact of flooding on people and the urban fabric but also for the purpose of planning flood adaptation options. From the perspective of network design and network improvement high-resolution hydraulic models offer the benefit of assessing the performance of individual or a group of storm drain inlets. In this context, storm drain inlets might be added or removed. This would allow, for instance, assessing the potential reduction of drainage efficiency of storm drain inlets as result of clogging or structural changes for instance [19].

1.3. CityCAT 1D/2D: Linking Storm Drain Inlets With Manholes

CityCAT automatically connects storm drain inlets with the nearest manhole using parametric relations to describe the flow through the inlet. The linking is based on the single-linking element (SLE) approach [1,20] where the Bernoulli equation and the energy loss equation are used to calculate the exchange of discharge between the surface and the subsurface. Importantly, this allows for the dimensions and design of various inlets to be included, as well as explicitly modelling the efficiency of the inlets that can be affected by clogging with debris or leaves. For the purpose of this study, uniform dimensions for all inlets (0.3 m × 0.3 m) and linking pipes (90 mm diameter) between the inlet and manhole were applied. CityCAT solves every time step the two-dimensional shallow water equations for the overland flow, the SLE equations for every link between the surface and the subsurface, and the mixed flow equations for the pipes.

1.4. Aim of Study

In order to access the full capabilities of CityCAT, a comprehensive record of storm drain inlet locations is required. Experience shows, however, that such records are not always available. For certain areas, no records exist at all, while other areas have only incomplete or outdated records. Consequently, missing storm drain inlet data prevent a full application of CityCAT. In order to fill this gap, the main aim of this study is to develop and evaluate an automated GIS (Geographic Information System) routine to generate synthetic storm drain inlet locations for the existing pipe networks. For evaluation purposes a detailed field survey was conducted in order to have a record of actual storm drain inlet locations for a catchment. From a practical perspective, the question is whether the routine can provide a straightforward and low cost alternative when compared to resource intensive field work when establishing a network of storm drain inlets.

Missing input data is often one of the biggest challenges when building flood models. Apart from storm drain inlet locations, this may also concern dimensions and locations of manholes and pipes. Taking the idea proposed in this work of using surrogate data and generic methods, the tool that is developed could potentially be extended in future work to assist in generating other crucial information that is required for high-resolution urban flood models.

The study will first outline the assumptions behind the initially developed GIS routine. Based on preliminary hydraulic model results that are obtained from CityCAT 1D/2D simulations, improvements to the GIS routine are presented. Finally, model results for a synthetically generated and surveyed storm drain inlet network are compared in order to validate the GIS routine and to study the sensitivity of the whole network drainage efficiency (for sub-surface and surface) to the density and placement of inlets.

2. Development of GIS Routine

2.1. Background

A variety of methods have been developed over the last decades for the design of sewer systems [21–25]. Although these methods vary, they are primarily aimed to provide assistance in the development of new sewer systems in the form of manholes and pipes. This study will focus on the design (location and density) of storm drain inlets for existing sewer systems.

The design and spacing of storm drain inlets in reality takes into account a large number of parameters [26,27] for instance: slope and cross-fall of the road, the storm drain inlet grating type and efficiency, surface roughness, flow width and velocity in the kerb channel, maintenance factors, design storm, and contributing catchment area per storm drain inlet. In this context [19], identify three major factors contributing to the inlet capacity and efficiency: lateral street slope, longitudinal street slope, and pavement roughness. If all of those variables were to be incorporated in a GIS routine, a significant amount of high quality input data would be required that is not usually available.

A further challenge in developing a generic GIS routine for locating storm drain inlets is the change of design criteria over time [28–30]. More recent design criteria shifted towards separated drainage systems and the need to accommodate larger runoff volumes. In this context, for example, in Scotland, newly built sustainable drainage systems have to accommodate storm events with return periods of up to 200 years [31]. Most installed sewer systems however are based on older design manuals, dating back several decades, which adopted return periods between 1 and 30 years [6]. Therefore, a generic tool to generate storm drain inlet locations for existing sewer systems has to find a balance in terms of the different design criteria over time.

2.2. Initial Assumptions behind Gis Routine

With the above considerations in mind, it has been decided to initially adopt a robust and simplified approach for the GIS routine that is developed. Furthermore, the routine should be universally applicable and therefore rely only on a minimum number of input data. Crucially however, the tool needs to incorporate information on the existing sub-surface system. The three assumptions behind the initial version of the GIS routine are therefore:

1. **All storm drain inlets are on roads and of the same grating type**: In reality storm drain inlets can be found in various places, but predominantly on roads—next to or underneath the kerb (Figure 1). Several studies were carried out investigating the impact of different locations, grating, and cover types on the drainage efficiency of storm drain inlets [19,32–34]. However, for simplification purposes, it is assumed that all storm drain inlets are located on roads and are of the same type. Hence, all synthetic storm drain inlets that are generated mimic the location shown in the left, Figure 1.

2. **All storm drain inlets are spaced at an equal distance**: The approach of an equidistant spacing of 50 m between storm drain inlets is obtained from [6].

3. **All storm drain inlets are close to the existing pipe network**: It is assumed that storm drain inlets are located at a certain distance to the nearest pipe. The pipe network in form of a polyline shapefile is used to conduct a spatial proximity analysis. A threshold distance of 20 m between storm drain inlet and pipe has been selected.

Based on those assumptions, two sets of input data are required: (1) a polygon shapefile representing the road network; and (2) a polyline shapefile representing the pipe network. With the above assumptions, no terrain information is accounted for at this stage. The integration of terrain information will be presented after the preliminary results.

2.3. Placement of Storm Drain Inlets

In the first step, the synthetic storm drain inlets are placed along a reference line. For this purpose, the road polygons are dissolved (i.e., no internal boundaries) into one single polygon (Figure 2a). Inside this polygon a buffer is created applying a distance of 0.25 m in order to assure that all of the storm drain inlets are placed inside the road. The buffer polygon is subsequently converted to a line feature, which is subsequently referred to as the reference line (Figure 2b). Finally, point features (i.e., storm drain inlets) are placed along the reference line at an equidistant spacing of 50 m (Figure 2b).

2.4. Alignment of Storm Drain Inlets with Pipe Network

The pipe network is applied next. As shown in Figure 2c, a 20 m buffer is created around the entire pipe network. Any storm drain inlet that is located outside the 20 m buffer will be discarded (Figure 2c). Synthetic storm drain inlets in the final layer therefore meet the following conditions: they are located inside the road polygon, 0.25 m inside the kerb and 50 m apart from each other, and they are within 20 m of the pipe network.

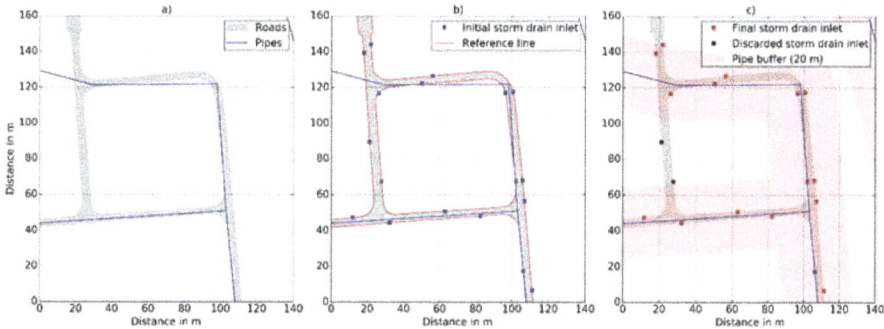

Figure 2. Generation of a synthetic storm drain inlet network: (**a**) Input data consisting of dissolved road polygon shapefile and pipe network polyline shapefile. (**b**) Creating reference line (0.25 m inside of road polygon shapefile) and placing points on reference line at an equidistant space of 50 m. (**c**) Storm drain inlets at a distance >20 m to nearest pipe segment are discarded.

3. Case Study

Having outlined the principles behind the initial GIS routine in general, a case study is conducted. For validation purposes, a field study was completed to survey the actual storm drain inlet network. The study area is located in central Scotland. The topography of the area is relatively flat. The drainage network, including pipes and manholes, were obtained from an Infoworks CS model, which was made available by Scottish Water. By incorporating additional drainage network elements from GIS records, the final drainage network that was applied consists of 294 manholes and 10,898 m of pipes. For the purpose of this study, only storm and combined drainage pipes were applied. Furthermore, no base flow or dry weather flow was taken into account.

3.1. Synthetic Storm Drain Inlet Locations: Applying the GIS Routine

Following the workflow of the GIS routine outlined, the first step was to obtain the road data. For this purpose, the OS (Ordnance Survey) MasterMap Topography layer was downloaded from Edina Digimap [35]. From the initial data set, all of the road polygons ('featureCod: 10172') were extracted. Applying the steps of the initial GIS routine, the network generated contained 376 synthetic storm drain inlets (Figure 3a).

3.2. Surveying of Actual Storm Drain Inlet Locations: Field Work

Approximately 13 km of roads were surveyed, requiring four full working days, including travelling time and post-processing of the data collected. The GPS equipment that was applied to record the storm drain inlet locations was the hand held device Leica GS15 with SMARTNET correction [36]. The network based Real Time Kinematic (RTK) function of the device combines satellite and GPRS signals to achieve greater accuracy. Overall, the positional accuracy observed during the field work was approximately +/−10 cm, which is thought to be sufficient for the purpose of this study. At a few storm drain inlet locations signal problems were experienced. Those locations were manually highlighted on a map and later added to the RTK-surveyed storm drain inlets in GIS.

If a parked vehicle made it impossible to survey the actual inlet, its location was taken at the closest distance possible and subsequently adjusted in GIS.

In a final step, the locations of the surveyed storm drain inlets were aligned with the below ground pipe network. Although a surveyed storm drain inlet would indicate the existence of below ground drainage features, a number of inlets were surveyed in areas without any GIS records of a pipe in close proximity. To avoid unrealistic long connections between those storm drain inlets and a manhole, and not to estimate any pipe dimensions and locations, it was decided to discard those storm drain inlets. This was done applying the same 20 m buffer that has been used for aligning the synthetic storm drain inlet locations (Figure 2c). As shown in Figure 3, the post-process surveyed network consists of 445 storm drain inlets.

3.3. Hydrodynamic Model for Case Study Area

Finally, the surveyed and synthetically generated storm drain inlet networks were applied in hydrodynamic simulation in the CityCAT 1D/2D software. The drainage network elements applied included the storm drain inlet locations, pipes, and manholes. The simulations were conducted using LiDAR (Light Detection and Ranging) terrain data with a resolution of 2 m × 2 m, and a uniform rectangular numerical grid was generated with $\Delta x = \Delta y = 2$ m. The catchment is made up of 525,554 cells, resulting in a total area of approximately 2.1 km^2. Buildings and green areas were extracted from MasterMap topography data. The 20 and 50 year return period storm event of 60 minutes duration applied (Figure 4) were generated using the Flood Estimation Handbook (FEH) procedure [37]. Surface roughness coefficients (Manning's n values) of 0.02 and 0.035 for impermeable, and permeable surfaces were applied, respectively. The model uses an adaptive time step algorithm in order to satisfy the Courant number condition. In this application, the time steps are fractions of a second and each simulation took approximately 29 h to complete on a server with an Intel Xeon processor at 2.6 GHz using 8 cores and DDR4 memory.

Based on interviews with the Local Council, Scottish Water, and residents during the field survey, it was found that the hydrodynamic model results identified most areas that had been affected from pluvial flooding in recent years. Detailed investigations into the simulated and observed inundation depths, however, were limited due to a lack of detailed data, and are therefore not presented in this study.

3.4. Preliminary Results

As shown in Figure 3, the initial GIS routine produced 376 storm drain inlets in comparison to the 445 surveyed ones. There are two reasons that are thought to be responsible for this difference. As described earlier, newly built area drainage networks are designed to accommodate larger storm runoff volume. The area highlighted in Figure 3b is a newly built area with a much greater density of storm drain inlets in comparison to the overall catchment. The second reason is the small scale terrain depression, which quickly results in an accumulation of surface water. Those areas are sometimes referred to as in-sag locations [26], which would see additional inlets installed to cope with the surplus of water.

The network drainage efficiency has been assessed by calculating the inflow volumes for the network at each time step in CityCAT. The initial network, with fewer storm drain inlets and not specifically accounting for ponding areas results in a clear under-representation of the captured flow (Qi) by the storm drain inlets (Figure 4). The flow captured (Qi) represents the portion of flow entering the storm drain inlets and is the difference between the flow approaching a storm drain inlet and the pass-over flow [19]. The graph in Figure 4 shows the total volume of water that is drained by all storm drain inlets for the synthetic and surveyed storm drain inlet networks for both of the storm events. Figure 4 not only shows an under-representation of the captured flow, but also a later onset of drainage when comparing the synthetic storm drain inlets with the actual ones.

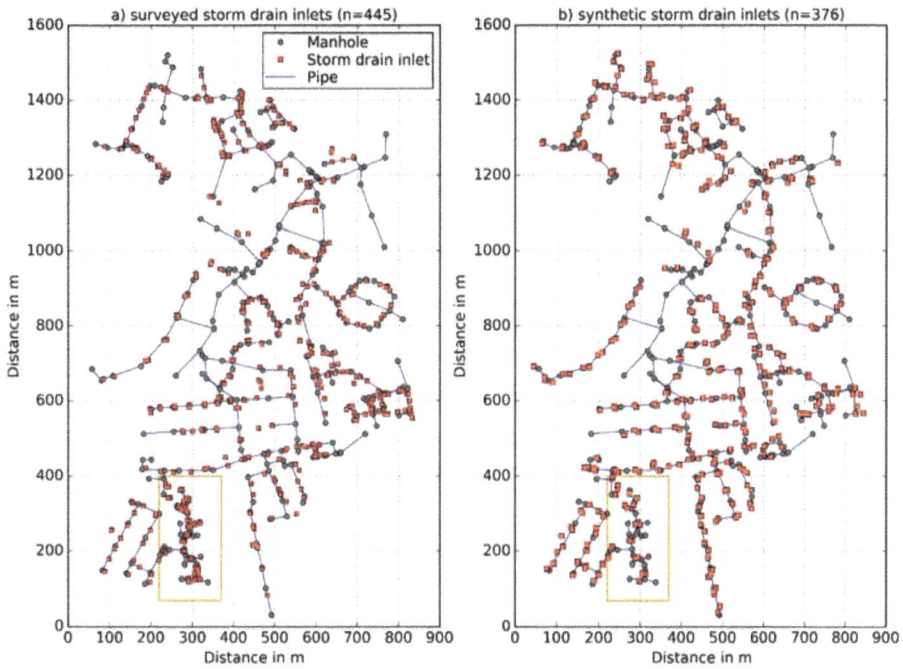

Figure 3. Surveyed storm drain inlet network (**a**) and synthetic storm drain inlet version 1 (**b**).

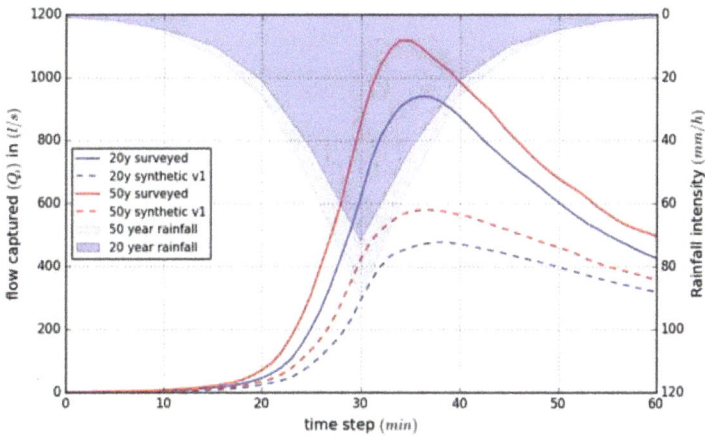

Figure 4. Comparison of flow captured (Qi) for surveyed and synthetic storm drain inlet network version 1 for a 20 and 50 year storm event.

4. Adaptations to Initial GIS Routine

Based on the preliminary results, it becomes evident that adaptations to the initial GIS routine are required. Those adaptations should aim to better approximate the total number of storm drain inlets, as well as the drainage performance of a synthetically generated storm drain inlet network. Terrain information is likely to be crucial, so the adaptions made focus on the following aims:

1. Increase the total number of synthetic storm drain inlets to match the actual number
2. Increase the drainage efficiency of synthetic storm drain inlets by re-distributing them applying terrain information
3. Increase the drainage capacity within surface water accumulation areas by adding more synthetic storm drain inlets

4.1. Storm Drain Inlet Density

In order to increase the total number of storm drain inlets, it was decided to specifically address areas that were built under more recent building standard. As outlined initially, the drainage network and its capacity within those areas is likely to have been designed to cope with larger runoff volumes when compared to older ones. It is therefore required to understand which areas across the catchment are relatively newly built. The spacing of storm drain inlets within those areas of interest (AOI) is subsequently reduced in order to increase the storm drain inlet density. For this purpose, the reference line inside the AOI was separated from the rest. Any existing synthetic storm drain inlet from the first network version that was found to be inside the AOI was deleted. Subsequently, new storm drain inlets were placed at a spacing of 20 m. Finally, the storm drain inlets from the AOI were merged with the remaining storm drain inlets from the first version.

4.2. Adjusting the Locations of Storm Drain Inlets

Apart from having too few storm drain inlets, the preliminary model results also suggested that the initially placed storm drain inlets are insufficient in terms of their drainage. It was therefore decided to re-distribute the initially placed storm drain inlets to lower elevated cells within the immediate surrounding of the storm drain inlets. First, a 3 m buffer was created around each storm drain inlet to extract the reference line. Around each of those 6 m long sections a second buffer area with a distance of 1.5 m was created (Figure 5b). Within each of those buffer areas the lowest terrain point was identified (Figure 5b). Finally, the shortest distance between the lowest terrain point and the reference line was calculated in order to identify the final location of the adjusted storm drain inlet (Figure 5c).

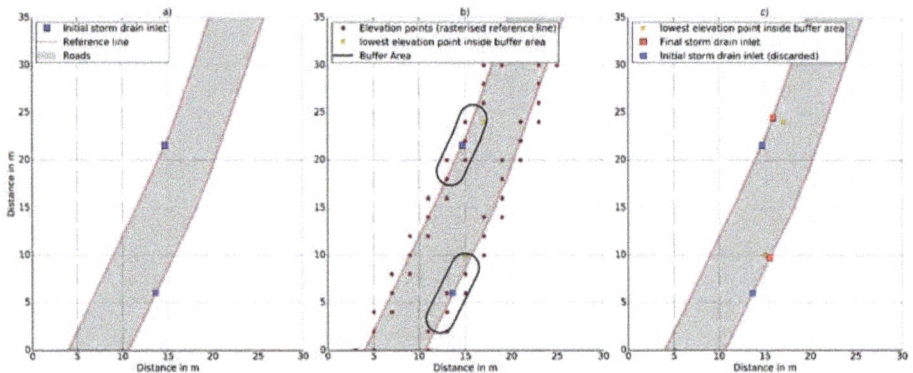

Figure 5. Re-distributing process of initial storm drain inlets (**a**) by applying terrain information to identify lower lying cells (**b**) and final, re-distributed inlet location with discarded, initial one (**c**).

Based on the criteria outlined above an adjusted storm drain inlet can be at a maximum distance of 4.5 m to its initial one. The more important change between the initial and adjusted storm drain inlet, however, is the difference in their elevation. A comparison between the terrain elevation at the initial location and the adjusted one found an average drop in elevation of 6.9 cm (Figure 6). The maximum drop in elevation between an initial and adjusted inlet is 0.91 m. At that location, the elevation of the initial storm drain inlet was affected by an embankment feature close to the road.

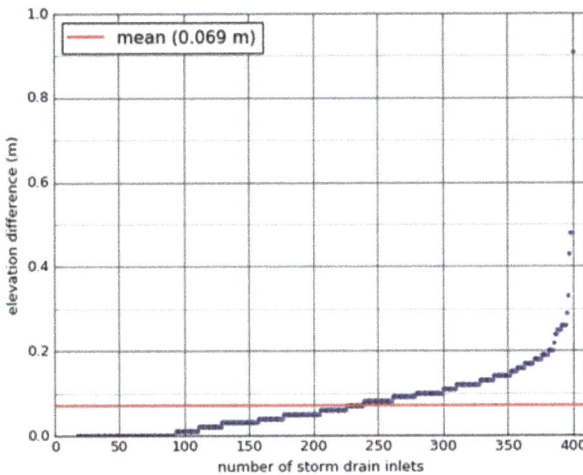

Figure 6. Difference of elevation between the initial storm drain inlet location and adjusted one.

4.3. Surface Water Accumulation Areas

The final improvement made to the initial GIS routine concerns surface water accumulation areas. In reality, those areas of local terrain depressions would likely have more storm drain inlets that are installed to cope with the surplus of water. To add additional storm drain inlets, areas of depression had to be identified first. For this purpose model results obtained from a CityCAT 2D (surface only) simulation were applied. The simulation was run for a 20-year return period on a 2 m LiDAR grid and included buildings as well as green areas. The results of the final time step (after 60 minutes) were subsequently used.

In a first step, all of the cells with an inundation depth $\geq$0.05 m were merged together to form continuous polygons (Figure 7a,b). Any polygon with an area < 200 m^2 and not intersecting with the reference line were discarded. For each polygon left the lowest terrain point inside the road polygon was identified (Figure 7b). The threshold values of 0.05 m and 200 m^2 are thought to be a reasonable combination to reflect the actual surface water ponding areas and not being misguided by potential erroneous terrain data. Subsequently, at the shortest distance between the lowest terrain point and the reference line, an additional storm drain inlet was placed (Figure 7b). Around this storm drain inlet, a buffer with a distance of 150% of the terrain resolution was created (Figure 7c). At the lower lying intersection point of this buffer line with the reference line a second additional storm drain inlet was added (Figure 7c). Adding a second storm drain inlet aims to reflect so called twin-gullies, which are commonly installed in sag locations [26].

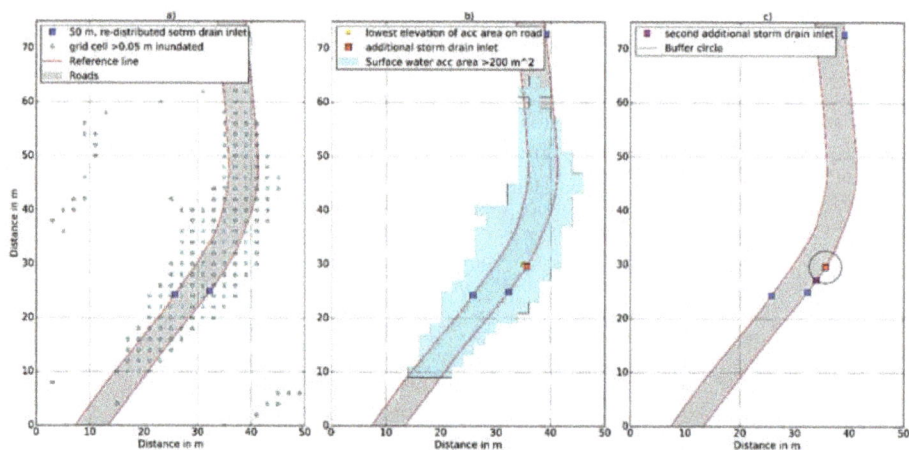

Figure 7. Placement of additional twin storm drain inlets in surface water accumulation areas. Extracted cells with inundation depth >0.05 m (**a**) are accumulated and converted to polygons (**b**). For each polygon >200 m^2 and intersecting with reference line the lowest elevation point inside the road polygon is identified and two additional inlets are placed (**b,c**). All four storm drain inlets shown in (**c**) are applied for hydrodynamic simulation in CityCAT 1D/2D.

5. Results and Discussion

5.1. Final Synthetic Storm Drain Inlet Network

Applying all of the adaptions outlined to the initially introduced case study area the final synthetic storm drain inlet network presented in Figure 8 consists of 443 storm drain inlets. In comparison to the 376 and 445 storm drain inlets of the initial synthetic network and surveyed one, respectively, the adaptations show a substantial improvement in terms of the number of storm drain inlets.

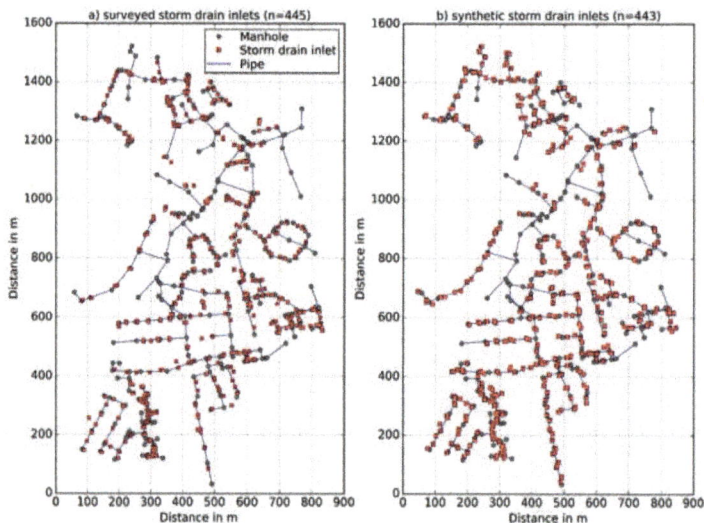

Figure 8. Surveyed storm drain inlet network (**a**) and synthetic storm drain inlet network final version (**b**).

In order to compare the hydraulic performance of the synthetic storm drain inlet network against the surveyed, two different sets of results are presented in the following looking at:

1. The surface/sub-surface interface by comparing the total volume of water entering all storm drain inlets (as shown previously in Figure 4).
2. The surface domain by comparing surface water inundation depth grids.

5.2. Surface/Sub-Surface Domain Interface: Water Volume Flow in Storm Drain Inlets

In comparison to the preliminary results (Figure 4), the final synthetic network model shows a significant improvement (Figure 9) in terms of the drainage efficiency when compared with the surveyed network model, in terms of both the total volume drained and the shape and timing of the inflow hydrograph (Qi). The results shown in Figure 9 highlight the significance of the adaptions that were made to the GIS routine and especially the incorporation of terrain information to re-position the inlets. As described earlier, the average elevation difference between the initial and adjusted storm drain inlet of 6.9 cm was relatively small, suggesting that together with the 42 storm drain inlets added in surface water accumulation areas, the system drainage performance is quite sensitive to small scale topographical changes and positioning. These issues are also highlighted in other studies [38,39].

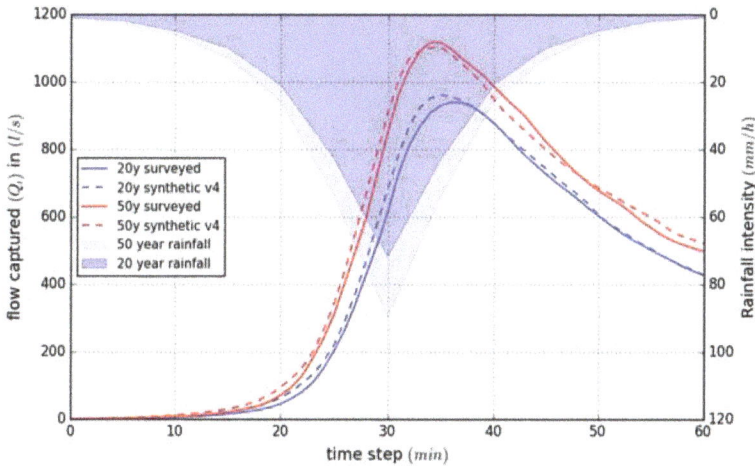

Figure 9. Comparison of flow captured (Qi) for surveyed and synthetic storm drain inlet network version 4 for a 20 and 50 year storm event.

5.3. Surface Water Domain: Inundation Depth on Grid

The final results that are presented investigate the impact of the different storm drain inlet networks on the surface water inundation depth. The map in Figure 10 was produced by subtracting the maximum surface water grid obtained for the simulation using the surveyed storm drain inlet network from that using the synthetic network. Only the results for the 20-year storm event are presented.

A negative difference indicates a greater surface water depth for the simulation based on the surveyed storm drain inlet network when compared to the one that is obtained from the simulation using the synthetic storm drain inlet network. Whereas, a positive difference means a greater surface water depth for the simulation based on the surveyed storm drain inlet network.

Overall, the differences in surface water depth mostly range between −0.01 and 0.01 m, respectively. Areas with a difference in surface water depth beyond −0.01 m and 0.01 m are scattered across the entire catchment. For a majority of those areas, the absolute difference in surface water depth is within 0.05 m. On a catchment level, the results shown in Figure 10 can be considered satisfactory in terms of the surface water drainage that is achieved by the synthetic storm drain inlet network. The results also underline the significance of storm drain inlets that are situated in surface water accumulation areas. At the same time, this highlights the critical aspect of over-estimating the number of synthetic storm drain inlets added in surface water accumulation areas.

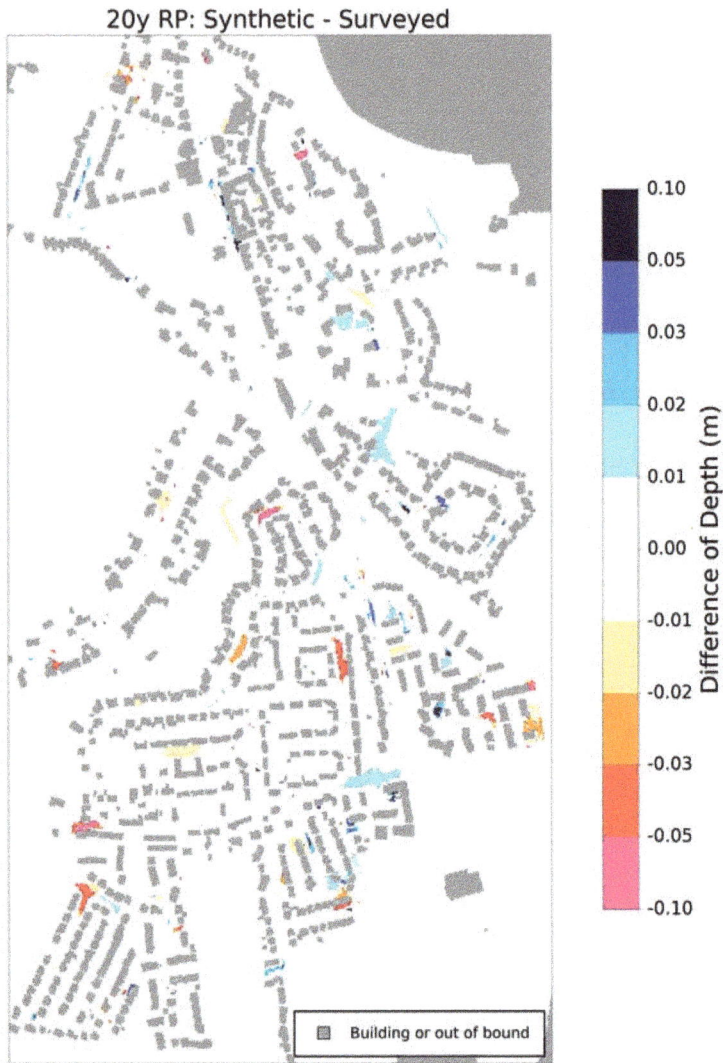

Figure 10. Comparison of maximum surface water depth for 20-year event. Depths indicate difference between simulations using final synthetic storm drain inlet network and surveyed one. If value (+): surface depth greater for simulation using synthetic storm drain inlets. If value (−): surface depth greater for simulation using surveyed storm drain inlets.

6. Conclusions

A GIS routine was developed, allowing the generation of a synthetic storm drain inlet network for the purpose of fully coupled 1D/2D urban flood modelling using CityCAT [17]. Simultaneously, actual storm drain inlet locations were surveyed during a field survey for validation and calibration purposes. Preliminary modelling results for a case study in Scotland using the simplest design assumptions based on a synthetic storm drain inlet every 50 m revealed an under-representation of the total flow captured when compared against a surveyed storm drain inlet network (Figure 4). Consequently, improvements to the GIS routine were made in order to account for higher densities of storm drain inlets in specific areas, but more importantly, to include terrain information. This was achieved in two ways. Firstly, the initially placed storm drain inlets were re-located to lower lying neighboring cells. Secondly, two additional storm drain inlets were added in areas of calculated surface water accumulation.

The updated model results showed significant improvements in terms of the flow that was captured (Figure 9) and the remaining surface water inundation depths on a catchment level (Figure 10). The GIS method developed therefore provides a reasonable and robust way of generating synthetic storm drain locations at relatively low costs when compared to field work. The results also stress the significance of having high-resolution terrain data since the re-distribution of storm drain inlet locations was conducted on a micro scale level. The resulting maximum horizontal and average vertical shift of 4.5 and 0.069 m, respectively, led to a considerable increase in the flow that was captured, and therefore highlight the sensitivity of the drainage efficiency to relatively small changes in the location and elevation of storm drain inlets.

Despite the results that were achieved, a synthetically generated storm drain inlet network should only be considered as a first iteration towards the final storm drain inlet network that is applied in a 1D/2D simulation. Particularly for critical locations, such as surface water accumulation areas, or when studying the impact of clogged or blocked storm drain inlets, it is necessary to have the exact location of a storm drain inlet. Also, when conducting detailed surface–sub-surface flow pathway analysis, the application of surveyed storm drain inlets is recommended. Having the capabilities of conducting such detailed analysis in CityCAT underlines at the same time the benefits of having a generic way of generating a synthetic network of storm drain inlets. As shown in this work, different densities of storm drain inlets can be generated for individual areas. Simultaneously, single storm drain inlets could be added or removed. Changing inlet densities and locations together with pipe dimensions could be applied in future work to evaluate not only the hydraulic, but also the cost implications, which was beyond the scope of this work.

Generally, more studies are required in order to validate the hydrodynamic results presented against measured flow data of actual storm events. This would allow for a wider and systematic sensitivity analysis addressing the location and spatial density of storm drain inlets, as well as the impact of their geometry and blockages. Critically, the GIS routine should also be applied and tested on urban catchments with different characteristics in terms of size, topography, etc. to avoid a potential over-calibration towards the catchment tested in this work. In general, further research in this area could focus on two things. Firstly, on improving the GIS routine that is developed in this work. This could range from including additional data or applying different means of placing the storm drain inlets by using more sophisticated statistical analysis tools. Secondly, moving away from the synthetic network to find different ways of recording the actual locations of storm drain inlets. Surveyed inlet locations are preferred in order to replicate the complex hydraulics of the actual urban drainage system. Methods to generate synthetic inlet locations are informed by design standards that may have changed over time and are often unique to specific locations. Furthermore, certain situations in reality might require adaptation of the design and planning standard in order to place storm drain inlets. Such locations are difficult to capture with a generic method for generating synthetic storm drain inlet locations. From a practical perspective, storm drain inlet locations could be collected as part of drainage maintenance work or other regular work that is carried out on roads. Alternatively, an automated process could be developed allowing for a detection of storm drain inlet locations based on photo-interpretation of the application of Google Street View images, similar to algorithms that are applied for face or license plate recognition [40].

Acknowledgments: This work has been funded by Scottish Water and the Engineering and Physical Science Research council (EPSRC) as part of grant 1368347 of Centre for Doctoral Training in Engineering for the Water Sector (STREAM, EP/L015412/1). Funding was also provided by two EPSRC projects, ITRC MISTRAL: Multi-scale Infrastructure Systems Analytics (EP/N017064/1) and Future Urban Flood Risk Management (EP/P004334/1). The authors gratefully acknowledge Dawn Lochhead, Bob Fleming, Dom McBennett (all Scottish Water) and Russell Stewart, Perth & Kinross Council for providing data and valuable expertise. Martin Robertson, Newcastle University provided invaluable technical support on the surveying aspects.

Author Contributions: R.B., V.G. and C.K. conceived the problem and established the concept behind the tool; R.B. developed the method; V.G. performed the modelling in CityCAT 1D/2D; R.B. analyzed the data; R.B. conducted the field survey of inlets; R.B., V.G. and C.K. wrote the paper.

Conflicts of Interest: The authors declare no conflict of interest.

References

1. Leandro, J.; Djordjević, S.; Chen, A.S.; Savić, D. The use of multiple-linking-element for connecting sewer and surface drainage networks. In Proceedings of the 32th Congress of IAHR Harmonizing the Demands of Art and Nature in Hydraulics, Venice, Italy, 1–6 July 2007; Volume 32, p. 204.
2. Djordjević, S.; Prodanović, D.; Maksimović, Č. An approach to simulation of dual drainage. *Water Sci. Technol.* **1999**, *39*, 95–103. [CrossRef]
3. Schmitt, T.G.; Thomas, M.; Ettrich, N. Analysis and modeling of flooding in urban drainage systems. *J. Hydrol.* **2004**, *299*, 300–311. [CrossRef]
4. Schmitt, T.G.; Thomas, M.; Ettrich, N. Assessment of urban flooding by dual drainage simulation model RisUrSim. *Water Sci. Technol.* **2005**, *52*, 257–264. [PubMed]
5. Leandro, J.; Lopes, P.; Carvalho, R.; Páscoa, P.; Martins, R.; Romagnoli, M. Numerical and experimental characterization of the 2D vertical average-velocity plane at the center-profile and qualitative air entrainment inside a gully for drainage and reverse flow. *Comput. Fluids* **2014**, *102*, 52–61. [CrossRef]
6. Butler, D.; Davies, J.W. *Urban Drainage*, 3rd ed.; Spon Press: Oxon, UK, 2011; ISBN 978-0-415-45526-8.
7. Tan, W. *Shallow Water Hydrodynamics: Mathematical Theory and Numerical Solution for a Two-Dimensional System of Shallow Water Equations*; Tan, W., Ed.; Elsevier Oceanography Series; Elsevier: Amsterdam, The Netherlands, 1992; ISBN 978-0-444-98751-8.
8. Toro, E.F. *Shock-Capturing Methods for Free-Surface Shallow Flows*; Wiley: Hoboken, NJ, USA, 2001; ISBN 978-0-471-98766-6.
9. Bourdarias, C.; Ersoy, M.; Gerbi, S. A mathematical model for unsteady mixed flows in closed water pipes. *Sci. China Math.* **2012**, *55*, 221–244. [CrossRef]
10. Innovyze Infoworks ICM. Available online: http://innovyze.com/products/infoworks_icm/ (accessed on 18 August 2017).
11. Preissmann, A. Propagation des intumescences dans les canaux et rivieres. In Proceedings of the First French Association for Computation, Grenoble, France, 14–16 September 1961; pp. 433–442.
12. Cunge, J.A.; Wegner, M. Intégration numérique des équations d'écoulement de barré de Saint-Venant par un schéma implicite de différences finies. *La Houille Blanche* **1964**, *1*, 33–39. [CrossRef]
13. Innovyze Modelling of Pressurised Pipes within InfoWorks ICM and CS. Available online: http://blog.innovyze.com/wp-content/uploads/2013/02/Modelling_of_Pressurised_Pipes_within_InfoWorks_ICM_and_CS.pdf (accessed on 18 August 2017).
14. Pina, R.D.; Ochoa-Rodriguez, S.; Simões, N.E.; Mijic, A.; Marques, A.S.; Maksimović, Č. Semi- vs. Fully-distributed urban stormwater models: Model set up and comparison with two real case studies. *Water* **2016**, *8*, 58. [CrossRef]
15. Maksimovic, C.; Prodanovic, D. Modelling of Urban Flooding—Breakthrough or Recycling of Outdated Concepts. In *Urban Drainage Modeling*; American Society of Civil Engineers: Reston, VA, USA, 2001; pp. 1–9.
16. Djordjevic, S.; Prodanovic, D.; Maksimovic, C.; Ivetic, M.; Savic, D. SIPSON—Simulation of Interaction between Pipe flow and Surface Overland flow in Networks. *Water Sci. Technol.* **2005**, *52*, 275–283. [PubMed]
17. Glenis, V.; McGough, A.S.; Kutija, V.; Kilsby, C.; Woodman, S. Flood modelling for cities using Cloud computing. *J. Cloud Comput. Adv. Syst. Appl.* **2013**, *2*, 14. [CrossRef]
18. Glenis, V.; Kutija, V.; Kilsby, C.G. City Catchment Analysis Tool—CityCAT. *Environ. Model. Softw.* **2017**, under review.

19. Despotovic, J.; Plavsic, J.; Stefanovic, N.; Pavlovic, D. Inefficiency of storm water inlets as a source of urban floods. *Water Sci. Technol.* **2005**, *51*, 139–145. [PubMed]
20. Leandro, J.; Chen, A.S.; Djordjević, S.; Savić, D.A. Comparison of 1D/1D and 1D/2D Coupled (Sewer/Surface) Hydraulic Models for Urban Flood Simulation. *J. Hydraul. Eng.* **2009**, *135*, 495–504. [CrossRef]
21. Argaman, Y.; Shamir, U.; Spivak, E. Design of Optimal Sewerage Systems. *J. Sanit. Eng. Div.* **1973**, *99*, 703–716.
22. Diogo, A.F.; Graveto, V.M. Optimal Layout of Sewer Systems: A Deterministic versus a Stochastic Model. *J. Hydraul. Eng.* **2006**, *132*, 927–943. [CrossRef]
23. Guo, Y.; Walters, G.A.; Khu, S.T.; Keedwell, E. A novel cellular automata based approach to storm sewer design. *Eng. Optim.* **2007**, *39*, 345–364. [CrossRef]
24. Möderl, M.; Butler, D.; Rauch, W. A stochastic approach for automatic generation of urban drainage systems. *Water Sci. Technol.* **2009**, *59*, 1137–1143. [CrossRef] [PubMed]
25. Guo, Y.; Walters, G.; Savic, D. Optimal design of storm sewer networks: Past, Present and Future. In Proceedings of the 11th International Conference on Urban Drainage, Edinburgh, UK, 31 August–5 September 2008.
26. Highways Agency. *Design Manual for Roads and Bridges—Spacing of Road Gullies*; Highways Agency: Guildford, UK, 2000; Volume 4.
27. Spaliviero, F.; May, R.W.P.; Escarameia, M. *Spacing of Road Gullies. Hydraulic Performance of BS EN 124 Gully Gratings and Kerb Inlets*; Report SR 533; HR Wallingford: Oxfordshire, UK, 2000.
28. Marsalek, J.; Barnwell, T.O.; Geiger, W.; Grottkert, M.; Huber, W.C.; Saul, A.J.; Schillingt, W.; Tornol, H.C. Urban Drainage Systems: Design and Operation. *Water Sci. Technol.* **1993**, *27*, 31–70.
29. Delleur, J.W. The Evolution of Urban Hydrology: Past, Present, and Future. *J. Hydraul. Eng.* **2003**, *129*, 563–573. [CrossRef]
30. Mailhot, A.; Duchesne, S. Design Criteria of Urban Drainage Infrastructures under Climate Change. *J. Water Resour. Plan. Manag.* **2010**, *136*, 201–208. [CrossRef]
31. Fletcher, T.D.; Shuster, W.; Hunt, W.F.; Ashley, R.; Butler, D.; Arthur, S.; Trowsdale, S.; Barraud, S.; Semadeni-Davies, A.; Bertrand-Krajewski, J.-L.; et al. SUDS, LID, BMPs, WSUD and more—The evolution and application of terminology surrounding urban drainage. *Urban Water J.* **2015**, *12*, 525–542. [CrossRef]
32. Almedeij, J.; Alsulaili, A.; Alhomoud, J. Assessment of grate sag inlets in a residential area based on return period and clogging factor. *J. Environ. Manag.* **2006**, *79*, 38–42. [CrossRef] [PubMed]
33. Gómez, M.; Russo, B. Hydraulic Efficiency of Continuous Transverse Grates for Paved Areas. *J. Irrig. Drain. Eng.* **2009**, *135*, 225–230. [CrossRef]
34. Gómez, M.; Russo, B. Methodology to estimate hydraulic efficiency of drain inlets. *Proc. ICE Water Manag.* **2011**, *164*, 81–90. [CrossRef]
35. OS MasterMap Topography Layer [GML2 Geospatial Data], Scale 1:1250, Updated December 2014, Ordnancce Survey GB. Using: EDINA Digimap Ordnance Survey Service. Available online: https://digimap.edina.ac.uk/ (accessed on 1 June 2015).
36. Leica Geosystems. Leica Viva GS15 Data Sheet. Available online: http://w3.leica-geosystems.com/downloads123/zz/gpsgis/viva%20gs15/brochures-datasheet/leica_viva_gs15_ds_en.pdf (accessed on 19 July 2017).
37. Reed, D.W.; Robson, A.J. *Flood Estimation Handbook*; Centre for Ecology and Hydrology: Lancaster, UK, 1999.
38. Aronica, G.T.; Lanza, L.G. Drainage efficiency in urban areas: A case study. *Hydrol. Process.* **2005**, *19*, 1105–1119. [CrossRef]
39. Palla, A.; Colli, M.; Candela, A.; Aronica, G.T.; Lanza, L.G. Pluvial flooding in urban areas: The role of surface drainage efficiency. *J. Flood Risk Manag.* **2016**. [CrossRef]
40. Frome, A.; Cheung, G.; Abdulkader, A.; Zennaro, M.; Wu, B.; Bissacco, A.; Adam, H.; Neven, H.; Vincent, L. Large-scale privacy protection in Google Street View. In Proceedings of the 2009 IEEE 12th International Conference on Computer Vision, Kyoto, Japan, 29 September–2 October 2009; pp. 2373–2380. [CrossRef]

water

water

MDPI

Article

Urban Estuarine Beaches and Urban Water Cycle Seepage: The Influence of Temporal Scales

Sérgia Costa-Dias [1,2], Ana Machado [1,2], Catarina Teixeira [1,2] (iD) and Adriano A. Bordalo [1,2,*] (iD)

1   CIIMAR—UP, Interdisciplinary Centre of Marine and Environmental Research of the University of Porto, Novo Edifício do Terminal de Cruzeiros do Porto de Leixões, Avenida General Norton de Matos, S/N, 4450-208 Matosinhos, Portugal; scdias@icbas.up.pt (S.C.-D.); ammachado@icbas.up.pt (A.M.); catarina@icbas.up.pt (C.T.)
2   ICBAS—UP, Institute of Biomedical Sciences of Abel Salazar—University of Porto, Rua Jorge Viterbo Ferreira 228, 4050-313 Porto, Portugal
*   Correspondence: bordalo@icbas.up.pt; Tel.: +351-220-428-195

Received: 4 October 2017; Accepted: 6 February 2018; Published: 9 February 2018

Abstract: Temperate estuarine beaches are an asset to coastal cities. Being located within the transition zone where the river meets the sea can provide several environmental benefits such as warm water temperature during the summer, flat waters, protection from coastal upwelling-induced morning fog, as well as additional recreational and cultural values. In this study we address a major question—can the urban water cycle impair the water quality dynamics during a bathing season in a temperate Atlantic estuary (Douro, Northwest Portugal)? Water quality was assessed according to the EU legal criteria at different time scales. No daily, weekly, or monthly patterns for microbiological descriptors were found, which rather followed the hourly tidal dynamics. Quality decreased during high tide, affecting potentially 800+ beach-users during mid-summer weekends (4 m^2 per person). Low water quality was transported upstream from highly populated urban areas. Therefore, the understanding of the dynamics of estuarine systems is essential to adapt the standard official approach, and the obtained results can be used to draw policy recommendations to improve the sampling strategy, aiming for more accurate assessment of the water quality to reduce the risk hazard of estuarine beaches.

Keywords: urban; beach; bathing; water quality; estuary; management

1. Introduction

Tourism in coastal areas reached its peak in recent decades, contributing to national economies as well as to the wellbeing of local communities [1]. In addition, when temperatures rise, going to the beach is a top choice for many people with tangible effects on both physical and mental health. If properly managed, a bathing area may be an important source of revenue [2], also providing a range of environmental services beneficial to society [3].

Coastal cities may have oceanic beaches nearby to complement their recreational offers. However, throughout history, estuaries have been strategically chosen for human settlements [4], and if estuarine margins are not unreachable, they provide alternative beaches right at hand, even in no-swim areas, i.e., those not surveyed in terms of water quality assessment according to the legal criteria. Therefore, beach water quality is of particular concern in the coastal zone.

Within the European Union, the current Bathing Water Directive (BWD) [5] regulates the water quality assessment, and swimming interdiction may be imposed or lifted, accordingly. This depends on the outcome of the microbiological assessment that, inherently to the methodologies, does not provide real-time values. Moreover, the mandatory advertisement of water quality for a particular beach may be delayed due to bureaucratic practice, making the utility of such information questionable, introducing confusion to the public, and undermining their confidence.

In line with the US Environmental Protection Agency approach to gauge the level of contamination and to infer potential health risks [6], the BWD determines the use of *Escherichia coli* and intestinal enterococci as indicators. Two decades before, EPA [7] recommended the use of intestinal enterococci as the sole indicator for ocean water bacterial monitoring. Recently, Fewtrell & Kay [8] reviewed epidemiological studies and quantitative microbial risk assessments of infection risk from recreational water use. Again, the selection of enterococci and/or *E. coli* as the most suitable water quality indicators was questioned, and the use of models and more locally customized approaches have been proposed, not in line with the current European BWD. Therefore, the way water is tested has strong implications for beach closure and restrictions, i.e., on the use and management. This issue is even more complex when considering the highly dynamic water quality of urban tidal beaches with deficient wastewater treatment, and multiple urban tributaries, as in the case of the Douro estuary.

The current sampling strategy proposed in the BWD [5] does not account for short-term but important variability, such as day-to-day, tide-to-tide, or even morning-afternoon, that occurs in tidal beaches exposed to urban run-off. Moreover, in the directive, only extraordinary short-term pollution events are referenced—which may lead to samples being disregarded if the scheduled collection is performed during those episodes, but not when systematic or periodic events occur. Indeed, levels of contamination affecting the overall water quality can vary substantially on a temporal scale of minutes to hours or days [9–11]. Illness of swimmers has been related to bacterial indicator concentrations measured on the same day but not the day before [12], calling the attention for the need of a better performing beach water quality assessment approach.

All around the world, modern approaches to city organization intend to implement the concept of the "urban water cycle", aiming for environmental, economic, and social sustainability in the use of water sources, while considering water supply and demand management options (for a review see [13]). Nonetheless, the common focus seems to be the supply, the water "entering" the cycle. The water "leaving" the cycle is seldom sufficiently monitored, and its fate is often overlooked. Therefore, if an urban water cycle is not properly closed, wastewater may be released into the environment, impacting the water quality of the receiving water body, including its recreational use.

In this work, we focused on an estuarine beach located in a European metropolitan area with c.a. 2 million inhabitants and deficient sewage treatment [14]. The objective was to assess different variability scales (spatial—within the beach; temporal—morning/afternoon, daily, weekly, and monthly), in the water quality during a bathing season. The ultimate goal was to optimize a sampling strategy to improve the significance of the results for the evaluation of water quality of the urban beaches, providing realistic assurance of bathing safety to the public.

2. Materials and Methods

2.1. Area Description

The work was performed at Zebreiros beach (Figure 1), within the Gondomar municipality, on the upper Douro estuary (41.07786° N, −8.51630° W). This particular beach is located in the large urban area of Porto (NW Portugal), and is highly popular during summer. Infrastructures and accessibilities have been considerably improved by local authorities, with parking, toilets, bars, and first aid assistance by lifeguards, in line with the Blue Flag requirements [15]. The beach was officially recognized as a bathing area in 2016, but the delimitation of the water front with interconnected buoys with a chord lacked any scientific background.

The estuary has semi-diurnal tides (range up to 4 m), and river flow towards the sea—end member depends greatly on the regime of a large hydroelectric power dam on the upper estuarine limit (2 km from the studied area, 21.5 km from the river mouth). Typically, during the summer, daily freshwater inputs range from 0 to 416 m³ s⁻¹ (average ± SE = 139 ± 15). Treated and partially treated sewage from 8 plants is discharged into the estuary (Figure 1).

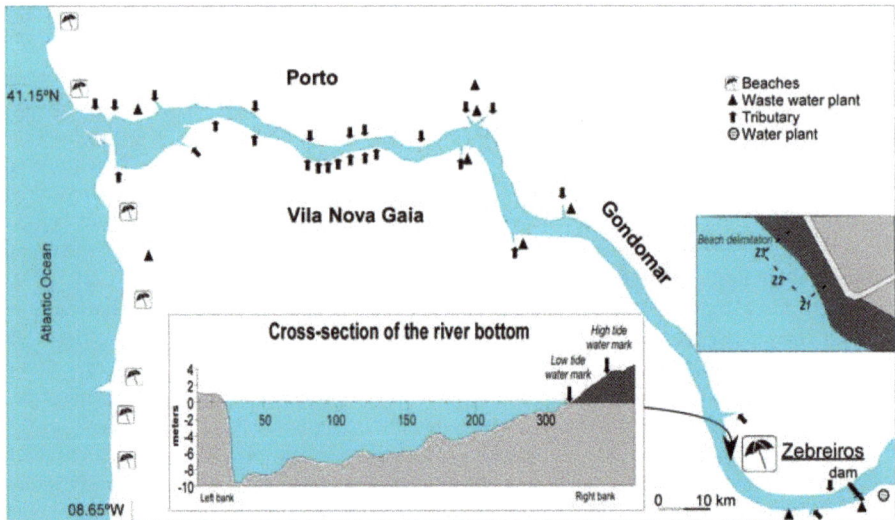

Figure 1. Zebreiros beach within the Douro estuary (NW Portugal). Inserts—cross section of the river bottom relative to the beach; and sampling location with beach delimitation.

2.2. Study Design

No detailed historical data on water quality were available. As such, in order to characterize Zebreiros beach and to evaluate its condition according to the relevant European directive for bathing waters [5], an intensive sampling program was implemented during the 4-month 2016-bathing season—officially between 15 June and 15 September.

The sampling program was designed to incorporate different time scale water quality variations, including hourly (during 25 h, 20–21 July), daily (through 8 days, 18–25 July), weekly (during 5 weeks, 4 July–2 August), and monthly (through 4 months, 7 June–6 September. With the exception of the hourly sampling surveys, all samples were collected during the morning low tide.

Within the official delimited beach area, three survey sites were established, namely Z1 (upstream beach limit), Z2 (middle of the beach), and Z3 (downstream beach limit) (Figure 1). The hourly sampling survey was performed at Z2 location only, due to logistic limitations.

During each sampling survey, measurements of subsurface (0–30 cm) key environmental parameters (temperature, conductivity, salinity, dissolved oxygen, turbidity, and pH) were performed with a multiparameter YSI 6000 probe. Tidal height was measured with a pressure sensor (YSI) and was compared to official tidal tables. Water samples were collected with sterile sampling bottles for microbiological analysis, and kept refrigerated in ice chests until processing. Microbiological indicators for the hourly sampling were collected for Z2 only.

Monthly sampling events also included the morphological characterization of the beach, by taking beach profiles with the Emery method [16]. Bathymetric profiles of the river area contiguous to Zebreiros beach were performed with a sonar, in August 2016, resulting in a depth chart of a specific stretch of the upper estuary.

Additionally, the number of beachgoers was recorded as a measure of beach use during all sampling surveys, and the available area per person calculated.

2.3. Analytical Procedures

In the laboratory, water samples were concentrated onto a sterile membrane of 0.45 μm pore size and 47 mm diameter (Schleicher Schull ME 25/21 ST). *Escherichia coli* were assessed on ChromoCult® Coliform agar (Merck, Darmstadt, Germany), and typical dark-blue to violet colonies were counted after 24 h incubation at 37 °C. Intestinal enterococci (IE) were assessed on Slanetz & Bartley agar (Oxoid Ltd., Hants, England), and typical reddish-brown colonies were counted after 48 h incubation at 44.5 °C [10]. The methods used were in accordance with ISO reference methods required by the BWD.

2.4. Data Treatment

The temporal and spatial variability of the main environmental parameters was mapped with Surfer 11.0 software (Golden Software Inc., Golden, CO, USA), using kriging (linear variogram model) as the gridding method.

Data were tested for normality and homogeneity of variances using Shapiro-Wilk and Levene's tests, respectively. *E. coli* and IE values were normalized by logarithm ($\log_{10}$) transformation prior to statistical analysis. Differences between time scales and sampling sites were analyzed by analysis of variance (ANOVA). Whenever significant differences were detected, a *post hoc* Tukey honestly significant difference (HSD) multicomparison test was performed. Time and space variability of *E. coli* was assessed by Kruskal-Wallis test since parametric assumptions were not met. Statistical analysis was performed at the 95% confidence level ($p < 0.05$) using STATISTICA 13.0 (StatSoft, Inc., Palo Alto, CA, USA).

3. Results

3.1. Beach and Bathing Water Abiotic Characterization

Bathymetric data were adjusted to low tide and were coordinated with beach morphology to produce cross-sections of the river bottom and Zebreiros beach surface. An example of the middle of the beach is presented in Figure 1 (insert). The deeper area on the left bank of the river corresponds to the navigational channel, artificially maintained by periodical dredging. The physical profile of the beach was revealed to be dynamic over time, but mainly in the fringe influenced by tides, with sand moving generally upstream (data not shown).

The daily, weekly, and monthly variability of selected water environmental parameters is presented in Figure 2.

The mean temperature was rather stable during the bathing season, ranging from a mean value of 23.12 °C ($\pm$0.21) in June to 24.77 °C ($\pm$0.03) in August, whereas the mean pH ranged from 7.28 ($\pm$0.04) to 7.63 ($\pm$0.03). The lowest oxygen saturation mean value occurred in August (7.1 $\pm$ 0.12 mg/L), with values over 7.88 mg/L in the other months. Average turbidity was 6 $\pm$ 0.8 NTU with the highest values (19.5 $\pm$ 3.0 NTU) in August. Water conductivity was rather stable, 300 $\pm$ 3 μS cm^{-1}, denoting the low water tidal conditions.

3.2. Water Quality Indicators

In accordance with the BWD, the number of cfu per 100 mL of *E. coli* and intestinal enterococci was used for microbiological water assessment, and eventually worsened from the upper (Z1) towards the downstream area of the bathing zone (Z3) in spite of the short distance (Figure 3). Throughout the study, no clear pattern for daily, weekly, or monthly time scales was found, but the hourly sampling seemed to follow the tidal cycle (Figure 4).

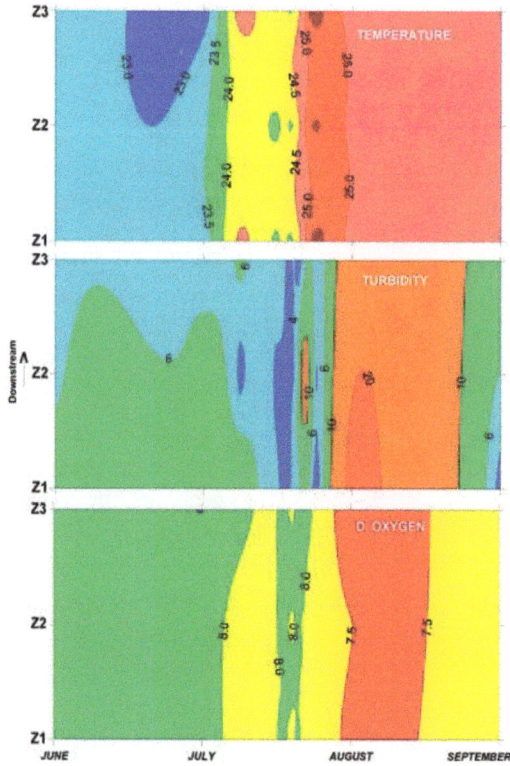

Figure 2. Spatial (Z1 through Z3) and temporal (June through September) distribution of key environmental descriptors: temperature (°C), turbidity (NTU) and dissolved oxygen (mg L^{-1}) at Zebreiros beach during the 2016 bathing season.

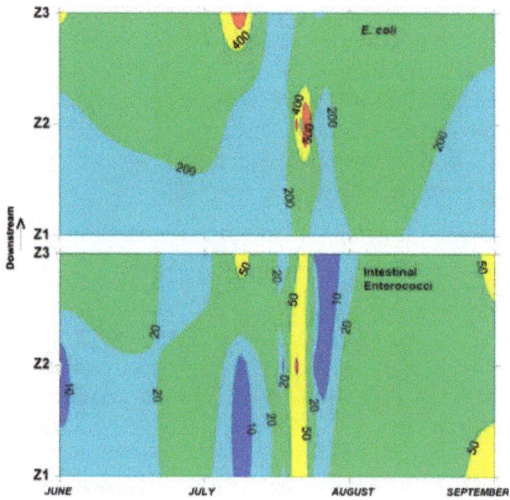

Figure 3. Spatial (Z1 through Z3) and temporal (June through September) distribution of *E. coli* and intestinal enterococci (cfu per 100 mL).

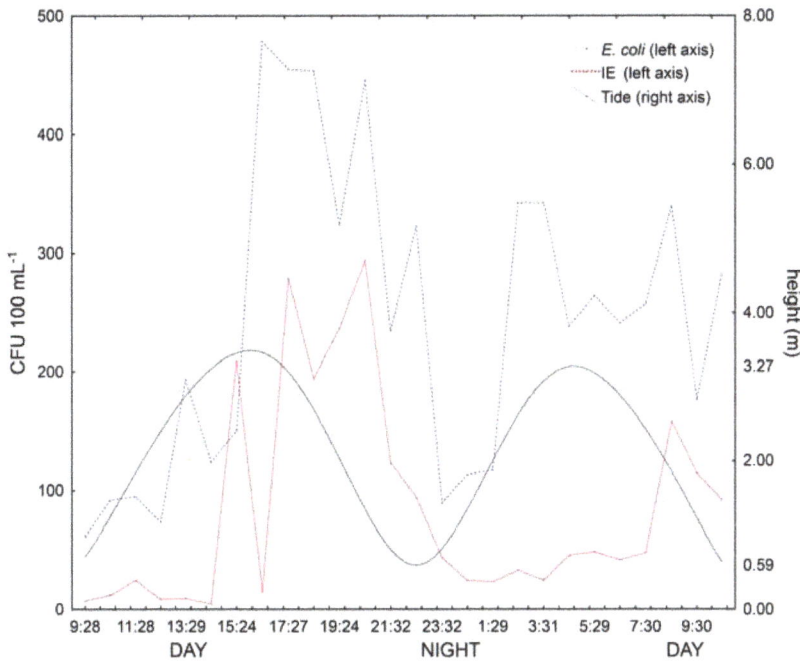

Figure 4. Hourly variability of the tidal height during two consecutive tidal cycles (m) and the distribution of *E. coli* and intestinal enterococci (IE) (cfu 100 mL^{-1}) at Zebreiros beach (20–21 July 2016).

It should be noted that microbial indicators tended to increase at mid low and high tide, when the water velocity in the Douro estuary tended to increase (data not shown), downstream and upstream, respectively. *E. coli* values were not significantly different between time scales, or sites ($p > 0.05$). With respect to the IE, significant ($p < 0.05$) differences could be identified between hourly and weekly surveys.

Nonetheless, variance within each temporal scale allows further interpretations. Hourly to monthly scales are represented in Figure 5, with emphasis on the BWD benchmarks. In spite of the overall median values being in the range *Excellent* (percentile 90 < 250 for *E. coli*; percentile 90 < 100 for IE), according to the European legal criteria, the variability was important, down to *Good* (percentile 95 < 500 for *E. coli*; percentile 95 < 200 for IE) and *Sufficient* (percentile 90 < 500 for *E. coli*; percentile 90 < 185 for IE).

3.3. Beach Use

The number of sun seekers using the beach varied greatly during the summer season and with the time of day. We recorded a few beachgoers in the early hours of a mid-week morning to 800+ on a weekend afternoon. Considering the total usable area of the beach with sand (within and outside the delimitation zone), this corresponded to an approximated value of 4 m^2 per user during peak periods, attesting the popularity of Zebreiros estuarine bathing waters.

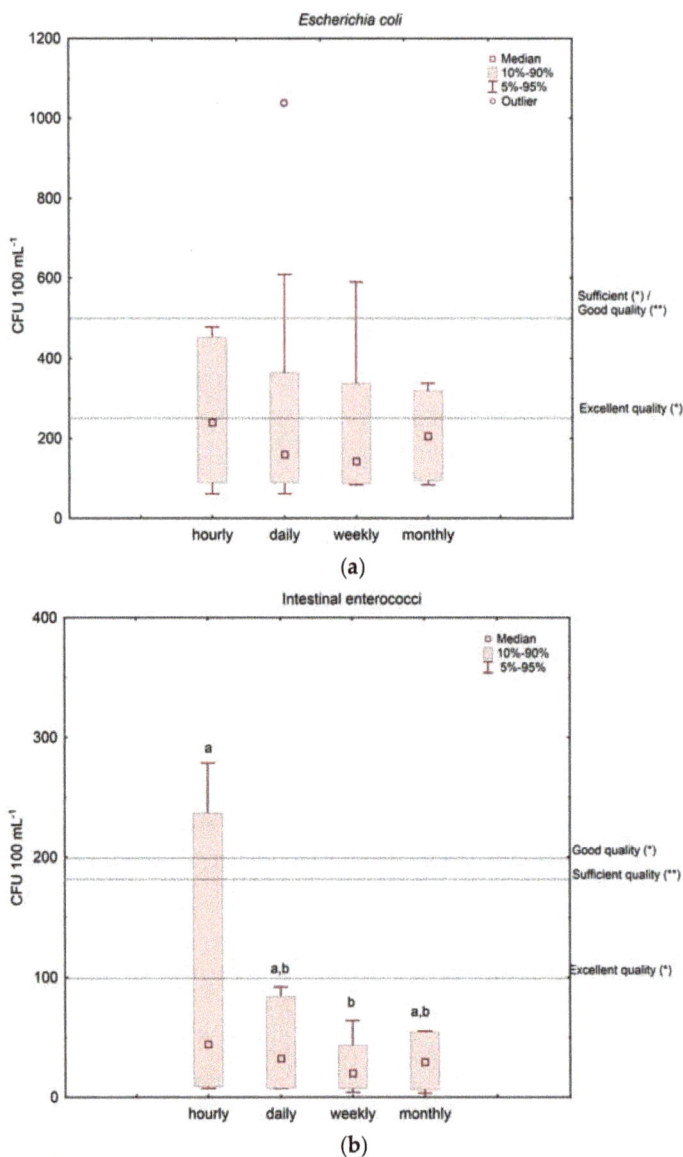

Figure 5. Temporal scale variances for the microbiological indicators: (**a**) *E. coli*; (**b**) Intestinal enterococci, measured at Z2 BWD benchmarks are given (Sufficient/Good/Excellent quality; assessment based on * percentile 95; ** percentile 90). Letters correspond to different homogeneous groups derived from ANOVA.

4. Discussion

4.1. Bathing Water Quality

The values recorded for the abiotic parameters in Zebreiros bathing waters are within the acceptable range for recreational purposes. Nonetheless, sampling in August coincided with a period of systemic severe fires in the surrounding unplanted forests, which might explain the noticeable decrease in water abiotic parameter quality, namely, dissolved oxygen and turbidity, owing to the deposition of ash and other debris. Concomitantly, the number of beachgoers also increased, contributing to the re-suspension of bottom fine particles and turbidity, a process that tends to increase IE in the water [8], but not noticed at Zebreiros (see below).

Regarding the water quality indicators according to the European legal criteria, Zebreiros beach would fall into different water quality categories according to the time scale of choice and the microbial indicator—IE or *E. coli*. As such, samples collected during the morning low tide would lead to a better water quality classification using IE as an indicator, falling within the category of *Excellent*, at any daily, weekly, or monthly scale; and a classification of *Inadequate* for bathing when hourly data are considered. Therefore, water quality tended to change faster in a short time period, being rather conservative on a day-to-day up to month-to-month framework. For the other indicator, *E. coli*, estimated values fell within classifications of *Sufficient* for daily and weekly scales, while hourly and monthly scales were classified as *Good*.

In coastal/ocean beaches, the main sources of fecal pollution with major influence on health risks to beach-goers are (i) storm water run-off, particularly when combined systems are still in use; (ii) untreated sewage point-discharge; (iii) fecal polluted rivers, and (iv) poorly treated sewage outflow. Bird droppings—a common endogenous source of bacteria should also be considered (e.g., [17]). Indeed, the removal of birds, such as gulls, may dramatically improve the water quality of ocean beaches [18]. In the Douro estuary, besides gulls, a myriad of other birds inhabit its shores all year round due to the intrinsic protection associated with the environment and the abundant feeding grounds, such as pigeons, ducks, waders, egrets, and sea crows. Their role as local sources of fecal materials is unknown.

The concentration of indicator bacteria has been used for decades to measure recreational water safety [19]. The current BWD [5] removed most tangible indicators except *E. coli* and intestinal enterococci in order to safeguard public health of beach goers. In this study, which was not designed to identify relationships between the concentration of bacterial indicators and illness among beach goers, IE seems to correlate better with such factors as tides—inferred from the 25-h sampling survey, than *E. coli*, probably due to the fact that enterococci have a larger decay rate with respect to beach water (e.g., [20]), i.e., survive longer. Indeed, the hourly variation of bacterial indicators seemed to act as a surrogate of the tidal cycle (Figure 4). Thus, the hourly scale, and implicitly the moment of the tidal cycle, might reveal the arrival of fecal spiked water masses from the highly urbanized downstream areas during high tide.

The official surveys performed by the Portuguese Environmental Agency in 2017 [21] showed that the worst cases of contamination, that led to the subsequent closure of Zebreiros beach, occurred around the peak of high tide. For operational reasons, all surveys were accomplished between 9:00 and 11:20 a.m., and the coincidence was unintentional. On the contrary, in 2016, only 7 surveys were carried out, all during low tide between 8:00 and 8:57 a.m., yielding better results.

Management of urban beaches may be problematic owing to the fact that different government and municipal agencies are involved, and jurisdictional limits are not always clear. The situation in the studied beach is further complicated by the fact that is under port jurisdiction. Moreover, no water quality surveys are carried out year-round, and bacteriological assessment of beach water is only performed selectively at the onset of the bathing season, and at least monthly later on (15 June to 15 September).

4.2. Perceived Quality

Roca et al. [22] compiled the recreational carrying capacity thresholds proposed by the literature, as the minimum recommended sand availability per user (m^2/user). Recommended values vary with beach type, and increase in accordance with accommodation costs, but the value of 4 m^2 per user is presented as the minimum acceptable limit. Although not the most comfortable for beach enjoyment, the value calculated in this study for Zebreiros beach during the peak season barely reaches that minimum limit. Moreover, the slope at the bottom, particularly during the high tide (Figure 1), concomitantly with the dramatic decrease of the dry area as water rises, reduces the available space. Nevertheless, since Zebreiros beach has supporting facilities and is mainly surrounded by semi-wild green areas within an urban zone, beachgoers seem not to be deterred, turning those bathing waters a popular spot during the summer time, regardless of the overall quality.

4.3. The Importance of Recreational Waters

In recent years, attention has been placed on the goods and services ecosystems provide, but in many cases, it is very difficult to infer their economic value. According to Alves et al. [23], non-market economic values hold historical, cultural, social, emotional, or ecological significance, and are economically important, attracting domestic and international visitors to a region.

When beaches are mainly used by locals—as in Zebreiros case for the time been, private investment is limited and municipalities have most of the financial burden, in order to contribute to the community wellbeing and to the regional appeal. On the other hand, the positive or negative perception of tourism by host populations determines the degree of community participation and support for tourism development [24], which must be considered for more effective management.

To the same extent, eventual health risks in the use of recreational waters due to inadequate or insufficient wastewater treatment will potentially harm users but furthermore will damage the external image of the region. In our case, fecal pollution is generated downstream—from the same municipality (Gondomar), and from two others within the same metropolitan area, and brought upstream with the rising of the tide.

As such, deficiencies in the urban water cycle that might jeopardize the water quality for recreational uses, and particularly bathing, come with severe risks at multiple levels, that should be accounted when managing bathing waters. With that in mind, the obtained results raise two pertinent questions: (i) sampling program timeframe, and (ii) urban water cycle leaks. The first one is linked to the meaning and the eventual limited value of sampling if closely undertaken according to the European legal criteria (BWD). Indeed, the possibility to drastically reduce the sampling frequency to just 1 sample/month during the 4-month bathing season seems inadequate to estuarine water quality dynamics. The second has profound management implications at the metropolitan area scale, since the quality decreases substantially during high tide due to the upstream movement of contaminated water from adjacent highly urbanized areas. Dealing with both is crucial to ascertain the risk posed by direct and indirect contact with bathing water, and to provide beachgoers with scientifically sound information in order to make better and informed decisions when dealing with the choice of when and where to go for estuarine bathing.

5. Conclusions

- In the Douro estuary, water quality varies on an hourly scale, depending on the tide, rather than on a daily, weekly, or monthly scale during the bathing season.
- Intestinal enterococci seem to be a better indicator of estuarine water quality than *E. coli*, particularly during short-term events (hour-to-hour scale).
- The water quality tended to worsen during high tide due to the tidal excursion of more contaminated surface water from downstream highly urbanized areas, where sewage is inadequately treated and dozens of small contaminated urban tributaries discharge their flows.
- The present EU sampling strategy seems inadequate for tidal environments.

Acknowledgments: The authors acknowledge the Hydrobiology laboratory staff that supported field and/or laboratory work; the Mayor of Gondomar for logistical support; and J. Vieira for kindly providing boat transportation. This work was implemented in the Framework of the Structured Program of R&D&I INNOVMAR—Innovation and Sustainability in the Management and Exploitation of Marine Resources (Reference NORTE-01-0145-FEDER-000035), namely within the Research Line ECOSERVICES, supported by the Northern Portugal Regional Operational Programme (NORTE2020), through the European Regional Development Fund (ERDF), which also covered open access publishing costs. C. Teixeira acknowledges FCT for a postdoctoral grant (ref. SFRH/BPD/110730/2015) through POCH, cofunded by MCTES and the European Social Fund.

Author Contributions: All authors conceived, designed and performed the study; and S.C.-D. and A.A.B. wrote the paper with contributions from all co-authors.

Conflicts of Interest: The authors declare no conflict of interest.

References

1. Mason, P. *Tourism Impacts, Planning and Management*, 3rd ed.; Routledge: New York, NY, USA, 2016; ISBN 978-1-315-78106-8.
2. Rabinovici, S.J.M.; Berknopf, R.L.; Whitman, R.L. Economic and health risk trade-offs of swim closures at a Lake Michigan beach. *Environ. Sci. Technol.* **2004**, *38*, 2737–2745. [CrossRef] [PubMed]
3. Costanza, R.; d'Arge, R.; de Groot, R.; Farberk, S.; Grasso, M.; Hannon, B.; Limburg, K.; d Naeem, S.; O'Neill, R.; Paruelo, J.; et al. The value of the world's ecosystem services and natural capital. *Nature* **1997**, *387*, 253–260. [CrossRef]
4. Pinto, R.; Marques, J.C. Ecosystem services in estuarine systems: Implications for management. In *Ecosystem Services and River Basin Ecohydrology*; Chicharo, L., Müller, F., Fohrer, N., Eds.; Springer: Dordrecht, The Netherlands, 2015; pp. 319–341. ISBN 978-94-017-9845-7.
5. TEPatCotE, Union. Directive 2006/7/EC of the European Parliament and of the Council of 15 February 2006 concerning the management of bathing water quality and repealing Directive 76/160/EEC. *Off. J. Eur. Union* **2006**, *L64*, 37–51.
6. Wade, T.J.; Pai, N.; Eisenberg, J.N.; Colford, J.M., Jr. Do U.S. Environmental Protection Agency water quality guidelines for recreational waters prevent gastrointestinal illness? A systematic review and meta-analysis. *Environ. Health Perspect.* **2003**, *111*, 1102–1109. [CrossRef] [PubMed]
7. Environmental Protection Agency (EPA). *Bacteriological Ambient Water Quality Criteria for Marine and Freshwater Recreational Waters*; PB86-158-045; US EPA: Springfield, VA, USA, 1986.
8. Fewtrell, L.; Kay, D. Recreational Water and Infection: A Review of Recent Findings. *Curr. Environ. Health Rep.* **2015**, *2*, 85–94. [CrossRef] [PubMed]
9. Boehm, A.B.; Grant, S.B.; Kim, J.H.; Mowbray, S.L.; McGee, C.D.; Clark, C.D.; Foley, D.M.; Wellman, D.E. Decadal and Shorter Period Variability of Surf Zone Water Quality at Huntington Beach, California. *Environ. Sci. Technol.* **2002**, *36*, 3885–3892. [CrossRef] [PubMed]
10. Bordalo, A.A. Microbiological water quality in urban coastal beaches: The influence of water dynamics and optimization of the sampling strategy. *Water Res.* **2003**, *37*, 3233–3241. [CrossRef]
11. Smith, S.D.A.; Markic, A. Estimates of Marine Debris Accumulation on Beaches Are Strongly Affected by the Temporal Scale of Sampling. *PLoS ONE* **2013**, *8*, e83694. [CrossRef] [PubMed]
12. Colford, J.M., Jr.; Schiff, K.C.; Griffith, J.F.; Yau, V.; Arnold, B.F.; Wright, C.C.; Gruber, J.S.; Wade, T.J.; Burns, S.; Hayes, J.; et al. Using rapid indicators for Enterococcus to assess the risk of illness after exposure to urban runoff contaminated marine water. *Water Res.* **2012**, *46*, 2176–2186. [CrossRef] [PubMed]
13. Rathnayaka, K.; Malano, H.; Arora, M. Assessment of Sustainability of Urban Water Supply and Demand Management Options: A Comprehensive Approach. *Water* **2016**, *8*, 595. [CrossRef]
14. Amorim, E.; Ramos, S.; Bordalo, A.A. Relevance of temporal and spatial variability for monitoring the microbiological water quality in an urban bathing area. *Ocean Coast. Manag.* **2014**, *91*, 41–49. [CrossRef]
15. FEE—Foundation for Environmental Education. Blue Flag Beach Criteria and Explanatory Notes 2017. Available online: https://static1.squarespace.com/static/55371ebde4b0e49a1e2ee9f6/t/5899e01ac534a5036aecbeeb/1486479387823/Beach+Criteria+and+Explanatory+Notes.pdf (accessed on 2 October 2017).
16. Emery, K.O. A simple method of measuring beach profiles. *Limnol. Oceanogr.* **1961**, *6*, 90–93. [CrossRef]

17. Araújo, S.; Henriques, I.S.; Leandro, S.M.; Alves, A.; Pereira, A.; Correia, A. Gulls identified as major source of fecal pollution in coastal waters: A microbial source tracking study. *Sci. Total Environ.* **2014**, *470–471*, 84–91. [CrossRef] [PubMed]

18. Converse, R.R.; Kinzelman, J.L.; Sams, E.A.; Hudgens, E.; Dufour, A.P.; Ryu, H.; Santo-Domingo, J.W.; Kelty, C.A.; Shanks, O.C.; Siefring, S.D.; et al. Dramatic improvements in beach water quality following gull removal. *Environ. Sci. Technol.* **2012**, *46*, 10206–10213. [CrossRef] [PubMed]

19. Noble, R.T.; Moore, D.F.; Leecaster, M.K.; McGee, C.D.; Weisberg, S.B. Comparison of total coliform, fecal coliform, and enterococcus bacterial indicator response for ocean recreational water quality testing. *Water Res.* **2003**, *37*, 1637–1643. [CrossRef]

20. Yamahara, K.M.; Sassoubre, L.M.; Goodwin, K.D.; Boehm, A.B. Occurrence and persistence of human pathogens and indicator organisms in beach sands along the California coast. *Appl. Environ. Microbiol.* **2012**, *78*, 1733–1745. [CrossRef] [PubMed]

21. SNIRH—Sistema Nacional de Informação de Recursos Hídricos. Available online: http://www.snirh.pt/snirh/_dadossintese/zbalnear/janela/par_graficos.php?code_cee=PTCE7N&ano=2017#tabela (accessed on 2 October 2017). (In Portuguese)

22. Roca, E.; Riera, C.; Villares, M.; Fragell, R.; Junyent, R. A combined assessment of beach occupancy and public perceptions of beach quality: A case study in the Costa Brava, Spain. *Ocean Coast. Manag.* **2008**, *51*, 839–846. [CrossRef]

23. Alves, B.; Ballester, R.; Rigall-I-Torrent, R.; Ferreira, O.; Benavente, J. How feasible is coastal management? A social benefit analysis of a coastal destination in SW Spain. *Tour. Manag.* **2017**, *60*, 188–200. [CrossRef]

24. Rasoolimanesh, S.M.; Ringle, C.M.; Jaafar, M.; Ramayah, T. Urban vs. rural destinations: Residents' perceptions, community participation and support for tourism development. *Tour. Manag.* **2017**, *60*, 147–158. [CrossRef]

Article

Convertible Operation Techniques for Pump Stations Sharing Centralized Reservoirs for Improving Resilience in Urban Drainage Systems

Eui Hoon Lee [1] and Joong Hoon Kim [2,*]

[1] Research Center for Disaster Prevention Science and Technology, Korea University, Seoul 02841, Korea; hydrohydro@naver.com
[2] School of Civil, Environmental and Architectural Engineering, Korea University, Seoul 02841, Korea
* Correspondence: jaykim@korea.ac.kr; Tel.: +82-02-3290-3316

Received: 9 August 2017; Accepted: 30 October 2017; Published: 31 October 2017

Abstract: Pump stations prevent backwater effects from urban streams and safely drain rainwater in urban areas. Urbanization has increased the required capacity of centralized reservoirs and drainage pumps; yet, their respective designs are based on the runoff of the target watershed at the time of design. In Korea, additional pump stations are constructed to supplement the insufficient capacity of centralized reservoirs and drainage pumps. Two pump stations in the same drainage area share centralized reservoirs, and there are gates between them. Operation of the gates and drainage pumps is based on the water level in the connected centralized reservoirs. The convertible operation is based on changes in flow between two pump stations with different effluent streams in shared centralized reservoirs. Efficient distribution of inflow to both pump stations provides additional storage capacity in centralized reservoirs and rapid drainage. For a rainfall event in 2010, flooding volumes for current and convertible operations were 58,750 and 7507 m^3, respectively. For an event in 2011, the corresponding figures were 3697 and 471 m^3. This shows that resilience increased by 0.10829 and 0.00756, respectively, for the two events. Accordingly, a new technique to operate multiple pump stations for reducing urban inundation is proposed.

Keywords: convertible operation; centralized reservoirs; pump operation; gate operation

---

1. Introduction

Structural measures, such as the installation of drainage facilities, and nonstructural measures, such as the operation of such facilities, have been developed and applied to prevent water inundation in urban areas. Structural measures are not only costly and time-consuming, but also need to be complemented by nonstructural measures; these are cheaper and easier to implement because urban areas may not be able to accommodate additional drainage facilities. As urbanization increases, runoff also increases and additional drainage facilities, such as pump stations, are constructed to drain the urban area. Appropriate operation of urban drainage systems, including additional facilities, is required to effectively prevent urban inundation.

The gates in centralized reservoirs are usually closed, and are operated in special cases, such as when receiving inflow from stream. The convertible operation of gates and pumps is based on the flow change in urban drainage systems. If all inflow is driven to a single pump station, drainage efficiency may reduce and the drainage area can be inundated. If there are two pump stations in a single drainage area, it is more effective to distribute inflow to both pump stations as this can improve drainage efficiency and provides additional capacity in centralized reservoirs, which enables more rapid drainage.

Many researchers have studied the real-time control (RTC) of urban drainage facilities as a nonstructural measure. These studies considered different types of operation, such as individual operation, cooperative operation, and convertible operation between centralized reservoirs with drainage pumps and gates in pump stations. In addition, these studies considered a range of operating time periods including yearly, daily, hourly, and per minute operation. The first studies have minutely and individual operations for urban drainage facilities. The optimal control of urban drainage systems has been investigated using a telemetry/telecontrol system [1]. Real-time operation of urban wastewater systems in Quebec was studied and this led to a significant improvement in their performance [2]. A real-time control strategy for the sewer system of Vienna was investigated, which was used to develop a hydrodynamic sewer system [3]. Optimal real-time control of the Quebec urban drainage system was studied to reduce the frequency and volumes of combined sewer overflows [4]. A forecast-based operation method to minimize flood damage in urban areas was used to predict the flow rate in storm sewer pipes and river water levels [5]. Real time control of urban drainage systems was investigated, with the additional efforts needed by this approach were highlighted through a comparison with conventionally operated systems [6]. The real-time operation of a sewer network was optimized using a multi-goal objective function [7]. An optimized adaptive network-based fuzzy inference system was recommended for the intelligent real-time operation of a pump station in an urban drainage system [8]. Proactive pump operation and capacity expansion in an urban drainage system was conducted to improve system resilience [9].

A cooperative operation and a convertible operation differ due to the flow path used. A cooperative operation includes a centralized reservoir located at the downstream end of a drainage area and a decentralized reservoir located at the upstream end. The cooperative operation refers to the concurrent operation of different drainage facilities in an urban drainage area. Only drainage pumps in centralized and decentralized reservoirs were used for the cooperative operation in a previous study [10]. A convertible operation requires two centralized reservoirs, linked by gates that control the rate of inflow; these can be operated to provide additional drainage capacity. Two pump stations can share several centralized reservoirs in the same drainage area.

Many studies on RTC of urban drainage facilities have been conducted since 2000, and some researchers have studied long-term operation across different time periods, including yearly, daily, and hourly operation. Initial research was carried out on screening urban wastewater systems for RTC potential [11]. Optimal real-time operation of multipurpose urban reservoirs was applied to the target area in Singapore [12]. The operation of detention facilities for flood control and multipurpose urban reservoirs in Singapore was also studied [13]. These studies only considered individual operation for the yearly, daily, and hourly operating time periods; cooperative operation and convertible operation across these periods were not investigated. The operations in these studies had a variety of purposes, but involved operation of only one hydraulic facility. In addition, these operations are not appropriate in some urban drainage areas because of the long operating times required.

Studies of cooperative operation of urban drainage facilities with per-minute operating times have been conducted. Inundation in an urban drainage system was simulated with several pump stations having different drainage areas [14]. Modelling and real-time control of an urban wastewater system was conducted with an integrated mathematical model [15]. Cooperative operation between centralized and decentralized reservoirs in an urban drainage system was proposed for reducing urban inundation [10]. Cooperative operations were possible in these studies, but the inflow volume to each drainage facility was not modified, and the direction of inflow could not be changed. This means that the capacity of drainage facilities, such as centralized reservoirs, to prevent urban inundation was not considered.

No studies regarding convertible operations between centralized reservoirs in one drainage area have been undertaken. The convertible operation comprises the gate operation in centralized reservoirs and the pump operation in pump stations. In this study, the pump operation in the convertible operation was used to maintain low water levels in the centralized reservoirs. Early operation of the pumps also enabled rapid drainage in the target drainage area, which reduced the backwater effect from the water level of the centralized reservoirs. The gate operation in the convertible operation was used to distribute the inflow efficiently to the centralized reservoirs, reducing urban inundation and providing additional capacity in centralized reservoirs. The gate operation could change the flow direction in the centralized reservoirs, effectively distributing the inflow between the two pump stations. In addition, resilience in the target watershed was examined by applying system resilience identified in a previous study to the convertible operations [10].

2. Materials and Methods

2.1. Overview

This study consisted of four parts: First, synthetic rainfall data were generated for the rainfall runoff simulation. Second, monitoring nodes were selected for the convertible operation in the centralized reservoirs; these nodes were chosen by searching for the first flooding node and the maximum flooding node. Third, the operating method for the convertible operation in centralized reservoirs was investigated and the convertible operation was applied to the target area. Finally, the results of the convertible operation were compared with the results of the current operation to examine the effects of the convertible operation and to assess system resilience. A flowchart of this study is shown in Figure 1.

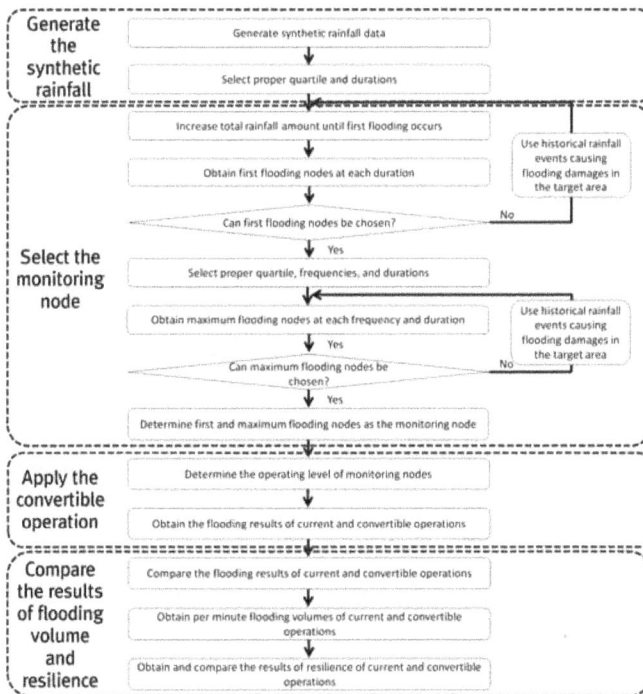

Figure 1. Flowchart of this study.

2.2. Generation of Synthetic Rainfall Data for Selecting Monitoring Nodes

In this study, monitoring nodes were selected by searching for the first and maximum flooding nodes. This search was based on rainfall runoff simulations using rainfall input data; in Korea, most drainage facilities are designed using synthetic rainfall data based on the Huff distribution [16]. The Huff distribution consists of four quartiles, split according to the peak time of the synthetic rainfall data. If the total time of the synthetic rainfall data is assumed to range between 0% and 100%, then the peak time of each quartile is divided into four regions. The peak values of the first, second, third, and fourth quartiles in the Huff distribution are located between 0% and 25%, 25% and 50%, 50% and 75%, and 75 and 100%, respectively. Each region has an appropriate quartile; the third quartile is appropriate to Korea [17]. The cumulative regression equation of the third quartile for the Huff distribution in Seoul is as follows [18]:

$$y = 37.835x^6 - 106.21x^5 + 105.18x^4 - 44.549x^3 + 9.1084x^2 - 0.3603x + 0.0005 \qquad (1)$$

where y represents the ratio of cumulative rainfall, and x is the ratio of the total rainfall duration. The process of generating synthetic rainfall data via the Huff distribution consists of three steps [9,10,19,20]: The first step is the generation of a cumulative rainfall distribution using the cumulative regression equation. The second step is the conversion from the cumulative rainfall distribution to the separated rainfall distribution. The final step is the process of applying the total rainfall amount according to the selected frequency and rainfall duration to the separated rainfall distribution. This process for generating the synthetic rainfall data via the Huff distribution is shown in Figure 2.

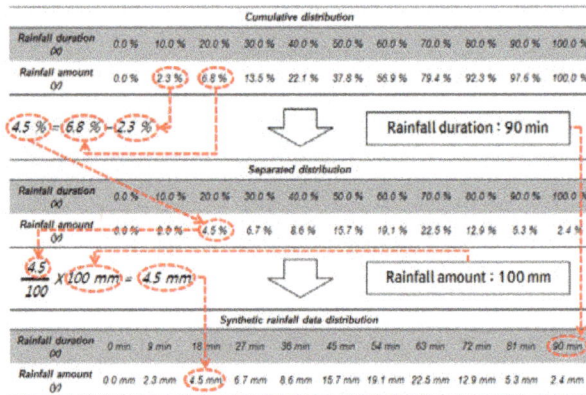

Figure 2. Process for generating the synthetic rainfall data via the Huff distribution (third quartile, 100 mm, 90 min).

2.3. Selecting Monitoring Nodes for the Convertible Operation

Monitoring nodes are important because the convertible operation of centralized reservoirs in pump stations is based upon the level of the monitoring nodes. Selection of the monitoring nodes is based on two methods: finding (1) the first flooding nodes; and (2) the maximum flooding nodes. These methods have been used in previous studies [9,10,19] into the operation of urban drainage facilities.

For the first flooding nodes, three durations were selected by considering the time of concentration. For example, 30, 60, and 90 min durations were chosen if the time of concentration was 30 min. When searching for the first flooding nodes, the initial total amount of synthetic rainfall in the study area was set at 1 mm and this was then increased in 1 mm increments until first flooding occurred at each duration. First flooding nodes at each duration were then selected as the monitoring nodes. If the first flooding nodes were different, the distribution of historical rainfall data was used following the same process. To identify the maximum flooding nodes, the synthetic rainfall data were generated at a minimum of three frequencies and for three durations. Rainfall runoff simulations for all rainfall data were conducted and the maximum flooding nodes at each simulation were chosen; the results were then reviewed and the node with the most frequent occurrence was then selected. If the maximum flooding nodes were different from each other and it was difficult to select appropriate monitoring nodes using the synthetic rainfall data, historical rainfall data were used to identify the monitoring nodes. Historical rainfall events in 2010 (21 September 2010) and 2011 (27 July 2011) were therefore applied for selection of the maximum flooding nodes. Figure 3 shows the detailed process used to identify the first flooding nodes and maximum flooding nodes used in this study.

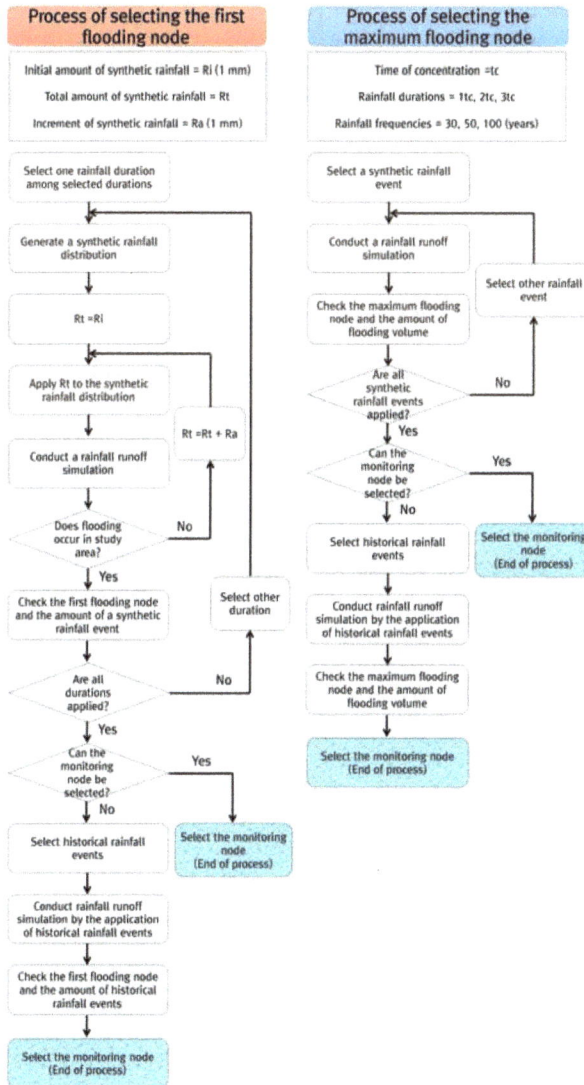

Figure 3. Process for selecting the first and maximum flooding nodes.

2.4. Convertible Operation of Centralized Reservoirs in Pump Stations

The general operation of centralized reservoirs in pump stations, such as one centralized reservoir in one pump station, has been recommended in a previous study [10]. Gate operation was not considered in that study because gates were not present in the target watershed. However, in this study, the convertible operation of multiple centralized reservoirs in multiple pump stations with drainage pumps and gates was examined.

Distribution of water through gate operation in multiple centralized reservoirs is very important for reducing flooding from backwater effects and to obtain additional storage. The flow in centralized reservoirs can be driven to one side by gravity if it is not controlled by drainage pumps or gates. The operation of drainage pumps is based on the level of monitoring nodes, and the operation of gates is based on the level of each centralized reservoir. In Korea, most urban sewer systems are combined sewer systems, and the sewer system in this study was a combined sewer system. Up to 3Q (three times the dry weather flow) is captured by an intercepting sewer in centralized reservoirs and sent to a Waste Water Treatment Plant (WWTP) located outside of the study area. Inflow over 3Q is discharged to urban watercourses by pump stations (Figure 4).

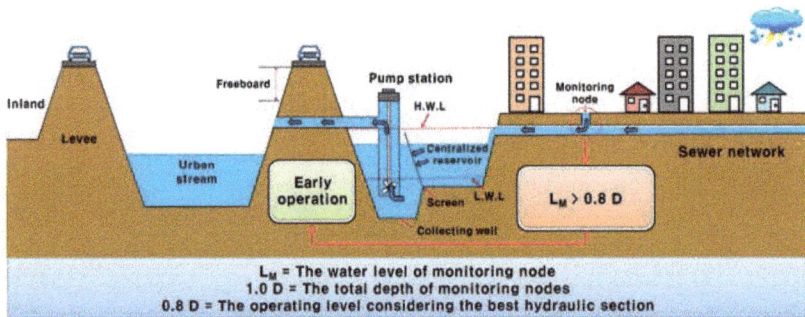

Figure 4. Pump operation of centralized reservoirs in pump stations.

In Korea, the water level of centralized reservoirs in pump stations is a standard used for operating drainage pumps. The initial operating level of the drainage pumps in centralized reservoirs was calculated as the sum of three factors: (1) the required depth; (2) the screen head loss; and (3) the freeboard for mechanical operation. The required depth (1) was obtained by dividing the value of the required volume (1-1) by the average area in a centralized reservoir (1-2). The required volume (1-1) is the product of the pump discharge and the pump preparation time. In this study, the screen head loss ranged between 0.1 to 0.3 m, and the freeboard for mechanical operation was distributed from 0.0 to 0.2 m. Other operating levels are based only on the required depth. The equation used to calculate the initial operating level of the drainage pumps in the centralized reservoirs is shown below:

$$O_i = \frac{P_i \times T_p}{4V_r A_l} + H_s + F_m + B_{cr} \tag{2}$$

where O_i is the initial operating level of the drainage pumps in a centralized reservoir and P_i is the initial operating pump discharge. T_p is the preparation time of the initial operating pump and V_r is the required volume in the centralized reservoir. A_l is the average area at each elevation in CR and H_s is the screen head loss. F_m is the freeboard for mechanical operation and B_{cr} is the bed elevation of the centralized reservoir.

The operation of the gates is based on the water level at the discharge location, and the levels of the centralized reservoirs receiving water are considered because of the flow between centralized reservoirs. The schematics of the gate operation in centralized reservoirs referred to in this study are shown in Figure 5.

Current operation

| Inflow 2 | Centralized reservoir 2 | Centralized reservoir 4 | No pumping |
| Gate 1 (open) | Volume1 (V1) | ✖ |

Gate 3 (closed) ✖   Gate 2 (closed) ✖

| Inflow 1 | Centralized reservoir 3 | Centralized reservoir 1 | Pumping |
| Gate 4 (open) | Volume2 (V2) |

Current operation | Gates 1 and 4 are open / Gates 2 and 3 are closed | ➡ V1 < V2

New operation

| Inflow 2 | Centralized reservoir 2 | Centralized reservoir 4 | Pumping |
| Gate 1 | Volume1 (V1) |

Gate 3 ↖   Gate 2 ↘

| Inflow 1 | Centralized reservoir 3 | Centralized reservoir 1 | Pumping |
| Gate 4 | Volume2 (V2) |

New operation | Gates 1, 2, 3, and 4 are operated (open or closed) | ➡ V1 = V2

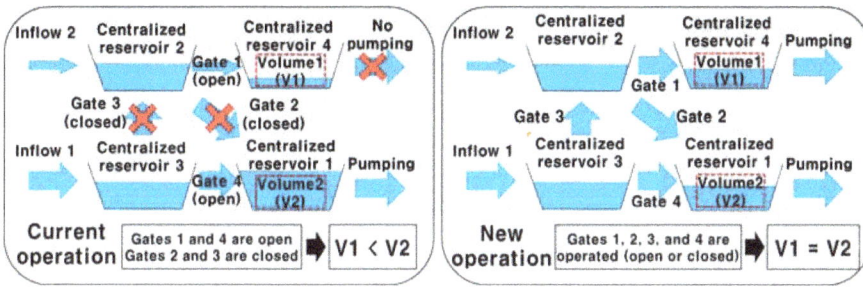

Figure 5. Schematics of the gate operation in centralized reservoirs.

In the current operation, inflows 1 and 2 are divided, and there is no gate operation. If inflow 1 is large and inflow 2 is small, the centralized reservoirs receiving inflow 1 have high levels, while the centralized reservoirs receiving inflow 2 have low levels. When the water is pushed to one side, the levels in the centralized reservoirs increase and the levels of the upstream conduits also increase via backwater effects. In the new operation, inflows 1 and 2 flow into different reservoirs and are distributed through gate operation based on the levels of the centralized reservoirs receiving the water. If inflow 1 is large and inflow 2 is small, gate operation can distribute the water appropriately.

The convertible operation is the operation between the centralized reservoirs that changes the flow path of the inflow. For example, inflow 1 in Figure 5 has several flow paths, including: (1) centralized reservoir 3 to centralized reservoir 2 to centralized reservoir 4; (2) centralized reservoir 3 to centralized reservoir 2 to centralized reservoir 1; and (3) centralized reservoir 3 to centralized reservoir 1. If inflow 1 flows to centralized reservoir 4, it is discharged by pump station 2. However, if inflow 1 flows to centralized reservoir 1, it is discharged by pump station 1. Inflow 2 also has several flow paths.

The volume of runoff from the catchment was based on the amount of rainfall and the catchment area. Two catchments have different runoff rates because rainfall could be distributed unevenly or because the size of the catchment area is different. Additionally, runoff from the catchment is not uniform because concentration time can very between catchments. In the current operation as shown in Figure 5, one pump station discharges inflow into centralized reservoir 4, while the other pump station continues to discharge inflow into centralized reservoir 1 because gates 2 and 3 are closed.

2.5. Determination of the Operating Method for Pumps and Gates in the Decentralized Reservoirs

Historical flooding occurred twice within the study area, once in 2010 and again in 2011. The level of pump operation in the convertible operation was calculated prior to the rainfall runoff simulation. The preparation time was generally considered to take between 5 and 30 min. The average area in the centralized reservoir was published in an official report for the pump station. The screen head loss ranged from 0.1 to 0.3 m, while the freeboard for mechanical operation was between 0.0 and 0.2 m. Other operating levels excluding the initial operating level were determined by the required depth only. Table 1 shows the pump operating levels of the centralized reservoirs in the target watershed.

To determine the appropriate operating level, both the best hydraulic section and local regulation were considered [21]. The appropriate range of operating level based on the best hydraulic section was 0.8 to 1.0 D; this is because the flow rate of a conduit at a depth from 0.8 to 1.0 D was larger than the flow rate when a conduit is filled with water if D represents the total depth of node. In Korea, the large conduits in urban drainage systems should have additional capacity, ranging from between 25% and 50%. Using local regulation, the appropriate range was 0.67 to 0.8 D [19]. Therefore, the level meeting both conditions was 0.8 D and this was the standard applied to this study.

Table 1. Current and new operations of the drainage pumps in the centralized reservoirs.

Gaebong 1 Pump Station				Gaebong 2 Pump Station			
Elevation of Current Operation in Gaebong 1 (m)	Pump Discharge of Current Operation (m³/s)	Elevation of New Operation in Gaebong 1 (m)	Pump Discharge of New Operation (m³/s)	Elevation of Current Operation in Gaebong 2 (m)	Pump Discharge of Current Operation (m³/s)	Elevation of New Operation in Gaebong 2 (m)	Pump Discharge of New Operation (m³/s)
5.0	0.00	5.0	0.00	3.0	0.00	3.0	0.00
6.5	7.67	5.5	7.67	6.0	5.17	5.0	5.17
6.7	15.34	5.7	15.34	6.1	10.34	5.1	10.34
6.9	23.01	5.9	23.01	6.2	15.51	5.2	15.51
7.1	30.68	6.1	30.68	6.3	20.68	5.3	20.68
7.3	38.35	6.3	38.35	6.4	28.10	5.4	28.10
7.5	46.02	6.5	46.02	6.5	35.52	5.5	35.52
7.8	53.69	6.8	53.69	6.6	42.94	5.6	42.94
8.1	61.36	7.1	61.36	6.7	50.36	5.7	50.36
8.4	73.36	7.4	73.36	6.8	57.78	5.8	57.78
8.7	85.36	7.7	85.36	6.9	65.20	5.9	65.20
9.0	97.36	8.0	97.36	7.0	72.62	6.0	72.62
9.3	109.36	8.3	109.36	7.1	80.04	6.1	80.04
9.6	121.36	8.6	121.36	-	-	-	-
9.9	133.36	8.9	133.36	-	-	-	-
10.2	145.36	9.2	145.36	-	-	-	-
10.5	157.36	9.5	157.36	-	-	-	-

In gate operation, gates 2 and 4 were automatically closed if the stream gate was closed and the inflow gate was open. In this case, reservoir 1 received inflow from the Mokgam Stream and the inflow was discharged via drainage pumps in the Gaebong 1 pump station. All inflow from drainage areas 1, 2, and 3 passed to reservoir 4 and was discharged via the Gaebong 2 pump station. Additional gate operation was conducted when water levels in the Anyang Stream were high. Both the current and new operations included additional gate operation. For the 2010 and 2011 rainfall events, the stream gate was open and the inflow gate was closed. In this study, gates 1, 2, 3, and 4 were opened or closed in the new operation according to the level of each centralized reservoir (whereas gates 2 and 3 were closed in the current operation). Table 2 shows the comparison between the current and new operations.

Table 2. Comparison between the current and new operations with gates.

Inflow	Current Operation	New Operation	Comparison
Drainage area 1	Reservoir 4	Reservoir 4	Same
Drainage area 2	Reservoir 2 → Reservoir 4	Reservoir 2 → Reservoir 4 (Depth of reservoir 4 < Depth of reservoir 1) Reservoir 2 → Reservoir 1 (Depth of reservoir 4 > Depth of reservoir 1)	Different
Drainage area 3	Reservoir 3 → Reservoir 1	Reservoir 3 → Reservoir 2 (Depth of reservoir 2 < Depth of reservoir 1) Reservoir 3 → Reservoir 1 (Depth of reservoir 2 > Depth of reservoir 1)	Different

2.6. Comparison between the Current and Convertible Operations Considering System Resilience

One method used to evaluate structural and nonstructural measures is system resilience. System resilience in urban drainage systems can be defined as the ability to prepare for, and recover from, the malfunction of urban drainage facilities or flooding (failure) [10]. All systems, including urban drainage systems, have risk factors, and it is important to have the ability to recover quickly (resilience), even if the risk is connected to a disaster situation. In water resources engineering, disaster response is formulated through statistical analysis of past events, which is then used to enable prediction of future events. However, risk factors vary over time, and forecasting based on historical statistics can have limitations. Recently, system resilience has been used to evaluate various systems, including urban drainage systems.

A previous study [10] recommended the use of the resilience index, which considers concentration time in urban drainage systems. This resilience index is based on the results of per minute flooding volumes obtained from rainfall runoff simulations. Two factors—flooding volume and system resilience—were examined to compare the effects of operations. Figure 6 shows the concept of system resilience with failure and recovery over time.

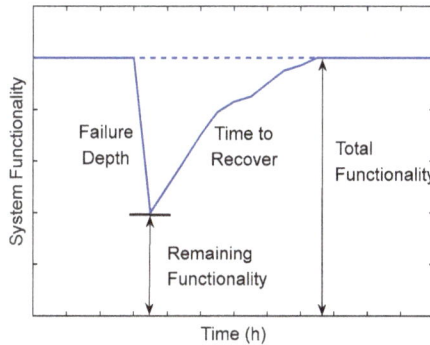

Figure 6. Concept of system resilience.

The resilience index in urban drainage systems is calculated using a performance evaluation function with per minute flooding volumes, concentration times, and unit drainage areas. The values of the performance evaluation function, as well as the resilience index, are distributed from 0 to 1, with 1 meaning there is no failure (flooding). The performance evaluation function is as follows:

$$u(T)_t = \max\left(0, \quad 1 - \frac{F_t}{\sum\limits_{t-t_c}^{t} R_t \times A_u}\right) \tag{3}$$

where $u(T)_t$ means the performance evaluation function at time t; F_t represents per minute flooding volume at time t (m^{-3}); t_c is the time of concentration; t is the current time; R_t is the rainfall volume at time t $(10^{-3}\ \mathrm{m})$; and A_u means the unit drainage area $(1\ \mathrm{ha} = 10^{-4}\ \mathrm{m}^2)$. The value of the performance evaluation function was negative if flooding occurred as a result of rainfall. In this case, it was calculated as 0 using Equation (3). The performance evaluation function was high if an urban drainage system showed the ability to recover from failure.

The resilience of the urban drainage system based on the performance evaluation function is shown in the following equation:

$$R_s = \frac{1}{T}\int_{T_0}^{T} u(T)_t\, dt \tag{4}$$

where R_s refers to the resilience of the urban drainage system; T_n represents the entire time; T_0 is the initial time; and $u(T)_t$ is the performance evaluation function at the current time. The results of each operation (the current and convertible operations) were compared and analyzed according to historical rainfall events.

2.7. Study Area

Gyeonggi province is the area surrounding Seoul (Seoul is the capital of Korea and has a population of 13 million people) and has a population of approximately 12.4 million people. Seoul and Gyeonggi province are traversed by the Han River. The Han River has many tributaries, including the Anyang Stream, which in turn has many tributaries, such as the Mokgam Stream. In this study, the area near the Gaebong 1 and Gaebong 2 pump stations, which are located in the lower reaches of the Mokgam Stream, was selected as the target watershed; its drainage area is 1086 ha. Figure 7 shows the location of the target watershed [21].

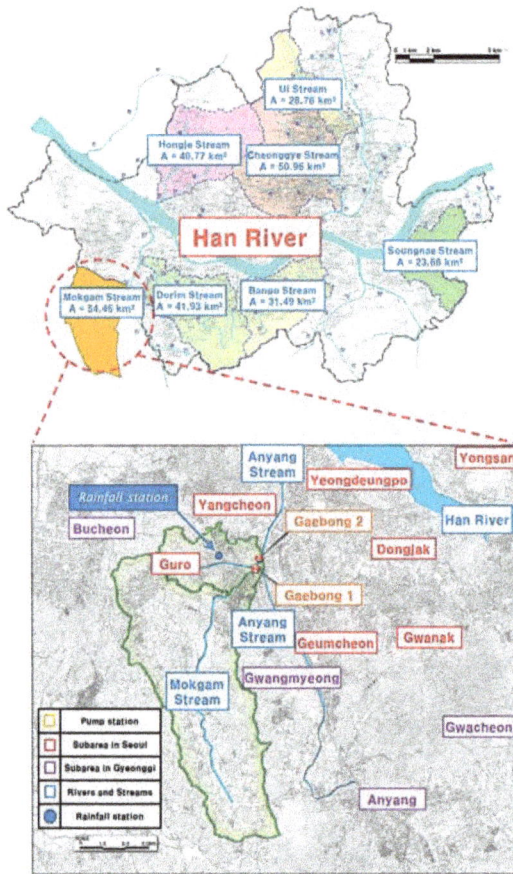

Figure 7. Location of the target watershed.

The target watershed had four centralized reservoirs and an inflow conduit between reservoirs 2 and 4 (Figure 8). The drainage system also had two pump stations, 711 sub-catchments, 618 nodes, and 630 links. The current operation of drainage facilities in the target watershed consisted of two steps. In the first step, the inflow gate from the Mokgam Stream and gate 3, located between reservoirs 2 and 3, were closed while the other gates were open. Inflow from drainage areas 1 and 2 was discharged via the Gaebong 2 pump station, including natural discharge gate 2, while inflow from drainage area 3 was discharged via the Gaebong 1 pump station, including natural discharge gate 1. In the second step, the stream gate was closed to prevent backwater effects from the Anyang Stream, and the inflow gate was open if the water level of the Mokgam Stream was high. Gates 2 and 4 were closed, and inflow from all drainage areas passed to reservoir 4 in the Gaebong 2 pump station. Reservoir 1 received the inflow from the Mokgam Stream [22].

A storm water management model (SWMM) 5.0 was selected as the rainfall runoff model and used to simulate the target watershed [23]. The sewer network and drainage facilities in the target watershed were based on Geographic Information System (GIS) data for Seoul provided by the Seoul Metropolitan Government. Figure 9 shows the sewer network of the target watershed.

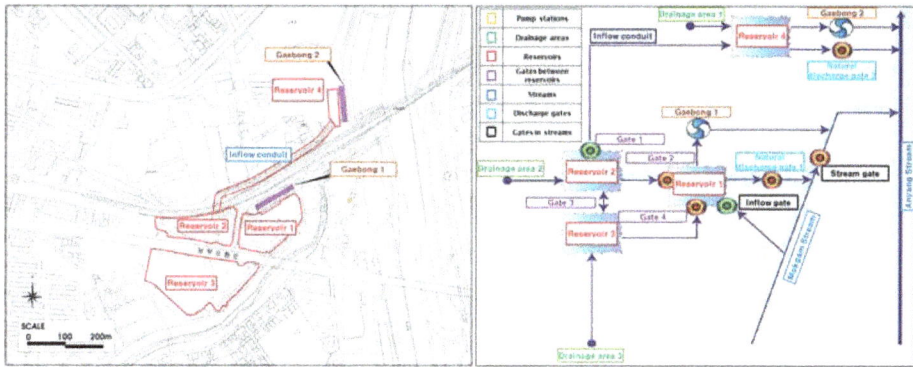

Figure 8. Drainage facilities of the target watershed.

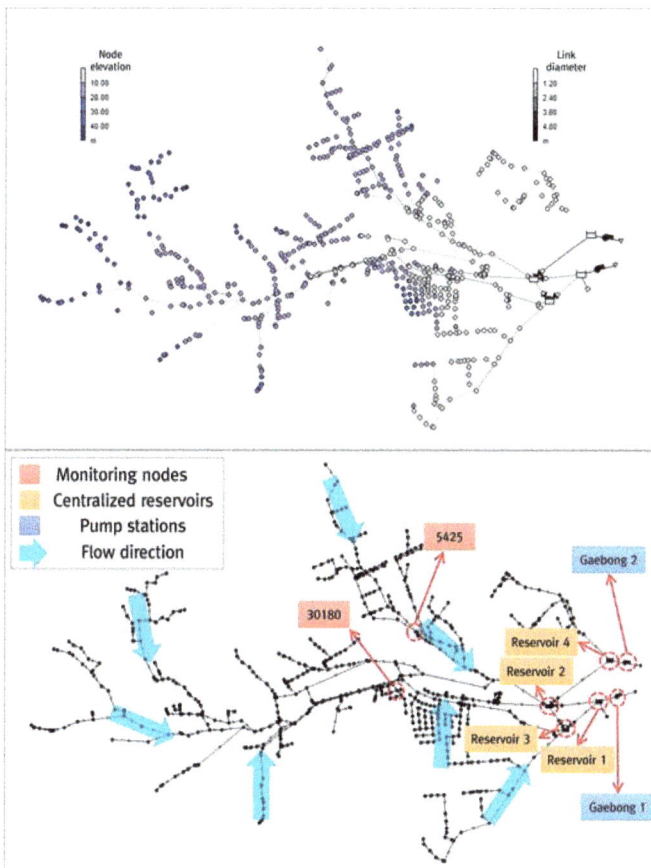

Figure 9. Sewer network of the target watershed.

Information for the two pump stations, Gaebong 1 and Gaebong 2, consisted of high water level (HWL), low water level (LWL), initial operating level, and drainage pump capacity. Information for the centralized reservoirs consisted of HWL, LWL, area, and effective storage capacity. This information was necessary for the operation of the drainage facilities and is shown in Table 3.

Table 3. Information of the drainage facilities in the target watershed.

Status	Gaebong 1		Gaebong 2	
High water level (m)	10.9		8.0	
Low water level (m)	6.0		3.0	
Initial operating level (m)	6.5		4.0	
Drainage pump capacity (m^3/min)	9440		4800	
Status	**Reservoir 1**	**Reservoir 2**	**Reservoir 3**	**Reservoir 4**
High water level (m)	10.90	8.00	8.00	8.00
Low water level (m)	6.00	3.00	6.00	6.00
Area (m^2)	17,920	17,920	14,590	28,457
Effective storage capacity (m^3)	51,700	31,858	25,408	32,234

3. Results

3.1. Selection of Monitoring Nodes for Pump Operation

Selection of the monitoring nodes was the first process for pump operation of the centralized reservoirs. First and maximum flooding nodes were selected as the monitoring nodes. First flooding nodes generally occurred in branch conduits. Branch conduits were located downstream of the node where the product of the runoff coefficient (C) and the drainage area (A) was smaller than 0.12 km^2 (CA < 0.12 km^2). However, it was difficult for branch conduits to represent the entire watershed because they had small drainage areas; hence, main conduits were selected for monitoring nodes. Main conduits were located downstream of the node where the product of C and A was larger than 0.12 km^2 (CA $\geq$ 0.12 km^2) [10]. The results of the monitoring node selection for the target watershed are shown in Table 4.

Table 4. Results of monitoring node selection.

Duration (min)	First Flooding Node (Rainfall Amount)	Event (Year)	Maximum Flooding Node (Flooding Volume)
30	30180 (43 mm)	2010	5425 (16,630 m^3)
60	30180 (48 mm)	2011	5425 (3409 m^3)
90	02473 (68 mm)	-	-

Nodes 30180 and 5425 were selected as monitoring nodes for pump operation of centralized reservoirs in convertible operation. These two nodes were located where the flow rate increased due to the steep slope of the upper conduit (5425) or the meeting of several conduits (30180). Their locations are shown in Figure 10.

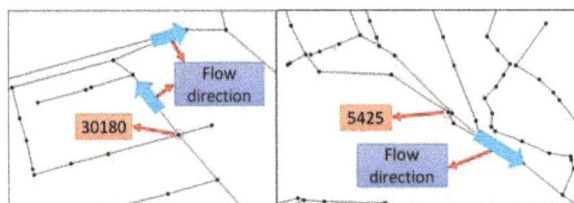

Figure 10. Monitoring node locations in the target watershed.

3.2. Results of Flooding Volume for the Current and Convertible Operations

The new convertible operation combined pump operation and gate operation, preventing backwater effects based on the levels of the centralized reservoirs. The new gate operation provided additional storage capacity in the centralized reservoirs. The total flooding volumes for the current and new operations were 58,750 m^3 and 7507 m^3, respectively, for the 2010 event. For the 2011 event, the total flooding volumes for the current and new operations were 3697 m^3 and 471 m^3, respectively. The flooding volumes over time for the current and new operations are shown in Figure 11.

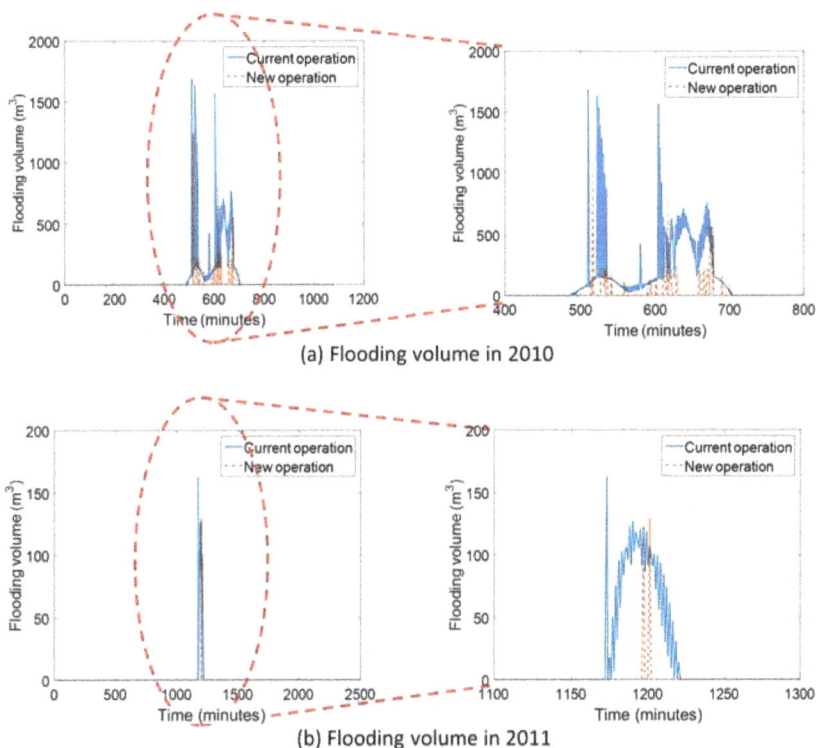

(a) Flooding volume in 2010

(b) Flooding volume in 2011

Figure 11. Flooding volumes over time for the current and new operations in (**a**) 2010 and (**b**) 2011.

As shown in Figure 11a, the flooding volume for the 2010 event for the current operation occurred from 500 to 700 min. The peak flooding volume for the current operation was approximately 1700 m^3 at 550 min, and the number of peak flooding volumes over 1500 m^3 was three. Two peak flooding volumes occurred consecutively, while one appeared independently. The duration of the flooding for the 2011 event for the new operation was similar to that of the current operation. The peak flooding volume for the 2010 event for the new operation was approximately 1200 m^3 at 570 min, and no other flooding volumes were over 1000 m^3. The results for the 2010 event clearly showed the difference between the two modes of operation. Two peak flooding volumes occurred continuously because the centralized reservoirs had no additional capacity under the current operation. In contrast to the current operation, the new operation showed a single peak flooding volume, with no additional flooding volume because the centralized reservoirs had additional capacity.

In Figure 11b, the flooding volume for the 2011 event for the current operation occurred from 1150 to 1220 min. The peak flooding volume for the current operation was approximately 160 m^3 at 1150 min, and the number of peak flooding volumes over 120 m^3 was two. Two peak flooding volumes occurred continuously and the pattern was similar to that of the 2010 event. The peak flooding volume for the 2011 event for the new operation was approximately 130 m³ at 1200 min, with no other flooding volumes over 120 m³. The results for the 2011 event also showed other differences, such as prevention of backwater effects and provision of additional capacity in the centralized reservoirs. The difference between the two operations was examined when the historical rainfall events were applied. Figure 12 shows the results of pump discharge over time for the current and new operations in the 2010 and 2011 events.

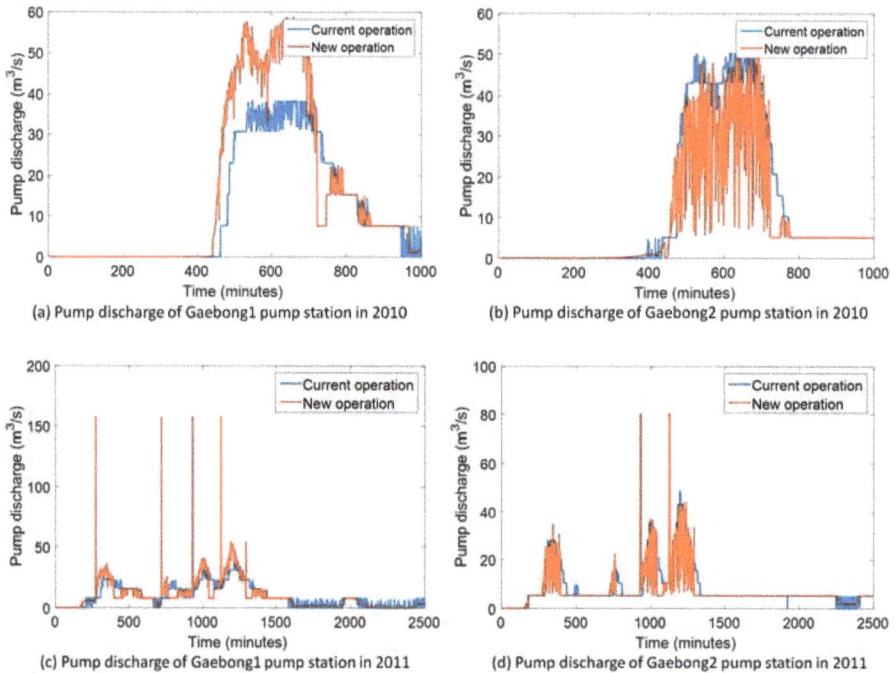

(a) Pump discharge of Gaebong1 pump station in 2010

(b) Pump discharge of Gaebong2 pump station in 2010

(c) Pump discharge of Gaebong1 pump station in 2011

(d) Pump discharge of Gaebong2 pump station in 2011

Figure 12. Pump discharge over time for the current and new operations.

As shown in Figure 12a, the peak value of pump discharge in the new operation was higher than the peak value of pump discharge in the current operation. In Figure 12b, the peak value of pump discharge in the new operation is similar to the peak value of pump discharge in the current operation. In Figure 12a,b, the peak values of pump discharge in both operations occurred from 500 min to 700 min; this is because flooding in the 2010 event, as shown in Figure 12a occurred from 500 min to 700 min. In Figure 12c,d, the peak value of pump discharge in the new operation is higher than the peak value of pump discharge in the current operation. Figure 12c shows four peak values and Figure 12d shows two peak values. All peak values of pump discharge in the new operation occurred before 1200 min, although flooding occurred at 1200 min in 2011. This means that the new operation maintained low water levels in the centralized reservoirs as a preemptive response for reducing flooding in the study area.

Flooding at the monitoring nodes was caused by multiple factors, including the backwater effects due to the high water level in centralized reservoirs and insufficient capacity in the sewer conduit downstream of the flooded node. To solve this problem, capacity of the sewer conduit should be increased as a structural measure, requiring cost and time. The purpose of this study was to investigate a new method of operation as a potential nonstructural measure. However, the reduction of back water effects by the convertible operation cannot prevent urban inundation completely. This approach would demonstrate better results if sewer conduit capacity and the convertible operation were both carried out.

3.3. Results of System Resilience for the Current and Convertible Operations

Figure 13 shows the results of system resilience analysis for the current and new operations, and highlights the increase in resilience for the 2010 and 2011 events.

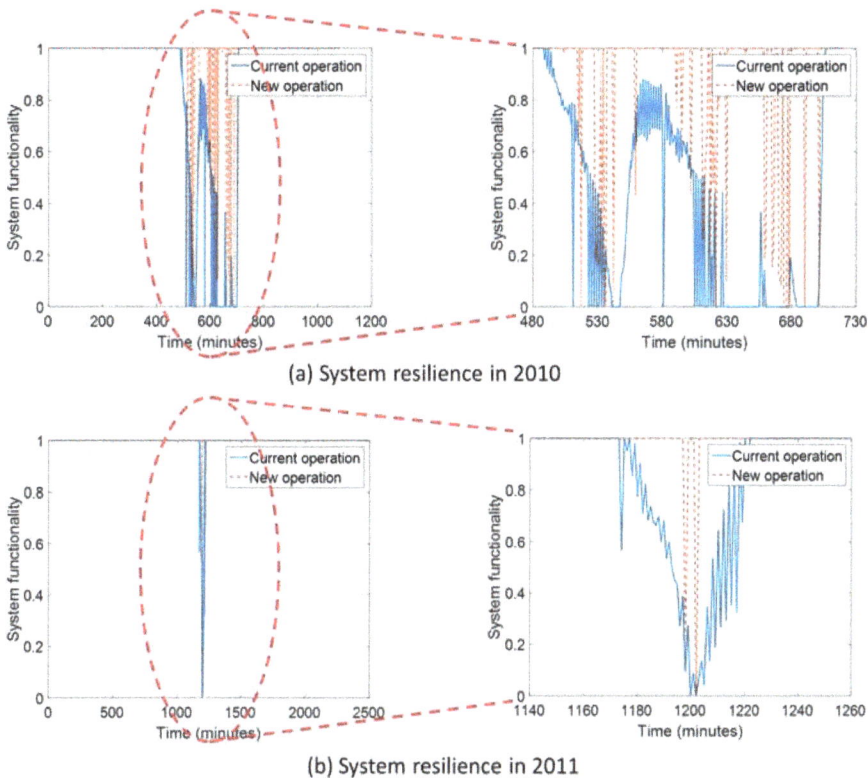

(a) System resilience in 2010

(b) System resilience in 2011

Figure 13. System resilience for the current and new operations in (**a**) 2010 and (**b**) 2011.

For the 2010 event, the system resilience for the current operation was 0.86529 and the system resilience for the new operation was 0.97358. Therefore, the increase of system resilience for the 2010 event was 0.10829. For the 2011 event, the system resilience for the current operation was 0.99174 and the system resilience for the new operation was 0.99930. The increase of system resilience in 2011 event was 0.00756. The values of system resilience for the 2010 event were lower than those for the 2011 event because rainfall duration for the 2011 event was longer than that for the 2010 event, and rainfall intensity for the 2010 event was higher than that for the 2011 event. If system resilience was low, the potential of system failure was high. Therefore, failure for the 2010 event was larger than the failure for the 2011 event, and the reduction in the failure for the new operation for the 2010 event was larger than that for the 2011 event.

4. Conclusions

This study consisted of four parts: (1) the generation of synthetic rainfall data; (2) selection of monitoring nodes; (3) determination of operating level for drainage pumps and gates; and (4) comparison of flooding volume for the current and new operations, and examination of system resilience. The results of this study lead to the following conclusions.

- The new convertible operation of centralized reservoirs consisted of pump operation for rapid drainage and gate operation for additional storage capacity.
- Pump operation was conducted by the levels of monitoring nodes, which were obtained by finding first and maximum flooding nodes using synthetic rainfall data via the Huff distribution and historical rainfall data.
- Gate operation converted the flow of each drainage area based on the levels of centralized reservoirs and generated the effective runoff distribution for continuous flooding mitigation.
- The new convertible operation could reduce flooding volume and system failure, which are indicators for evaluation of urban drainage systems.

In this study, gate operation in convertible operation changed the drainage flow in each sub-area, and pump operation in convertible operation was used to induce rapid drainage and secure additional capacity in centralized reservoirs. Convertible operation represents a practical method that could be applied to urban drainage areas with pump stations that share centralized reservoirs. In future studies, cooperative operation between pump stations that share discharge streams and various drainage facilities, including underground drainage tunnels, should be investigated.

Acknowledgments: This research was supported by a grant (17AWMP-B066744-05) from the Advanced Water Management Research Program funded by the Ministry of Land, Infrastructure, and Transport of the Korean government and The National Research Foundation (NRF) of Korea, funded by the Korean government (MSIP) (No. 2016R1A2A1A05005306).

Author Contributions: Eui Hoon Lee carried out the survey of previous studies and wrote the draft of the manuscript. Eui Hoon Lee revised the draft until the final version of the manuscript and simulated the results. Eui Hoon Lee and Joong Hoon Kim conceived the original idea of the proposed method.

Conflicts of Interest: The authors declare no conflict of interest.

References

1. Cembrano, G.; Quevedo, J.; Salamero, M.; Puig, V.; Figueras, J.; Martí, J. Optimal control of urban drainage systems. A case study. *Control Eng. Pract.* **2004**, *12*, 1–9. [CrossRef]
2. Schütze, M.; Campisano, A.; Colas, H.; Schilling, W.; Vanrolleghem, P.A. Real time control of urban wastewater systems—Where do we stand today? *J. Hydrol.* **2004**, *299*, 335–348. [CrossRef]
3. Fuchs, L.; Beeneken, T. Development and implementation of a real-time control strategy for the sewer system of the city of Vienna. *Water Sci. Technol.* **2005**, *52*, 187–194. [PubMed]
4. Pleau, M.; Colas, H.; Lavallée, P.; Pelletier, G.; Bonin, R. Global optimal real-time control of the Quebec urban drainage system. *Environ. Model. Softw.* **2005**, *20*, 401–413. [CrossRef]
5. Tamoto, N.; Endo, J.; Yoshimoto, K.; Yoshida, T.; Sakakibara, T. Forecast-based Operation Method in Minimizing Flood Damage in Urban Areas. In Proceedings of the 11th International Conference on Urban Drainage, Edinburgh, Scotland, UK, 31 August–5 September 2008; p. 31.
6. Beeneken, T.; Erbe, V.; Messmer, A.; Reder, C.; Rohlfing, R.; Scheer, M.; Schuetze, M.; Schumacher, B.; Weilandt, M.; Weyand, M. Real time control (rtc) of urban drainage systems—A discussion of the additional efforts compared to conventionally operated systems. *Urban Water J.* **2013**, *10*, 293–299. [CrossRef]
7. Fiorelli, D.; Schutz, G.; Klepiszewski, K.; Regneri, M.; Seiffert, S. Optimised real time operation of a sewer network using a multi-goal objective function. *Urban Water J.* **2013**, *10*, 342–353. [CrossRef]
8. Hsu, N.S.; Huang, C.L.; Wei, C.C. Intelligent real-time operation of a pumping station for an urban drainage system. *J. Hydrol.* **2013**, *489*, 85–97. [CrossRef]
9. Lee, E.H.; Lee, Y.S.; Joo, J.G.; Jung, D.; Kim, J.H. Investigating the impact of proactive pump operation and capacity expansion on urban drainage system resilience. *J. Water Resour. Plan. Manag.* **2017**, *143*. [CrossRef]
10. Lee, E.H.; Lee, Y.S.; Joo, J.G.; Jung, D.; Kim, J.H. Flood reduction in urban drainage systems: Cooperative operation of centralized and decentralized reservoirs. *Water* **2016**, *8*, 469. [CrossRef]

11. Zacharof, A.I.; Butler, D.; Schütze, M.; Beck, M.B. Screening for real-time control potential of urban wastewater systems. *J. Hydrol.* **2004**, *299*, 349–362. [CrossRef]

12. Galelli, S.; Goedbloed, A.; Schwanenberg, D.; van Overloop, P.J. Optimal real-time operation of multipurpose urban reservoirs: Case study in Singapore. *J. Water Resour. Plan. Manag.* **2012**, *140*, 511–523. [CrossRef]

13. Raimondi, A.; Becciu, G. On pre-filling probability of flood control detention facilities. *Urban Water J.* **2015**, *12*, 344–351. [CrossRef]

14. Hsu, M.H.; Chen, S.H.; Chang, T.J. Inundation simulation for urban drainage basin with storm sewer system. *J. Hydrol.* **2000**, *234*, 21–37. [CrossRef]

15. Vanrolleghem, P.A.; Benedetti, L.; Meirlaen, J. Modelling and real-time control of the integrated urban wastewater system. *Environ. Model. Softw.* **2005**, *20*, 427–442. [CrossRef]

16. Huff, F.A. Time distribution of rainfall in heavy storms. *Water Resour. Res.* **1967**, *3*, 1007–1019. [CrossRef]

17. Yoon, Y.N.; Jung, J.H.; Ryu, J.H. Introduction of design flood estimation. *J. Korea Water Resour. Assoc.* **2013**, *46*, 55–68.

18. Korea Precipitation Frequency Data Server. Available online: http://www.k-idf.re.kr (accessed on 20 April 2017).

19. Lee, E.H.; Kim, J.H. Design and operation of decentralized reservoirs in urban drainage systems. *Water* **2017**, *9*, 246. [CrossRef]

20. Lee, E.H.; Kim, J.H. Development of resilience index based on flooding damage in urban areas. *Water* **2017**, *9*, 428. [CrossRef]

21. Korea Water and Wastewater Works Association. *Standard on Sewer Facility*; Ministry of Environment: Seoul, Korea, 2011.

22. Seoul Metropolitan Government. *Report on Design and Expansion of Gaebong1 Pump Station*; Seoul Metropolitan Government: Seoul, Korea, 2009.

23. United States Environmental Protection Agency. *Storm Water Management Model User's Manual Version 5.0*; EPA: Washington, DC, USA, 2010.

water

MDPI

Article

Use and Utility: Exploring the Diversity and Design of Water Models at the Science-Policy Interface

Natalie Chong [1,*] (ID), Peter M. Bach [2,3,4] (ID), Régis Moilleron [1], Céline Bonhomme [1] and José-Frédéric Deroubaix [1]

[1]   Laboratoire Eau Environnement et Systèmes Urbains (LEESU), École Nationale des Ponts et Chaussées, University of Paris East, 6-8 Avenue Blaise Pascal, 77455 Champs-sur-Marne, France; regis.moilleron@enpc.fr (R.M.); celine.bonhomme@enpc.fr (C.B.); j.deroubaix@enpc.fr (J.-F.D.)
[2]   Swiss Federal Institute of Aquatic Science & Technology (Eawag), Überlandstrasse 133, 8600 Dübendorf, Switzerland, Switzerland; peter.bach@monash.edu
[3]   Institute of Environmental Engineering, ETH Zürich, 8093 Zürich, Switzerland
[4]   Monash Infrastructure, Department of Civil Engineering, 23 College Walk, Monash University, Clayton VIC 3800, Australia
*   Correspondence: natalie.chong@enpc.fr; Tel.: +33-01-64-15-39-50

Received: 31 October 2017; Accepted: 4 December 2017; Published: 19 December 2017

Abstract: Effort to narrow the gap between the production and use of scientific knowledge for environmental decision-making is gaining traction, yet in practice, supply and demand remains largely unbalanced. A qualitative study based on empirical analysis offers a novel approach to exploring key factors, focussing on seven water models in the context of two organisations at the science-policy interface: the PIREN-Seine in France and the CRC for Water Sensitive Cities in Australia. Tentative linkages drawn from these examples identify: (1) objective and expertise; (2) knowledge and tools; and (3) support structures as main drivers influencing the production of scientific knowledge which, in turn, affect the use and utility of modelling tools. Further insight is gained by highlighting the wide spectrum of uses and utilities existing in practice, suggesting that such 'boundary organisations' facilitate interactions and exchanges that give added value to scientific knowledge. Coordinated strategies that integrate inter-, extra-, and intra-boundary activities, framed through collaborative scenario building and the use of interactive modelling platforms, may offer ways to enhance the use and utility of scientific knowledge (and its tools) to better support water resources management, policy and planning decisions, thus promoting a more cohesive relationship between science and policy.

Keywords: boundary organisation; environmental decision-making; integrated modelling; knowledge brokering; model usability; strategic planning

---

1. Introduction

The trade-off between scientific complexity and 'usability' of scientific knowledge and tools to support management, policy and planning decisions is a fundamental question at the heart of the science-policy interface. Similar to all areas of environmental decision-making, water resources managers must make decisions under high system complexity and uncertainty, which demands effective integration of useful and relevant scientific information [1]. In this context, 'useful' scientific knowledge possesses a utilitarian function by clarifying and expanding different options for decision makers to achieve desired outcomes [2,3] and must also be perceived as credible (reliable and of high quality), relevant (context-specific) and legitimate (transparent and objective) [4–8].

The myriad of challenges and opportunities affecting usability have been well documented in the literature, summarized by Lemos et al. [9] as a function of the interconnected factors of fit, interplay and interaction. Elucidating the complexity of these dynamics requires a departure from the traditional 'linear' model of research use, where scientists produce scientific information, viewed as objective or neutral facts, which are then transmitted to a passive audience [10]. Instead, use and utility should be understood as the product of a complex and nuanced relationship comprised of mediated interactions between the various actors involved. Accordingly, the science-policy interface represents a set of social processes between scientists and decision makers, which facilitates the exchange and co-construction of knowledge to support decision-making [11], while also taking into account the complex, iterative, and selective nature of the decision making process [12].

A growing body of work dedicated to the subject defines the science-policy interface in terms of 'boundaries' [6,13,14], which 'demarcate the socially constructed and negotiated borders between science and policy, between disciplines, across nations and across multiple levels' ([5] (p.1)). 'Boundary organisations,' are intermediary organisations straddling the frontiers of science and policy through the co-production of shared interests, knowledge and tools [6], that can facilitate and/or hinder communication, collaboration and collective action [5]. Touted by some as promoting the best of both worlds, others remain cautious of how we distinguish 'science' from 'non-science' within these arrangements, otherwise known as the 'boundary problem' [14]. In the same vein, Jasanoff [13,15] argues that scientific claims are socially constructed through various social influences and constraints, which can place unusual strains on science when applied to real-world situations. While concerns over the bureaucratisation or standardisation of science are certainly valid, Guston [6] maintains that boundary organisations can help to avoid these issues, by having one foot in science and the other in policy, thereby keeping one another in check.

Within this discourse, modelling tools can be considered 'boundary objects' [16,17], which serve to deepen scientific understanding, while concurrently supporting key management, policy and planning decisions [1,18]. Their dual function as a research and an operational tool has enabled practitioners to navigate the complexities of water resources management and planning, which demands not only a nuanced understanding of dynamic environmental processes but also the ability to negotiate trade-offs between a multitude of social, economic, political and ecological interests among competing stakeholders. On the other hand, models have different forms and functions, not all of which are equal in terms of: (1) their *use*, i.e., 'the method or manner of employing or applying something'; and (2) their *utility*, i.e., its 'fitness for some purpose or worth to some end' or 'something useful or designed for use' [19].

While considerable efforts have been made to bridge the gap between the production and use of scientific knowledge in decision making [5,9,20–26], many authors continue to highlight a mismatch between supply and demand [1,2,10,12,20,27–29], suggesting the need for further insight into the production as well as the use and utility of such knowledge and tools in practice. Much of the existing literature on creating 'usable' science focuses on opportunities and challenges without delving into what exactly this information is *used* or *useful for* in the context of environmental decision making. To date, discussion on model complexity vs. usability has been largely based on the notion that a model's use and utility is contingent upon its 'usability' (e.g., user-friendly interface, simplified processes and outputs, etc.). However, this overlooks the multitude of uses (ranging from direct to indirect), which exist in practice. Here, we distinguish 'utility' from what others have referred to as 'usability' [9] in an effort to incorporate this diversity. Within this literature, boundary organisations have been identified as an effective strategy for producing knowledge that is both useful and usable for decision making [9,27,30–32], yet there is still a lack of empirical data to reinforce this hypothesis. In an effort to address these gaps, this paper aims to provide further insight by using a novel approach based on empirical analysis to explore the boundary organisation hypothesis: the way an organisation or a (set of) tools is structured can help or hinder the production of scientific knowledge that is perceived as valuable for the implementation (or elaboration) of public policies. We explore this hypothesis, focussing on the use and utility of modelling tools within the context of two interdisciplinary research programs whose core activities are rooted in research-industry collaboration (public or private):

the PIREN-Seine (Programme Interdisciplinaire de Recherche sur l'Environnement de la Seine) in France and the CRC (Cooperative Research Centre) for Water Sensitive Cities in Australia.

The choice of these examples derived from a desire to compare two exemplary experiences, which share the overall objective of improving collaboration and exchange at the science-policy interface. Specifically, both aim to address challenges of water resources management, policy and planning through the advancement of scientific knowledge and the development of modelling tools in partnership with various stakeholders. These challenges include technical factors, such as model complexity, uncertainty and the availability and reliability of data, as well as socio-economic factors such as institutional barriers and paradigms, competing objectives, time and resource constraints and lack of effective communication and understanding. However, they approach these challenges using strategies that are fundamentally different: one being more 'research-oriented,' while the other is more 'industry-oriented.' On one hand, the PIREN-Seine in France favours models with more scientific rigour at the cost of usability for industry partners. On the other, the CRC for Water Sensitive Cities in Australia is developing modelling tools designed for industry use, though it remains to be seen whether they will be readily adopted. The breadth and diversity of modelling tools represented in both examples provides a sufficient dataset with which to draw from, while the openness and transparency of these programs allowed for the collection of empirical data, which can be considered an adequate representation of reality. As both programs use modelling tools developed (or partially developed) outside of their defined 'boundaries,' we are also able to go beyond the two case studies to explore the legacy of seven water models across two countries. Finally, the diversity of modelling tools found in both examples represents different stages of model development and use, thereby giving further insight into current and potential use and utility.

Through an empirical analysis of the PIREN-Seine in France and the CRC for Water Sensitive Cities in Australia, this paper aims to narrow the gap between the production and use of scientific knowledge by exploring the nuanced relationship between the use and utility of modelling tools within boundary organisations at the science-policy interface. Section 2 presents the methods and materials used to inform this analysis, as well as the framework for discussion. Based on Grounded Theory (GT) [33], our approach is an exploration of the factors influencing the use and utility of modelling tools, using empirical data as a starting point. Through an historical perspective, Section 3 offers a comprehensive characterisation of the different strategies implemented by the two organisations in order to enhance utility. Brief descriptions of seven water models will be presented to provide context for the discussion that follows. Section 4 explores the links between the respective strategies and their effect on the use and utility of these models for decision-making. We deepen this discussion in Section 5, by characterising the different types of use and utility represented in the two examples. By delving into these specificities, we highlight the influence of model use (direct or indirect) on its utility and vice versa. Moving past the assumption that knowledge is *useful* only when it is *used*, we posit that the social value of this knowledge is also derived from the different types of interactions and exchanges existing between the complex, dynamic web of science-policy boundaries. Finally, we arrive at the conclusion that the use and utility of scientific knowledge (and its tools) could be enhanced through coordinated strategies which frame these inter-, intra- and extra-boundary exchanges and interactions through the co-construction of scenarios and the use of interactive modelling platforms.

2. Materials and Methods

We conducted a qualitative study using an approach based on Grounded Theory (GT), a general research methodology that derives theory through the systematic collection and analysis of data [33–35]. Rather than having an established framework or theory from the outset with which to test against research data, this method offers a more flexible, adaptive approach through an iterative process that involves: raising generative (but not static or confining) questions to guide research, identifying core theoretical concepts through the systematic collection and analysis of data, and developing tentative linkages between core concepts and data [35]. This approach allows for an exploration (and subsequent identification) of the factors influencing the use and utility of modelling tools through:

1. A characterisation of the strategies implemented by the two organisations and a description of the different types of modelling tools, which is used to explore the influence of these strategies and the potential use and utility embedded in the structure of the model (Section 3);
2. Systematic observation and analysis of the interactions and perceptions of the different producers and users of modelling tools, which allows us to form tentative linkages (Section 4) and;
3. A characterisation of the different model uses (ranging from direct to indirect), which is shown to inform their utility (and vice versa) (Section 5).

Our analysis draws primarily on systematic document analysis (e.g., activity reports, scientific literature produced by both programs), formal semi-structured interviews with researchers and practitioners from both countries, and observations during science-practice engagement activities. This provided a rich data set for comparing PIREN-Seine in France and CRC for Water Sensitive Cities in Australia.

2.1. Document Analysis

Document analysis focused on the work produced by the PIREN-Seine and the CRC for Water Sensitive Cities throughout the duration of each program, which included hundreds of peer-reviewed journal articles as well as grey literature such as periodic activity reports (over 700 reports from the PIREN-Seine and over 150 from the CRC for Water Sensitive Cities), synthesis documents and other communications. As a lot of the modelling in the Australian urban water sector also emerged from a long legacy of research and industry collaboration dating back to the 1990s, we were also cognisant of older documents prior to the commencement of the CRC for Water Sensitive Cities research program including those from its predecessors, the CRC for Catchment Hydrology, the CRC for Freshwater Ecology and the eWater CRC. Pertinent documents were identified by searching different combinations of the following keywords: 'decision making'; 'exploratory modelling'; 'management'; 'model'; 'modelling'; 'planning'; 'policy'; 'strategic planning'; 'water sensitive cities' and 'water sensitive urban design' on each program's website in addition to major search engines (Scopus, Web of Science, Google Scholar). Keywords were selected and narrowed down from initial searches based on relevance to the respective research program and included the names of specific 'operational' or industry partners (practitioners) and known modelling tools in order to obtain information about their use and application in practice.

Though this process was systematic, the permeable nature of the 'boundaries' between science and policy limited our ability to adequately define the models represented in this study. First, there is no clear consensus regarding ownership. While for some, it is a question of licencing and rights, for others, the model developer is considered the 'owner,' since they have the ability to change the code. Second, model development is typically a long process, where different actors may be involved in some capacity at various stages, contributing to its overall development and evolution. This work can be carried out under the auspices of the boundary organisation or it can be done through external contracts or exchanges. Third, these models are not entirely independent. That is to say, they often include modules or sub-modules that were developed outside of the program. In some instances, models were created in other research contexts and were subsequently developed, further elaborated and maintained by the program. With these limitations in mind, we refer to 'PIREN support tools' or 'Water Sensitive City (WSC) tools' to distinguish modelling tools that were developed, used and supported within these two contexts to conduct research associated with the respective program. To capture (as much as possible) the breadth and diversity of modelling tools represented in both cases, we took a broad definition of 'model' to mean any model, modelling tool, or part of a modelling tool mentioned in the documents produced by either program, that was either developed or used at one time or another by a researcher of that program. Under this definition, a model can also refer to a sub-model or module that can simulate biophysical or chemical processes using mathematical equations and numeric calculations.

2.2. Semi-Structured Interviews and Observation of Engagement

Since what is written and officially communicated is not necessarily what is said and done in practice; observation and semi-structured exploratory interviews were implemented to support initial findings in both France and Australia. A total of 36 and 21 interviews were conducted in France and Australia respectively with researchers (including modellers and non-modellers) and practitioners (including modellers, water authorities, consultants, regulating authorities and government officials) who were either previously or are currently involved (both directly or indirectly) with modelling activities within these two contexts. Interviews were semi-structured, based on a general question guide (provided in Annex A) that focussed on themes relating to: (1) the development and use of modelling tools; (2) the relationship between researchers and partners; (3) the regional context; as well as (4) the objectives and themes of the respective research program. Questions were adapted to individual participants according to their role and involvement in modelling activities, the program, or their position. Interviews were open-ended and lasted anywhere from 1 to 4 h with an average duration of 1.5 h. Interviews were transcribed and coded according to the four themes listed above. Anecdotal observations were used as secondary data, which was collected throughout 2015–2017, during numerous meetings, seminars and conferences organised by the PIREN-Seine in France. This included two general assembly and annual planning meetings organised to reflect on the year's work and co-define upcoming program objectives. In Australia, anecdotal observation was limited to seminars and conferences organized by the CRC for Water Sensitive Cities from May to August 2017, which included one major national conference in Perth and two workshops.

3. Retracing the History of (Co-) Production in France and Australia

The PIREN-Seine and the CRC for Water Sensitive Cities provide a platform for researchers and practitioners to collectively address some of the key issues of water resources management, policy and planning, using different strategies to achieve a common objective. As its name suggests, the PIREN-Seine focuses on the Seine River basin in France, while the CRC for Water Sensitive Cities extends its focus across cities to include the Yarra, Swan-Canning and Brisbane river basins in Australia, which represent notable examples of historically significant catchments facing serious issues of water quality and quantity due to increasing anthropogenic pressures caused by rapid urbanisation, population growth and climate change.

The PIREN-Seine has adopted a territorial perspective of the Seine River basin, with a desire to understand the ecological functioning of the entire watershed in relation to human activities [36,37]. Most of the research is centred on issues of water quality, though water quantity concerns are also explored (mostly from a quality perspective), particularly in light of recent major flood events. While industry collaboration is considered an essential part of the program, the intrinsic desire to maintain scientific integrity is reflected in the knowledge and tools produced, which have traditionally leaned towards academic pursuits as a primary function and responding to operational demands as secondary. As a result, modelling tools are primarily seen as 'research tools,' which have been used to support management and planning decisions, though the tools themselves have only been adopted by industry partners in exceptional cases of mutual interest and investment.

In contrast, the CRC for Water Sensitive Cities focuses on issues of urban water management in cities throughout Australia and abroad in pursuit of sustainability, resilience and liveability [38,39]. This has been partially motivated by extreme weather conditions experienced within the region, such as the Millennium Drought [40–42], which lasted more than a decade and has shifted the primary focus towards issues of water supply security (e.g., seawater desalination, rain water harvesting) even though water quality remains a serious concern, particularly for recreation and consumption [43,44]. Direct uptake of research into practice being the main objective, a large part of this work has been devoted to adoption pathways and socio-technical transitions, resulting in tools that lean towards practical application as a primary function. While this has proved successful in some cases, leading to wide-scale adoption of one example (i.e., the MUSIC model) that we feature in this study, it remains to be seen whether the new generation of modelling tools will be able to generate the same appeal. The contrasting strategies and diversity of models at different stages of development represented by these examples makes for a fruitful comparison for exploring how organisational configurations and context-specific drivers may influence the production of knowledge and tools within these spaces and what that means in terms of use and utility. A summary of the two research programs is presented in Table 1 below.

Table 1. Summary of Research Programs.

	PIREN-Seine	**CRC for Water Sensitive Cities**
Duration	(1989–)	(2012–2021)
Level/Scale	Territory; Basin	Urban; City
Interest	Seine River basin	Cities in Australia and abroad
Research Priority	Quality/Quantity	Quantity/Quality
Main Objective	To produce research to better understand river system functioning that can also support decisions	To produce research and tools for industry use to achieve water sensitive cities
Types of Actors	National research institutes, universities, mixed research groups, research laboratories, public institutions, regulating authorities	Universities, public utilities, governments (local, state), regulating authorities, capacity-building organisations, consulting companies, software companies

As the PIREN-Seine and the CRC for Water Sensitive Cities both position themselves at the science-policy interface, a comparison between the two presents a mutual learning opportunity: the CRC for Water Sensitive Cities can benefit from the nearly 30 years of experience from the PIREN-Seine, while the PIREN-Seine can gain insight from an international perspective. Additionally, this analysis can provide guidance for similar examples on a wider scale: The Seine River basin is facing strong anthropogenic pressures that are characteristic of many large watersheds, while Australia can be considered a 'litmus test' for other countries as it continues to face extreme weather conditions that may soon become the norm under climate change.

3.1. The PIREN-Seine, France

Established in 1989, the PIREN-Seine (PIREN) is an interdisciplinary research program in France comprised of 22 research teams and 140 researchers from a range of academic backgrounds. The majority are rooted in the fields of hydrology, biology, chemistry, or engineering, while a growing number of geographers, agronomists, political economists, political scientists and sociologists have become involved. The main types of actors in relation to modelling activities are represented in Table 1. Notable industry partners include the Syndicat Interdépartemental pour l'Assainissement de l'Agglomération Parisienne (SIAAP), a public institution responsible for wastewater treatment and sanitation in the Paris region and the Agence de l'Eau Seine-Normandie (AESN), a public institution responsible for the management of water resources in the Seine-Normandy watershed, both of whom, are heavily involved in modelling activities within the PIREN-Seine.

Partnerships between universities, research units and research institutions not only provide a pool of expertise, they can also be a source of funding, either through specific projects that directly or indirectly contribute to the work of the PIREN-Seine or through in-kind contributions in the form of researchers who are paid by their own institutions or doctoral students and post-doctoral researchers who support them. As for industry partners, relationships are largely financial, allowing them direct access to the knowledge and tools produced by the PIREN-Seine. They also play an active role in the elaboration of the program's research objectives and, in some cases, the modelling tools as well. In the case of the regulating authority—the Direction Régionale et Interdépartementale de l'Environnement et de l'Énergie (DRIEE)—the relationship has an added regulatory element. While each organisation has a defined role within the basin, individual relationships are not clearly defined, as many researchers and industry partners have formal and informal relationships that extend beyond the 'borders' of the program. For example, several individuals who have previously obtained their doctoral degree under the supervision of PIREN-Seine researchers now represent industry partners. Furthermore, a model that may have been developed within the context of the PIREN-Seine may see further development outside of the program through external contracts with individual researchers, research teams or even external consultancies.

Over the past three decades, the objectives and research themes of PIREN have evolved in response to changing research and operational needs and emerging trends, while gradually incorporating new disciplines and perspectives [45], which also went hand-in-hand with the development and evolution of modelling tools. Phase 1 (1989–1992), emerged from the need to create dialogue and fundamental partnerships between researchers and water actors as a prerequisite for mobilising research that could address specific water quality concerns at a territorial scale. Initial objectives soon evolved towards obtaining a more global vision that encompasses the entire river basin, a mentality that echoed the 1992 Water Act [46] and the Master Plan for Water Development and Management (SDAGE) [47]. Whereas Phase 1 looked at the longitudinal dimension of the aquatic continuum (upstream–downstream), Phase 2 (1992–1996) turned its attention to transverse interactions between watercourses and riparian zones such as wetlands, as well as the urban water cycle and the fate of pollutants in the river system. It is within this phase where the perception of models began to change from being seen as strictly research tools to their consideration for decision support.

From 1998 to 2006, work in Phases 3 and 4 aimed to contextualize the hydrographic network within the different interactions and anthropogenic influences occurring within the watershed. A retrospective outlook was used to consider the historic and dynamic nature of the hydrological system, which in turn increased the capability of models to simulate and test prospective management and planning scenarios. Phase 5 (2007–2010) integrated public health risks posed by emerging micropollutants such as new molecules with little known effects, pharmaceuticals and pathogens. Territorial studies also investigated the impact of ecological engineering and the reform of the Common Agricultural Policy (CAP). Phase 6 (2011–2014) further expanded into 5 main research axes, which reflected the concerns and challenges jointly identified by researchers and industry partners. These include: (1) creation of agricultural scenarios according to water quality requirements; (2) identification of the role of wetlands; (3) a deepened understanding of water quality in the current climate; (4) a better understanding of the relationships between chemical pressures and ecological states; and (5) understanding dynamics of chemical pressure over a long duration.

The current phase, Phase 7 (2015–2020), focuses on gaining an in-depth understanding of the mechanisms that regulate water resources and climate change scenarios to support management strategies that are more adapted to the agricultural, environmental and urban issues facing the region. Scenario building has become increasingly popular, allowing researchers and industry partners to collectively envision and anticipate possible futures. This outlook is reflected in official discourse, which promotes a science-policy transfer through a newly dedicated transfer unit ('cellule de transfert'). At the same time, a shared mentality insists upon its foundation in research, aiming to provide knowledge and expertise that helps inform management and policy decisions without directly implicating itself in the role of a policy maker.

3.2. PIREN-Seine Models: From Aggregation to Integration

PIREN-Seine modelling tools have evolved in parallel to its research objectives, adapting to suit changing demands and/or being used with other models to answer specific questions or to provide a more global view of the functioning of the system. This has produced a variety of models, including hydrologic models, biogeochemical models, hydraulic models, agronomic models, economic models and a model that simulates the environmental impact on fish populations. The majority of these models address issues of water quality, particularly the transfer of nutrients or pollutants through different parts of the system. However, within a large river system such as the Seine, individual models are only capable of telling 'part of the story,' limited to a specific temporal and spatial scale. At the same time, increasingly strict requirements from regulations such as the European Water Framework Directive (EU-WFD) [48] are placing increasing pressure on researchers and decision makers to restore water bodies to 'good ecological status' [49–51], which demands a global vision of the system. These trends have resulted in change in trajectory from individual models responding to specific questions, to the adaptation or coupling of models to answer bigger questions, towards modelling chains and/or platforms that can be applied to the entire Seine system. Here, we present four main models (see Table 2) based on their history of development and use (directly and indirectly) by industry partners: ProSe, Seneque, MODCOU and STICS.

Table 2. Overview of PIREN-Seine Support Tools *.

Model	Type	Key References
ProSe	River quality model	Even et al. [52]; Garnier and Mouchel [53]
Seneque	Catchment quality model	Garnier and Mouchel [53]; Billen et al. [54]
MODCOU	Surface-groundwater model	Ledoux [55]; Ledoux et al. [56]
STICS	Agronomic model	Brisson et al. [57,58]

* Limited to the models presented in this paper.

3.2.1. ProSe

Short for *'Projet Seine,'* the model ProSe was developed by researchers at École des Mines ParisTech in collaboration with PIREN-Seine research teams, research institutions, universities and industry partners [52,59] within the context of the PIREN-Seine. Originally designed to study problems of water quality and chronic deoxygenation related to effluent discharges from wastewater treatment plants on downstream sectors of the river and accidental overflow of sewage networks during rainy events [60,61], it has also been applied to hydraulic problems and questions associated with the transport of particles [60].

The modular structure of ProSe allows for greater adaptability in simulating different scenarios, therefore its applicability is widespread. In recent years, ProSe has undergone several revisions (producing versions 1 to 4), increasing previous functionality in terms of knowledge gained as well as the ability to be coupled with other models. Although it is neither a standardized nor commercial model, simulations using ProSe are requested and sometimes required by regulating authorities such as the DRIEE to justify project proposals (SIAAP representative, 27 June 2016), and it is now widely considered a reference model for water quality of the Seine. The development and evolution of ProSe has been partially motivated by special interest from the SIAAP, who uses ProSe as a medium- to long-term management and planning tool (SIAAP representative, 29 November 2016). This has resulted in additional investment (time and resources), which extends outside of PIREN-Seine, either through ARMINES, a consultancy arm of École des Mines engineering school, or through a working group involving researchers and practitioners interested in adapting ProSe to meet operational demands (PIREN researcher, 28 April 2016). As such, ProSe is considered both a research and operational tool, even though the tool itself is one and the same (PIREN researcher, 16 June 2016). However, despite being frequently cited as an example of this dual functionality, the future of ProSe remains uncertain. Many original developers have either retired or expressed interest in moving on to other research projects (PIREN researcher, 16 June 2016), while its only current operational user (SIAAP) is moving towards artificial intelligence and real-time control methods and is considering replacing the model with statistical techniques for daily operations (SIAAP representative, 10 March 2016).

3.2.2. Seneque

Seneque, which stands for 'Seine en equation', was developed by the research team METIS—an interdisciplinary research unit at the University of Pierre and Marie Curie (UPMC)—in the context of the PIREN-Seine, though some of its components (i.e., RIVE) were developed outside of the program. Based on the concept of stream-order, Seneque simulates the transport of nutrients and the biogeochemical functioning of the hydrographic network using a simplified and idealised conceptualisation of the drainage network of large regional basins with a refined representation of in stream microbiological processes using the RIVE model [62,63]. Also referred to as Riverstrahler, Seneque is essentially the same model applied to the Seine River basin and coupled with a GIS interface [64]. The added functionality of a user-friendly interface has enhanced the user's ability to visualize and explore results in a way that is more accessible to non-specialist users. Since its creation, Seneque has undergone several revisions and has been applied to different situations in combination with other models [52,62,65,66].

Also considered to be an 'operational' model, Seneque has been appropriated directly by the AESN as a medium- to long-term planning tool, used for example, to evaluate the ecological state of the basin by amalgamating different datasets to construct 'snapshots' at different spatial and temporal scales. The model has since reverted back to a 'research' tool mostly due to a loss in internal expertise at the AESN (AESN representative, 8 June 2016). The most recent incarnation of the model, Pynuts, has allowed researchers more flexibility in terms of model development, to explore new research questions using updated technology without having to invest time and resources on interfacing. However, plans to add an interface are in the works to, once again, allow it to be appropriated directly by industry partners in the future.

3.2.3. MODCOU

The MODCOU model was developed by researchers at École des Mines ParisTech [55,56,67,68] to simulate the movement and circulation of surface and groundwater. MODCOU describes surface and groundwater flow at a daily time step: the surface model calculates the water balance between evaporation, runoff and infiltration, while the underground model calculates the transfer of water in aquifers and surface-groundwater exchanges [67,69].

Much of the work on MODCOU is concentrated on its integration with other models. For example, it is often coupled with other models such as STICS (presented next) [67,70] in order to obtain a more complete understanding of nitrate contamination and the influence of agricultural activity on surface and groundwater. To date, MODCOU has been effectively applied to predict surface and groundwater flows in many French basins with varying scales and hydrogeological settings [67,71]. Though it has remained as a research tool, studies requested by partners such as the AESN to assess the impact of climate change on water resources have used MODCOU to evaluate groundwater levels and monitor trends in nitrate and pesticide content.

3.2.4. STICS

The model STICS has been developed by the Institut National de Recherche Agronimique (INRA) since 1996 [57] in collaboration with large research and professional institutes [58]. It was not developed in the context of the PIREN-Seine but is considered here as a PIREN support tool since it is often used to conduct research within the context of the program. STICS (Simulateur mulTIdisciplinaire pour les Cultures Standard) is an agronomic model that simulates crop growth, soil water and nitrogen balances driven by daily climatic data [57,58,72]. Intended to simulate the evolution of water, carbon and nitrogen in the soil-plant system over one or more years successively [57,73], STICS was designed and developed with the dual objective of calculating agronomic variables (e.g., plant biomass, harvested yield, protein content of the grain, nitrogen balances of the crop) and environmental variables (e.g., flow of water and nitrate out of the root zone) [58,72]. Crop generality allows for adaptation to various crops, whereas robustness in the model allows the user to simulate various soil-climate conditions without considerable bias in the outputs.

Development of the model has focused on usability through collaboration between model developers and users in a way that allows users to participate in its evolution. Mostly considered a research tool, its conceptual modularity has allowed STICS to be chained with other models in order to understand the transfer of nitrates and pesticides into surface and groundwater [74]. These types of studies are often requested by partners such as the AESN, who are interested in monitoring the impact of agriculture on water quality. In this way, it can also be considered a decision-support tool, although researchers are charged with running the model and scenarios.

3.3. The CRC for Water Sensitive Cities, Australia

Established in 2012, the CRC for Water Sensitive Cities (CRCWSC) [75] is one of many Cooperative Research Centres in Australia, which are part of a government initiative to fund innovative research that can directly meet the needs of industry. CRCWSC involves over 200 researchers from various backgrounds (hydrology, biology, chemistry, engineering, economics and social sciences), from national and international universities and research institutions. Setting itself apart from other on-going CRCs, the CRC for Water Sensitive Cities builds upon the research base of previous CRCs (the CRC for Catchment Hydrology from 1992 to 2005, CRC for Freshwater Ecology from 1993 to 2005 and eWater CRC from 2005 to 2008) and focuses specifically on creating water sensitive cities [76,77], or sponge cities [78–80], guided primarily by three main principles: (1) Cities as water supply catchments; (2) Cities providing ecosystem services; and (3) Cities comprising water sensitive communities [77].

Main actors in relation to modelling activities are represented in Table 1. Some of these partnerships are financial in nature, either through direct funding to the program, funding for specific projects which contributes to the work of the CRC for Water Sensitive Cities, or through in-kind contributions of researchers paid by their home institutions. Most partners are directly involved in research support, either as researchers themselves, or 'beta-testers,' who test, apply, provide feedback, and play an essential role in disseminating the knowledge and tools on the ground. This network also includes associate partners, who may access the knowledge or tools and help test, apply, and disseminate this research without direct investment, and who may also contribute to capacity-building activities.

Whilst PIREN is a research program that is renegotiated every 4–5 years, CRCWSC runs for 9 years (2012–2021), as opposed to the average 5 years of other CRCs. Its research program comprises two parts: Tranche 1 (2012–2016), focused on research and Tranche 2 (2016–2021), focuses on adoption pathways and implementation of the research produced in addition to building new knowledge. Within the first tranche, four diverse programs in the areas of Society (Program A), Water Sensitive Urbanism (Program B), Future Technologies (Program C) and Adoption Pathways (Program D), have produced research outputs that have either fed directly into the development of new modelling tools or have applied, adopted, and expanded existing industry standard models in new contexts. In particular, Program D focussed on developing partnerships between relevant actors at all levels (from community to government), capacity building, and holistic decision-support tools. With the first tranche completed, this program has continued in an evolved form in Tranche 2.

3.4. Water Sensitive City (WSC) Models: New Tools for New Strategies

WSC (models or tools) have moved away from decision support based on deterministic or stochastic models towards integrated modelling platforms and visualisation—an evolution in strategic planning within a new era of 'deep uncertainty' [81,82] and greater collaboration [83,84]. Whereas running models individually can support management and policy decisions on a short- to medium-term, an integrated modelling approach allows for exploratory modelling and adaptive planning for an uncertain future [81]. In the context of CRCWSC research, models are complementary, meant for use at different parts of the workflow. Here, we focus on three models (see Table 3): MUSIC, WSC Toolkit, and DAnCE4Water. Two of these models began development well before the CRC for Water Sensitive City program began, but have since been extended or upgraded based on the latest research resulting from Tranche 1 and are currently used—or intend to be used—by industry partners.

Table 3. Overview of Water Sensitive City (WSC) Tools *.

Model	Type	Key References
MUSIC	Stormwater quality model	Wong et al. [85]; http://www.ewater.org.au/products/music/
Water Sensitive Toolkit	Infrastructure planning tool	https://watersensitivecities.org.au/solutions/water-sensitive-cities-toolkit/
DAnCE4Water	Cloud-based city modelling platform	Rauch et al. [86]; www.dance4water.org

* Limited to the models presented in this paper.

3.4.1. MUSIC

The Model for Urban Stormwater Improvement Conceptualisation (MUSIC) was developed in 2001 by the CRC for Catchment Hydrology (1992–2005), involving many past and current researchers of CRCWSC. This work continued after it merged with the CRC for Freshwater Ecology (1993–2005) to eventually form eWater, a government owned non-profit organisation (and CRCWSC industry partner) offering capacity building, technical support services and modelling tools to support integrated water resources management and governance. Developed with the objective of synthesizing research into an easy-to-use tool, MUSIC is a decision support system that allows water managers to evaluate stormwater management systems based on specific water quality objectives, as well as determine appropriate sizing of stormwater treatment facilities and associated infrastructure [85]. Its core feature is how it describes water quality behaviour through a first-order kinetic decay model (K-C* Model) of three key pollutants (suspended solids, phosphorus, nitrogen), and hydrodynamic behaviour within a stormwater treatment device through the continuously stirred tank reactor (CSTR) concept [85,87]. The current version of MUSIC (v6) has expanded and updated initial capabilities to a wider range of stormwater treatment devices and new performance indicators [88]. Through on-going research efforts and communication between eWater and CRCWSC, many improvements to MUSIC's capabilities and functionality have been made and its applicability to non-Australian cities like Singapore is being assessed.

As one of eWater's most widely adopted models, MUSIC has since become the industry standard across Australia for stormwater quality management and Water Sensitive Urban Design (WSUD). Early endorsement from two key industry partners in Melbourne and Brisbane, who were investigating ways of protecting receiving waters from urban stormwater pollution, heavily contributed to rapid adoption across many municipalities in Australian's east, particularly in the states of Victoria and Queensland [89]. Practitioners use MUSIC to design integrated stormwater management plans based on a specific catchment and to demonstrate compliance to local standards. It has also been used for CRCWSC research, contributing to the development of other tools such as the WSC Toolkit.

3.4.2. Water Sensitive Cities (WSC) Toolkit

Developed in Tranche 1 of the CRCWSC program, the WSC Toolkit synthesises key research outcomes into easy-to-use modules for assessing the benefits of WSUD. The model aims at supporting strategic planning, by focussing on evidence-based quantification of the benefits of urban green infrastructure (GI) initiatives in order to develop business cases that are both robust and water sensitive [90]. The model is capable of: (1) improving stream health impacts based on the effectiveness of WSUD in mitigating runoff volumes, frequency and pollutant concentrations [91–94]; (2) assessing changes in flow frequency and reduction of geomorphic impact on streams based on the stream erosion index [95] and; (3) mitigating the urban heat island effect through urban greening and retaining water in the landscape [96]. Other modules are still under development including a future climate module, which will draw from a database of future rainfall projections for major Australian cities and can be used independently or as input data for future climate scenarios [97,98]. An economic valuation module is also planned, to consider the likely willingness-to-pay of community members based on various improvements made to liveability and sustainability of the catchment.

The WSC Toolkit is currently in closed 'beta-testing' mode, with its adoption slowly taking place in select municipalities across Australia. Much of its momentum is currently driven by the need for quick and easy microclimate assessment tools that enable local municipalities to formulate a business case for funding more WSUD and green infrastructure projects. The ability of the WSC Toolkit to communicate directly with MUSIC is also a strategic choice and leverages the familiarity of an existing large user base.

3.4.3. DAnCE4Water

The DAnCE4Water model (Dynamic Adaptation for eNabling City Evolution for Water) began as part of the European Framework Program 7—'PREPARED enabling change' (www.prepared-fp7.eu) prior to the CRC for Water Sensitive Cities [86,99]. It was then adopted within Program A (Society) of the CRC for Water Sensitive Cities, where it evolved into a cloud-based city modelling platform. Aspiring to be an interactive, 'user-friendly' decision support tool for different water actors to explore future scenarios and evaluate different policy and action strategies, DAnCE4Water takes into account the interactions between urban water infrastructure, the urban environment, and social dynamics [86]. This is represented by three modules rooted in a central unit, or 'conductor,' which runs each scenario by storing, managing and providing required data to the relevant modules [100]. Formerly driven by a societal transitions model [101], DAnCE4Water now relies on the interplay between urban development and societal dynamics influenced by an economic willingness-to-pay framework. The urban development module, in particular, projects the changes of the urban environment down to the household level [102]. Various biophysical modules are used to simulate the impact of urban development on infrastructure, and include well-known hydraulic models such as EPANET [103] and EPA SWMM [104], as well as a link with MIKE URBAN for flood risk assessment [105].

While this modelling tool has great potential for strategic planning and adaption, its use and utility remain undetermined for the moment, as (at the time of writing) it is still under development and not yet fully operational due to its scale and broad city-scale scope. The underlying computational and web-based framework has, however, paved the way for smaller tools that are currently being trialled across Australia, such as the Water Sensitive Cities Index, which enables municipalities to benchmark how 'water sensitive' their local area is compared to their peers and the overarching vision of CRCWSC [106].

4. Influence of Organisational Configurations and Context-Specific Drivers

Our exploration of the strategies and modelling tools in the context of the PIREN-Seine in France and the CRC for Water Sensitive Cities in Australia provides insight into their effect on the use and utility of modelling tools in each example. Both PIREN and CRCWSC fit the criteria for 'boundary organisations': they straddle the boundary between two distinct worlds (i.e., science and policy) but are accountable to both, provide opportunity and sometimes incentives for the development and use of shared objects or 'boundary objects' [6,16,17] (e.g., modelling tools), and involve participation of actors from both sides of the boundary, as well as actors who play a mediating role [6]. In this way, they not only mobilise various stakeholders but also orient research and available tools towards achieving common goals, which in turn, informs the potential and/or intended use and utility of their scientific knowledge and tools. Here, we draw from systematic observation and analysis to explore tentative linkages, highlighting the role of organisational configurations and context-specific drivers on model use and utility in practice. These can be classified into three main categories: (1) objectives and expertise; (2) knowledge and tools; and (3) supporting structures.

4.1. Objective and Expertise

Empirical data suggests that the objective(s) of the program and the expertise of the individuals involved have a large influence on the scientific knowledge that is produced and subsequently, how that knowledge is used (if at all). On one hand, PIREN has a territorial focus with expertise on the Seine River basin, although some of the knowledge and tools have been applied to other basins (PIREN researcher, 2 May 2016). On the other, CRCWSC has an urban focus, which began in Australia in the early days of its Melbourne-based predecessor (the 'Cities as Water Supply Catchments' Project) and has now expanded abroad through the involvement of international partners. Although PIREN engages researchers from different disciplines, most have a background in natural sciences or engineering with a focus on water quality. Other than annual conferences and planning sessions, research teams mostly keep to themselves (PIREN researcher, 1 December 2016). The representation of social sciences is small but growing, moving from quantitative studies to more qualitative studies, which include historical trends, social dynamics and the production of science. Likewise, CRCWSC involves an interdisciplinary team, though there is a greater balance between the natural and social sciences, which is seen as both necessary and inseparable (CRCWSC researcher, 22 June 2017).

Differing perspectives on the relationship between science and policy is perhaps the biggest difference between the two programs: PIREN tends to favour research over policy, while CRCWSC specifically orients its research towards use and adoption. On one hand, PIREN prefers a more marked distinction, with the objective of providing expertise and support without taking an active role in policy (PIREN researcher, 29 June 2016), although this perspective is not necessarily shared among all individuals and the mentality is generally becoming more open. Even if researchers would like their work to be applicable in practice, policy issues are commonly perceived as something beyond their role and responsibility (PIREN researcher, 12 January 2017). While this allows them to maintain scientific objectivity, it may also limit their impact in terms of knowledge dissemination and practical application, or at least render it more difficult to ascertain. On the other hand, CRCWSC has a clear objective: to promote sustainability, resilience and liveability through WSUD and the water sensitive cities by directly engaging with local councils, regional and national governments and citizens. Taking an active role in connecting science and policy and specifically organising its research around its use and transfer in practice has resulted in direct impacts on policy and planning (e.g., regulation standards set by MUSIC) (CRCWSC researcher, 9 June 2017).

Increasingly blurred borders and long collaborative relationships (official and unofficial) have likely contributed to building trust, credibility and legitimacy; a sentiment that was expressed in some form or another by all interview participants. In both examples, the science-policy interface resembled the web of interactions described by Vogel et al. [107]: in PIREN, many practitioners came from the same academic training as researchers (AESN representative, 8 June 2016), while in CRCWSC, it was common for researchers and practitioners to have held positions on both sides of the boundary at different stages in their career (CRCWSC researcher/industry partner representative, 21 June 2017). While this also occurs in PIREN (some industry partners were previous students of PIREN researchers), the lines between research and practice in this example have traditionally been more distinct. Collaboration, co-production and co-development resulting from the multitude of official and unofficial interactions and exchanges (inter-, intra- and extra-boundary) create mutual understanding and communication, which subsequently promote feelings of trust among different actors. In both examples, all interview participants expressed 'trust' in the models, as far as models can be trusted, knowing they are only a representation of reality. Confidence is fostered through official interactions such as conferences, working groups, planning sessions and workshops, as well as unofficial interactions where practitioners can consult researchers even when they are 'off-the-clock' (SIAAP representative, 10 March 2017). Whereas blurring the borders may foster collaboration, understanding and trust, maintaining legitimacy may, in some cases, require the borders to be restored (even if only temporarily) in order to clearly distinguish science from policy. This allows scientific knowledge (e.g., model outputs) to maintain scientific objectivity, since it is produced by researchers using scientific tools, and is therefore presumed to be free from political bias (SIAAP representative, 29 November 2016).

4.2. Knowledge and Tools

One of the biggest differences between knowledge and tools that have emerged from the two examples is their definition of purpose. Whereas PIREN support tools tend to place research as their primary objective and (indirectly) policy and planning as secondary, WSC tools are designed to make the underpinning research available and actionable for practitioners to demonstrate compliance and show the multiple benefits of local water sensitive solutions to regulators, authorities and communities. On one hand, a wide range of PIREN support tools are considered useful for practitioners, yet these tools tend to be highly academic and sometimes difficult to translate directly into action. On the other, the 'user-friendly' design of WSC tools is meant to promote adoption by industry partners, though some are still too new to be fully evaluated for use and utility.

In some cases, models may be improperly used or stretched beyond their capabilities to answer questions that they were not designed to answer (Australian water utility representative, 8 August 2017). While this is a general concern among model developers (CRCWSC researcher, 20 June 2017), there is a general feeling of trust among water actors that models will not be intentionally abused (CRCWSC researcher, 25 July 2017). For PIREN, a higher level of trust is felt among practitioners who have modelling expertise or who were involved in the development process, owing to a better understanding of the objectives and limitations of the model (AESN representative, 8 June 2016). For the most part, uncertainties were not explicitly discussed between researchers and industry partners in either case; the onus is therefore placed on experts and technicians to transmit relevant information (PIREN industry partner representative, 7 March 2017). Industry partners who have internal modelling expertise may also run their own uncertainty analyses, motivated by the direct consequences of such uncertainties on their work (SIAAP representative, 3 March 2017). 'Acceptance' or explicit concerns over uncertainty is therefore linked to potential consequences (social, economic, environmental) of management and planning decisions that were based on modelling results.

Other tools might have to be simplified to enhance their use and utility. For example the Water Sensitive Cities Index [106], which is less of a model and more of a benchmarking tool (CRCWSC researcher, 20 June 2017) has found opportunities for application due to its simplicity. Conversely, a more critical view was expressed for some of the larger-scale strategic planning tools, which may be considered 'helpful but unnecessary', as it was opined that conventional methods such as cost-benefit analyses or SWOT analyses could deliver the same results (Government Representative, 31 July 2017). It is important to highlight that this view may stem from a previous controversial experience that the state of Victoria has had with the use of such large-scale 'black boxes' [108]. Although this case was frequently cited, interview participants in Australia still generally expressed high levels of trust in models due to the demand for greater transparency and communication following this incident (CRCWSC researcher, 20 June 2017).

In the case of PIREN, the lack of 'operational' models that partners can use themselves is a strategic choice, not only for reasons of objectivity but also due to time and resource constraints:

> "Tools are available if [partners] want to use them as is but they don't have the human resources and they don't finance the interfacing either … We think more in terms of services, where the user defines what they want to do or what they want to evaluate and we [researchers] will perform the simulations and deliver the results".
>
> (PIREN researcher, 29 June 2017)

In this way, providing services are considered to be a more efficient use of resources for both researchers and industry partners, none of whom are prepared to invest time and human resources for a model they may only require on occasion. However, there may be less of an incentive to provide these services in cases where industry demand does not pique scientific interest.

4.3. Support Structures

Support structures refer to the different configurations that can promote or reinforce scientific knowledge or tools. This includes financial structures, organisational configurations, technical support and regulatory measures. Lemos et al. [9] suggest that usability can be improved through strategies of value-adding, retailing, wholesaling and customisation. While these may exist to some extent in both examples, the limitations posed by their respective 'boundaries' (in objectives and expertise, knowledge and tools and support structures) may not leave enough room to fully incorporate these strategies unless it is made to be a deliberate aim. For PIREN, this necessitated external contracts and support structures through the creation of ARMINES, the consulting arm of École des Mines ParisTech (PIREN researcher, 29 June 2016). While ARMINES provides a lucrative side line activity, which tailors research to specific industry demands, it is usually the research (scientific knowledge) itself that is customised, rather than the tools. For example, an industry partner such as AESN may request a specific study to be conducted and only require the results. In France, retailing, wholesaling and customisation of modelling tools is often perceived as the work of consultants, not researchers. For CRCWSC, modelling work was also outsourced with the MUSIC model through support from eWater (CRCWSC researcher, 9 June 2017). The structure of eWater is more aligned to strategies of retailing, wholesaling and customisation of tools, resulting in higher adoption of their tools. On one hand, boundary organisations play an important role in putting key players together with support and tools oriented towards a common objective and on a much wider scale than other science-policy partnerships. On the other, their 'boundaries' may limit their ability to fully support effective strategies that promote use and utility alone. The 'best of both worlds' may, in fact, be found in coordinated strategies that combine interactions and exchanges inside, outside and between these 'boundaries.'

Within these structures, financing often plays a large role on what is or can be done. On one hand, PIREN benefits from an extended and, for the moment, indefinite duration, allowing them more freedom to explore a wider range of research questions over a longer time period. However, their research actions are limited by a fixed amount of public funding from industry partners, an amount that has not seen much increase over the years despite a growing number of researchers who are involved in the program. Additionally, the autonomy of researchers is also subject to external funding sources that may come from universities, national research projects, or European projects, which allows certain freedoms while posing other constraints. On the other hand, CRCWSC is working on a 9-year timeline with a fixed budget of public and private funding from industry partners, governments and companies. Compared to PIREN, they are working with a bigger budget on a smaller time frame, which has allowed them to focus on specific goals and meet targeted objectives. In-kind support is also a major contributor in both examples, by way of researchers and doctoral students. At the same time, CRCWSC could face major challenges on the impact and sustainability of their work, particularly regarding the refinement, maintenance and adoption of modelling tools once the program ends. This can partly be addressed with technical support structures, which include user guidelines, technical manuals, training workshops, capacity building, and user support in the form of collaboration between researchers and industry partners, which fosters mutual understanding, transparency, and trust. The WSC Toolkit, for example, has initiated some of these structures including a user manual and a series of national training workshops for some of its operational features, building upon the experience learnt in the development and adoption of MUSIC (CRCWSC researcher, 20 June 2017). Technical support also exists in PIREN, though more through official or unofficial collaboration between researchers and partners. In the case of ProSe, for example, a technical working group was created in parallel to PIREN, involving some of the same researchers and partners while remaining outside of its boundaries.

Another important supporting structure is regulation, as illustrated by the examples of MUSIC in Australia and ProSe in France, both of which are required (even if unofficially) by regulating authorities. As an industry recognized tool, MUSIC has helped standardize regulations (e.g., [109,110]) across different territories with shared water networks (Government representative, 26 June 2017). The use of MUSIC as a compliance tool has also supported its legitimacy, since it 'helps speed up the process' for project proposals (Australian water utility representative, 8 August 2017). Additionally, models that are used nation-wide undergo a government-recognised accreditation process, which enhances the perception of its validity (Government representative, 26 June 2017). However, despite accreditation and validity, cost can be a limiting factor, with licences ranging from AU$0 for a 21-day limited trial version to prices starting at AU$5000 for a multiple user licence [111] (www.ewater.org.au). In the case of ProSe, the fact that the SIAAP is the only operating partner capable of running the model independently gives them a better bargaining position, however; the requirement to use ProSe also limits their ability to explore other models that may be better adapted to their needs (SIAAP representative, 3 March 2017). Regulations and the demand for evidence-based decisions may also place pressure on science to answer non-scientific questions. For example, since Paris won the bid to host the summer Olympics in 2024, there has been increased pressure for scientists to improve the water quality in the Seine in order to make it swimmable. Although issues of water quality are of scientific interest, particularly for PIREN, some may consider specific requirements for recreational use (e.g., faecal contamination levels) to be outside of the interest or expertise of PIREN researchers.

5. Moving Beyond the 'Usability Approach'

Technological advancement, coupled with the production of expertise, has led to the development of a large number of modelling tools [1,49,112], which aim to address specific environmental questions at different temporal and spatial scales. In parallel, practitioners face increasing pressure to base management and policy decisions on scientific evidence and data [18,113,114]. In this context, it would seem natural for modelling tools to be adopted by managers and decision makers, yet this is still far from the norm [115]. While challenges posed by the lack of communication or expertise are often cited among the main driving factors influencing the adoption of models [1,49,112,116,117], much of the literature is based on the dichotomy of 'use' vs. 'non-use.' However, the tentative linkages explored in the previous section suggest a more complex and nuanced relationship between use and utility that stretch beyond the common understanding of 'usability,' where the value of scientific knowledge and tools is tied to its ability to be applied (or directly used) in practice [9]. Proponents of, what we refer to as the 'usability approach,' often speak about 'usability' without detailing *how* scientific knowledge is actually used and *what* it is used for in practice, which we argue, have consequences on its use and utility. Building on previous research and aiming to deepen 'usability approach' thinking, this section explores the myriad of uses and utilities represented in our two examples.

5.1. Use vs. Utility

The major utilities for WSC tools and PIREN-Seine support tools generally fall under three main categories: (1) Enlightenment; (2) Decision support; and (3) Negotiation support, which reinforces previous findings [113,118,119]. Enlightenment can refer to a general contribution to overall understanding, specific information used for daily management or medium to long-term planning, or to monitoring trends and emerging issues. Decision support refers to daily management, medium to long-term planning, or evaluating actions taken, as well as to anticipating future trends. Negotiation support can refer to justifying a project or proposal, a way of asserting a certain role or position among a network of actors, or a way of acquiring or maintaining bargaining power. These categories are typically not independent and often coincide. The utility of a model is further influenced by three factors: objective, relevance and knowledge/expertise [120]. Objective refers to the set of priorities that the user seeks to be satisfied by the model. In other words, what is asked of the model, what purpose will it serve and what can be done with the model or its results? Relevance refers to how closely the model simulations correspond to the issues at stake for the user. In other words, the capability of the model to respond to the specific needs of the user, as well as the importance given to what is modelled. Finally, knowledge/expertise relates to the background or training of the user and their experience with modelling activities. This includes their capacity to run the model independently, add or modify components, understand its functions and limitations, know what data is required, and effectively translate and/or interpret the results.

5.2. User Involvement

In addition to the various utilities listed above, model use was found to be better represented as a spectrum based on four levels of user involvement [120] ranging from:

- Direct++, which indicates total mastery of the model;
- Direct+, which refers to independent model use without being able to change the model itself;
- Direct, which refers to a good understanding of what is being modelled while retaining limited involvement in the modelling process; to
- Non-Direct, which refers to complete detachment from modelling activities.

In Direct++, users can run the model independently, have access to input data, run simulations and are capable of making changes to the model itself (to the code, parameters, etc.). Next, Direct+ users understand how the model works; they can run simulations by themselves and may participate in the development of a model but are not able to make changes to it themselves. Direct use refers to users who have a good understanding of what is modelled and may participate in the elaboration of scenarios but are not involved in the modelling process itself. This type of user typically requests studies from experts and prefers to use the results instead of investing in in-house modelling expertise. Finally, there is Non-Direct use, where users are removed from the modelling process but can still benefit indirectly, as the knowledge produced by models is diffused into the global domain. A general framework outlining the relationships between use and utility from our two examples is found in Figure 1 below.

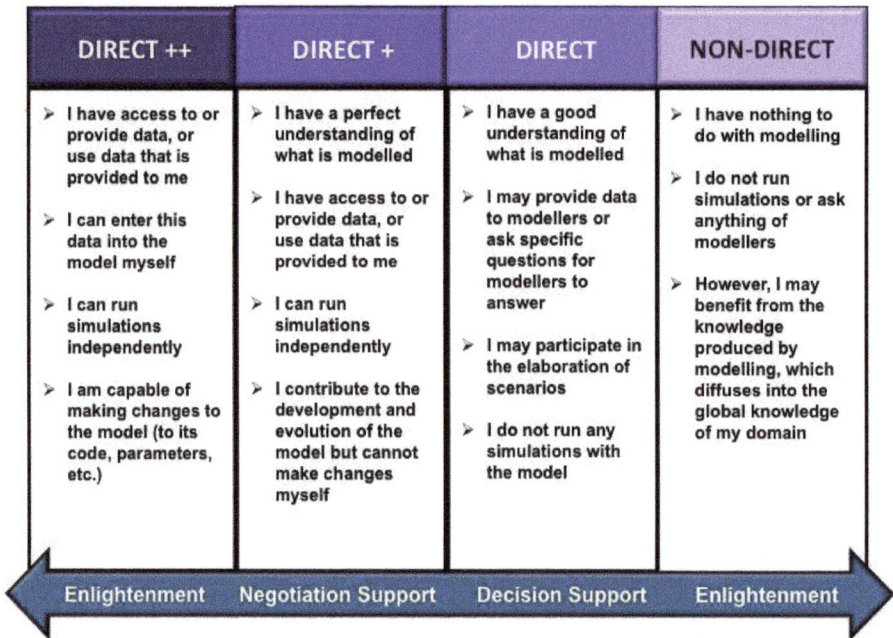

DIRECT ++	DIRECT +	DIRECT	NON-DIRECT
➤ I have access to or provide data, or use data that is provided to me ➤ I can enter this data into the model myself ➤ I can run simulations independently ➤ I am capable of making changes to the model (to its code, parameters, etc.)	➤ I have a perfect understanding of what is modelled ➤ I have access to or provide data, or use data that is provided to me ➤ I can run simulations independently ➤ I contribute to the development and evolution of the model but cannot make changes myself	➤ I have a good understanding of what is modelled ➤ I may provide data to modellers or ask specific questions for modellers to answer ➤ I may participate in the elaboration of scenarios ➤ I do not run any simulations with the model	➤ I have nothing to do with modelling ➤ I do not run simulations or ask anything of modellers ➤ However, I may benefit from the knowledge produced by modelling, which diffuses into the global knowledge of my domain
Enlightenment	Negotiation Support	Decision Support	Enlightenment

Figure 1. General Framework for Use and Utility Integrating User Involvement.

5.3. Integration and Application of Concepts

Of the numerous modelling tools that were either developed and/or used by PIREN over the past few decades, only two models (Seneque and ProSe) were identified as being used directly (at one time or another) by an operational partner, while one of the two (ProSe) is still in regular use today, suggesting greater 'non-use' of PIREN support tools. In retrospect, we could say that this is due to the fact that PIREN models are too academic and not 'user-friendly', rendering them less usable and therefore less useful. However, while this may be true in some cases, many of the partners interviewed maintained that the knowledge and tools produced by PIREN were integral to their work. We can explain this discrepancy by combining use, utility and user involvement into a general framework (Figure 1), which represents empirical findings from both examples. Within this framework, most users tend to fall on opposite ends of the spectrum: the majority of researchers involved in modelling activities are considered Direct++ users, while most operational partners are considered Non-Direct users, with the exception of the SIAAP who is a Direct+ user of the model ProSe. Although some models are occasionally used for decision and negotiation support, the main utility of PIREN support tools is for enlightenment, which explains why most partners find the tools useful even if they do not use them (directly). Although enlightenment is a fundamental utility of all types of uses, more prominent examples are found at the opposite ends of the spectrum in Direct++ and Non-Direct uses. For example, researchers make simulations with models (Direct++) to gain a deeper understanding of the transfer of micropollutants in the basin. While this information is relevant to operational partners, the science may not be at the point where it can be translated into action, or, similarly, the regulations may not have caught up with the science. Monitoring these research activities (Non-Direct) in the meantime will help to guide future planning by anticipating these emerging trends. On the flipside, CRCWSC aims to produce modelling tools that are adopted (directly) by water managers and decision makers. Using the general framework, we can say that most of the researchers are Direct++ users, while most industry partners are (or aim to be) Direct+ or Direct users. While models such as MUSIC have achieved this objective, it is too early to say whether newer tools such as DAnCE4Water or the WSC Toolkit will share the same success.

Compared to CRCWSC, the uses and utilities found within PIREN appear to be more varied. In both examples, Direct++ users tend to be researchers or model developers, while the knowledge they produce can be useful for researchers and industry partners of all user types for enlightenment. For example, in the case of PIREN, MODCOU is considered a research model (mostly Direct++ and Direct+ uses), yet the results are used by the AESN (mostly Direct or Non-Direct uses) to monitor and identify trends, which allows them to develop more adaptive climate change strategies (AESN representative, 8 June 2016). While most of the WSC models are aimed at Direct and Direct+ uses by industry partners, there is only one current instance of a Direct+ use within PIREN (case of the SIAAP who uses ProSe for enlightenment, decision and negotiation support). While most models serve an enlightenment function, the SIAAP also uses ProSe to support decisions (e.g., when sizing infrastructure and implementing new projects) as well as negotiation support, since they are required to justify proposals to the regulating authority using ProSe (SIAAP representative, 27 June 2016). Despite having in-house capacity to run the model independently and contributing to model development and data collection, practitioners are not able to change the code and must turn to researchers for specific requests (PIREN-Seine researcher, 28 April 2016). In Australia, MUSIC is a similar example of a Direct+ use by industry partners. As it has become the industry standard, using MUSIC to support decisions and justify proposals, though not always required, is beneficial (CRCWSC researcher, 9 June 2017). Direct uses are also common within the PIREN, in cases where partners ask for a specific study to be conducted. For example, when the AESN uses STICS-MODCOU to evaluate nitrates and pesticide flows in a specific aquifer (AESN representative, 8 June 2016).

While uncertainty related to modelling was rarely explicitly discussed, findings in both examples suggested that the 'acceptability' of uncertainty was implicitly informed by its use and utility. Direct and Direct+ users in PIREN were more concerned with quantifying uncertainty, as the stakes were relatively higher. An underestimation of pipe sizing by the SIAAP could, for example, directly contribute to major flooding in dense urban areas resulting in high economic, social and environmental costs. Failure to account for model uncertainty in these cases could also undermine project proposals based on modelling results, which in turn, undermines their negotiating power as it calls into question the expertise. On the other hand, the technical expertise required of these user types allows them to maintain trust in the model, by knowing what you can and cannot trust (SIAAP representative, 29 November 2016). Conversely, Non-Direct users may also maintain a high level of trust in the models despite a lack of technical expertise. In this case, trust is not in knowing *what* to trust (in the model) but rather, *whom* you can trust (experts) (DRIEE representative, 12 May 2016). For CRCWSC, uncertainty was considered 'more acceptable' (implicitly) in strategic planning tools such as DAnCE4Water. Since its intended use is to explore a range of possible future scenarios, the high level of associated uncertainty is a given (CRCWSC researcher, 14 June 2017).

Despite research and practice becoming increasingly collaborative processes, several studies continue to highlight the weak correlation between scientific production and use in practice. For example, through an empirical analysis of 20 scientific assessments co-produced by researchers and decision makers, Weichselgartner and Kasperson [27] revealed that decision makers did not sufficiently draw from available research-based knowledge, while at the same time, the knowledge produced by researchers was not sufficiently usable (directly). In another example, Holmes and Clark [28] analysed the studies conducted by the Environment Research Funders' Forum (ERFF) in the United Kingdom, pointing out that there was still significant lag time between current practice and guidance. Similarly, in their assessment of management practices in the Columbia River Basin, Callahan et al. [29] found that climate forecasts were significantly underutilised by managers despite their potential to support their ability to manage water resources in the face of increased climate variability.

The general framework of use and utility provided in Figure 1 extends the concept of use and utility from the strict dichotomy common to 'usability approach' thinking to a spectrum of uses and utilities that are found in examples such as PIREN and CRCWSC. Maintaining this dichotomy could lead some to develop solutions that are counteractive to their objective of increasing the adoption of modelling tools by practitioners and decision-makers. For example, a simplified model with a user-friendly interface may seem like a logical solution to overcome issues of communication and lack of expertise between researchers and practitioners. However, it may be of little use to a practitioner who requires a complex model to answer specific questions, but does not want to invest the time and resources towards in-house expertise. A better understanding of the nuanced relationship between use and utility can therefore support the development of tools that are more adapted to the needs of practitioners and decision-makers, according to what is needed (the model itself or the results), how they are used (level of user involvement), and what they are used for (justify proposals, monitor trends, etc.). While there is no one-size-fits-all solution (nor do we advocate for one), our analysis may help identify key points to consider when assessing the use and utility of modelling tools to better support water resources management, policy, and planning decisions.

Furthermore, discussion on how to produce 'usable' science could benefit from more in-depth analyses of specific examples. The fundamental difference between the CRC for Water Sensitive Cities and the PIREN-Seine is their objective and approach, which has resulted in different tools with different purposes. On one hand, the CRC for Water Sensitive Cities has taken a more market or policy driven approach, resulting in the production of more 'operational' tools, as well as active involvement from developers, water actors, local councils, and state governments (CRCWSC researcher, 9 June 2017). Not only does this promote research that is directly 'usable' for policy, it also establishes a target audience and a built-in user base (Australian water utility representative, interview 27 July 2017). In addition to decision support, WSC tools are designed with the specific (and arguably political) objective of achieving water sensitive cities in mind. In the case of MUSIC, its development and use as a compliance tool further entrenches the intimate relationship between science and policy, by creating both supply and demand (CRCWSC researcher, 9 June 2017). On the other hand, the PIREN-Seine has traditionally focused on the production of research and research tools as a primary objective to enlighten policy and planning decisions (PIREN researcher, 29 June 2016). Whereas CRCWSC takes an active role in policy, PIREN prefers the role of policy supporter rather than direct advisor, resulting in mostly 'research' tools and knowledge that is often difficult to translate to action and with an impact on policy that is not as easily quantifiable. However, the example of ProSe illustrates how a 'research' tool can also be 'operational' when mutual interest and supporting structures are strategically aligned (SIAAP representative, 27 June 2016).

While some commonalities can be extrapolated, our analysis of the specific organisational configurations and context-dependent drivers supports findings of previous authors [10,12] who stress the importance of moving beyond the traditional 'linear' model of research use, and advocate for a better account of the complex and nuanced interactions which take place at the science-policy interface. Vogel et al. [107] suggest we begin by reimagining these relationships in terms of 'spider webs', which are 'composed of nodes and a multitude of ephemeral linkages' (p. 360). Commonly held perceptions concerning the production of 'usable' knowledge and tools for management and policy tends to oversimplify the problem [2,10,12,107], which, in turn, limits opportunities for overcoming this fundamental challenge. Attempts to tackle this issue would therefore benefit from reframing the discussion to include and embrace the diversity that exists in modelling, which will not only provide a more informed understanding, but also help guide the development of knowledge and tools that are more adapted to different user needs. While the debate over scientific complexity vs. usability is still valid for specific models, it does not always need to be a trade-off. Instead, we can think of models as having different forms and functions, which can be used to complement one another or at different stages of the workflow to support different levels of planning and action (CRCWSC researcher, 6 June 2017; CRCWSC researcher, 20 June 2017). For example, deterministic models or real-time control for short to medium term management and planning and modelling platforms for longer-term planning and strategic thinking.

While each program uses a different approach, both are moving towards the idea of co-construction through collaborative scenario building and the use of modelling chains and/or integrated modelling platforms to address industry demands. On one hand, this is a logical choice, as scenario building and strategic modelling can support more robust and adaptive strategies (CRCWSC researcher, 14 June 2017). On the other, focus on 'co-construction' over 'co-production' may be considered a strategic choice, since a strict focus on the co-production of modelling tools requiring researchers and practitioners to invest heavily in time and resources, may not end up being very productive (PIREN researcher, 29 June 2016). Therefore, changing the discourse to the concept of co-construction of scenarios rather than co-production of models may allow for a more effective collaborative exchange as well as a more efficient use of resources. Partners may still be involved in the development of modelling tools, by providing feedback or as 'beta-testers' but the technical development (e.g., changing the code, adding parameters) resides with the researchers, who have the technical expertise. This way, each side plays to its strengths, while enhanced communication and understanding can be facilitated through interactive spaces such as workshops, seminars and working groups [10,107].

Finally, a more efficient science-policy relationship may benefit from a shifted focus from knowledge transfer to knowledge brokering [121–123], which helps ensure appropriate translation of research findings and facilitates the creation, sharing, and use of knowledge [124]. Knowledge brokers have played a key role in the dissemination of the work of the CRC for Water Sensitive Cities, helping to bring together different stakeholders towards the same objectives and increasing their impact on policy (CRCWSC researcher, 25 June 2017). In both cases, knowledge brokering would enhance the use and utility of modelling tools by helping developers understand user needs and helping users understand the objectives and limitations of the model. Modelling chains and platforms may be considered effective 'boundary objects,' by linking different modules together to tackle questions that are relevant to both research and policy. The same can be said of scenario building through strategic thinking exercises facilitated by these tools. In this context, the use and utility of the model itself is less of a concern, since the purpose is not to produce a specific outcome but rather to co-conceptualize and envision a range of possible outcomes. This allows different actors to come together and explore different strategies in a more neutral setting. Whether it is used directly or indirectly, collaborative scenario building and the use of modelling chains and/or platforms may prove to be a more effective path to enhancing the use and usability of scientific knowledge in practice.

6. Conclusions

Science and policy have become increasingly interdependent and science-policy collaborations more common, yet clear pathways for producing 'usable' scientific knowledge and tools remain uncertain. A novel approach based on an empirical analysis was used in the context of two boundary organisations in France and Australia to explore the tentative links between program strategy and the use and utility of modelling tools. Organisational configurations and context-specific drivers of: (1) objective and expertise; (2) knowledge and tools and; (3) support structures were identified as primary factors. Empirical findings highlighted a complex and nuanced relationship between use and utility, which suggests the need to go beyond 'usability approach' thinking. Further insight was also given into the role played by boundary organisations in bringing together relevant actors, facilitating formal and informal exchanges and building capacity, credibility, salience and legitimacy, suggesting that knowledge brokering and coordinated strategies which effectively integrate inter-, extra-, and intra-boundary activities would likely enhance use and utility. An exploration of the layered complexities between use and utility also suggests that added social value is created through mediated interactions and exchanges, which are facilitated by boundary organisations. The trend towards collaborative scenario building and the use of modelling chains and/or interactive modelling platforms offers ways of framing these interactions to better support management, policy and planning decisions. In this way, models may become a tool for communication and mediation between various actors, serving as a common reference point for co-conceptualising robust and adaptive strategies towards a shared vision of water resources management.

Acknowledgments: This study was co-funded by the PIREN-Seine program, École Nationale des Ponts et Chaussées, University of Paris East and LABEX Futurs Urbains in France, in cooperation with the Cooperative Research Centre for Water Sensitive Cities in Australia. The authors would like to extend their gratitude to all researchers and partners who graciously contributed valuable time, insight and knowledge to this research.

Author Contributions: N.C. and P.M.B drafted and edited the manuscript, with P.M.B. contributing heavily to the data and analysis on the Australian side. N.C. conducted interviews and provided anecdotal observation; R.M., C.B. and J.F.D. helped to formulate the theories and frameworks used and contributed to the analysis of the data on the French side.

Conflicts of Interest: The authors declare no conflict of interest.

Appendix A. Interview Question Guide

Background/History
What is your involvement in the PIREN-Seine/CRC for Water Sensitive Cities?
How did you get involved?
How long have you been involved?
How did the program get started? (Ex. Demand from researchers, industry or government?)
What is your background/training/experience?
How would you describe the relationship between researchers and partners in the program?
Do you think science should play a role in influencing policy?
How is the program funded?
Who finances it?
How much funding does the program have in total?
How much does each partner contribute?
What are the financial obligations from both sides?
In general, do you think there's a large gap between research and policy?
How does the program help to overcome this?
What could be improved?

Models: Development, Evolution, Use

Were you involved in the development of any modelling tools?
Which ones?
How were you involved? (Ex. did you develop the code, a module, provide feedback, etc.)
Who was involved in the development? (Ex. research teams, universities, institutions, partners, etc.)
How were the different actors involved? (Ex. funding, feedback, research, etc.)
What was the reason/need for developing this model?
Were there other models that existed at the time that could have done the same thing? If so, why develop a new model instead of using the existing one?
What were the main challenges in developing this model?
How has the model evolved? (Ex. different modules, more functionality, etc.)
What are the advantages/limits of the model?
Who uses the model?
Which actors? (Ex. Specific researchers, partners)
How do you use the model?
What does the model allow you to do, that you could not do (or not as easily do) without?
Do you run the model yourself or do you use the results?
What are some of the challenges in using this model?
Would you say it is easy to use for someone without training/expertise in modelling?
Would you prefer to be able to use the model yourself or just use the results?
Is the model used outside of the context of this program?
Do the outputs of the model meet the needs/demands of the user? If not, what could be improved?
Would you say it's more of a research model or an operational model?
What do you consider to be a 'research' or 'operational' model?
What type of user is the model designed for?
What type of use is the model designed for?
Can you think of any models that were developed within the context of the program but were not used or forgotten over time?
Would you say there's a big industry demand for modelling tools?
What types of tools are they looking for? (Ex. deterministic models, planning and visualisation tools, etc.)

Trust/Uncertainty

What do you need in order to 'trust' a model?
How is uncertainty taken into account in the modelling process/decision-making process?
Do partners ask for specific information on uncertainty?
What is considered to be an 'acceptable' level of uncertainty and how is this determined?
Can you think of a time where modelling results or the model itself were put into question?
Does the lack of available/reliable data pose a problem for you in trusting the model?
Would you say there is generally a lot of trust in modelling?
Would you prefer to have a model with a high level of associated uncertainty or to not have a model at all?

Scenarios

What simulations/scenarios were made with this model?
Who is involved in the construction of a scenario?
How do you determine which scenarios to test?
Out of an infinite number of possible future scenarios, how do you decide on the plausible scenarios to test?

Role of Modelling in Decision-Making

When are models used/their results taken into account in the decision-making process?
Besides modelling, what other factors influence the final decision?
Do you use this model more for daily management, or long-term planning?
Is it required by the regulating authority to use this model?
Can you give me specific examples of when the model (or its results) was used to make a decision?
Do you think that the knowledge/tools produced by this program have a big influence on policy in the country?

References

1. Liu, Y.; Gupta, H.; Springer, E.; Wagener, T. Linking science with environmental decision making: Experiences from an integrated modeling approach to supporting sustainable water resources management. *Environ. Model. Softw.* **2008**, *23*, 846–858. [CrossRef]
2. McNie, E.C. Reconciling the supply of scientific information with user demands: An analysis of the problem and review of the literature. *Environ. Sci. Policy* **2007**, *10*, 17–38. [CrossRef]
3. Haas, P. When Does Power Listen to Truth? A Constructivist Approach to the Policy Process. Available online: http://www.ingentaconnect.com/content/routledg/rjpp/2004/00000011/00000004/art00001 (accessed on 30 October 2017).
4. Cash, D.; Clark, W.C. From Science to Policy: Assessing the Assessment Process. *John F. Kennedy School of Government Faculty Research Working Papers Series* **2001**. [CrossRef]
5. Cash, D.; Clark, W.C.; Alcock, F.; Dickson, N.; Eckley, N.; Jäger, J. Salience, Credibility, Legitimacy and Boundaries: Linking Research, Assessment and Decision Making. *SSRN Electron. J.* **2003**. [CrossRef]
6. Guston, D. Boundary Organizations in Environmental Policy and Science: An Introduction. *Sci. Technol. Hum. Values* **2001**, *26*, 399–408. [CrossRef]
7. Sarkki, S.; Niemela, J.; Tinch, R.; van den Hove, S.; Watt, A.; Young, J. Balancing credibility, relevance and legitimacy: A critical assessment of trade-offs in science-policy interfaces. *Sci. Public Policy* **2014**, *41*, 194–206. [CrossRef]
8. White, D.D.; Wutich, A.; Larson, K.L.; Gober, P.; Lant, T.; Senneville, C. Credibility, salience and legitimacy of boundary objects: Water managers' assessment of a simulation model in an immersive decision theater. *Sci. Public Policy* **2010**, *37*, 219–232. [CrossRef]
9. Lemos, M.C.; Kirchhoff, C.; Ramprasad, V. Narrowing the Climate Information Usability Gap. *Nat. Clim. Chang.* **2012**, *2*, 789–794. [CrossRef]
10. Nutley, S.M.; Walter, I.; Davies, H.T.O. *Using Evidence: How Research Can Inform Public Services*; Policy Press: Bristol, England, 2007; ISBN 978-1-86134-664-3.
11. van den Hove, S. A rationale for science–policy interfaces. *Futures* **2007**, *39*, 807–826. [CrossRef]
12. Young, J.C.; Waylen, K.A.; Sarkki, S.; Albon, S.; Bainbridge, I.; Balian, E.; Davidson, J.; Edwards, D.; Fairley, R.; Margerison, C.; et al. Improving the science-policy dialogue to meet the challenges of biodiversity conservation: having conversations rather than talking at one-another. *Biodivers. Conserv.* **2014**, *23*, 387–404. [CrossRef]
13. Jasanoff, S.S. Contested Boundaries in Policy-Relevant Science. *Soc. Stud. Sci.* **1987**, *17*, 195–230. [CrossRef]
14. Gieryn, T.F. *Boundaries of Science*; Springer: Berlin, Germany, 1995; pp. 293–332. [CrossRef]
15. Jasanoff, S. Procedural choices in regulatory science. *Technol. Soc.* **1995**, *17*, 279–293. [CrossRef]
16. Star, S.L.; Griesemer, J.R. Institutional Ecology, "Translations" and Boundary Objects: Amateurs and Professionals in Berkeley's Museum of Vertebrate Zoology, 1907–39. *Soc. Stud. Sci.* **1989**, *19*, 387–420. [CrossRef]
17. Leigh Star, S. This is Not a Boundary Object: Reflections on the Origin of a Concept. *Sci. Technol. Hum. Values* **2010**, *35*, 601–617. [CrossRef]
18. Argent, R.M.; Perraud, J.-M.; Rahman, J.M.; Grayson, R.B.; Podger, G.M. A new approach to water quality modelling and environmental decision support systems. *Environ. Model. Softw.* **2009**, *24*, 809–818. [CrossRef]
19. Dictionary by Merriam-Webster: America's Most-Trusted Online Dictionary. Available online: https://www.merriam-webster.com/ (accessed on 30 October 2017).
20. Sarewitz, D.; Pielke, R.A. The neglected heart of science policy: Reconciling supply of and demand for science. *Environ. Sci. Policy* **2007**, *10*, 5–16. [CrossRef]
21. Jacobs, K. *Connecting Science, Policy and Decision-making: A Handbook for Researchers and Science Agencies*; National Oceanic and Atmospheric Administration, Office of Global Programs: Silver Spring, MD, USA, 2002.
22. Xu, Y.-P.; Booij, M.J.; Mynett, A.E. An appropriateness framework for the Dutch Meuse decision support system. *Environ. Model. Softw.* **2007**, *22*, 1667–1678. [CrossRef]
23. Castelletti, A.; Soncini-Sessa, R. A procedural approach to strengthening integration and participation in water resource planning. *Environ. Model. Softw.* **2006**, *21*, 1455–1470. [CrossRef]
24. Brown Gaddis, E.J.; Vladich, H.; Voinov, A. Participatory modeling and the dilemma of diffuse nitrogen management in a residential watershed. *Environ. Model. Softw.* **2007**, *22*, 619–629. [CrossRef]

25. Dilling, L.; Lemos, M.C. Creating usable science: Opportunities and constraints for climate knowledge use and their implications for science policy. *Glob. Environ. Chang.* **2011**, *21*, 680–689. [CrossRef]
26. Jacobs, K.; Garfin, G.; Lenart, M. More than just Talk: Connecting science and decision making. *Environ. Sci. Policy Sustain. Dev.* **2005**, *47*, 6–21. [CrossRef]
27. Weichselgartner, J.; Kasperson, R. Barriers in the science-policy-practice interface: Toward a knowledge-action-system in global environmental change research. *Glob. Environ. Chang.* **2010**, *20*, 266–277. [CrossRef]
28. Holmes, J.; Clark, R. Enhancing the use of science in environmental policy-making and regulation. *Environ. Sci. Policy* **2008**, *11*, 702–711. [CrossRef]
29. Callahan, B.; Miles, E.; Fluharty, D. Policy implications of climate forecasts for water resources management in the Pacific Northwest. *Policy Sci.* **1999**, *32*, 269–293. [CrossRef]
30. Buizer, J.; Jacobs, K.; Cash, D. Making short-term climate forecasts useful: Linking science and action. *Proc. Natl. Acad. Sci. USA* **2016**, *113*, 4597–4602. [CrossRef] [PubMed]
31. Cash, D.W.; Clark, W.C.; Alcock, F.; Dickson, N.M.; Eckley, N.; Guston, D.H.; Jäger, J.; Mitchell, R.B. Knowledge systems for sustainable development. *Proc. Natl. Acad. Sci. USA* **2003**, *100*, 8086–8091. [CrossRef] [PubMed]
32. Boezeman, D.; Vink, M.; Leroy, P. The Dutch Delta Committee as a Boundary Organization. *Environ. Sci. Policy* **2013**, *27*, 162–171. [CrossRef]
33. Glaser, B.G.; Strauss, A.L.; Strutzel, E. The Discovery of Grounded Theory; Strategies for Qualitative Research. *Nurs. Res.* **1968**, *17*, 364. [CrossRef]
34. Glaser, B.G.; Holton, J. Remodeling Grounded Theory. *Forum Qual. Sozialforschung* **2004**, *5*. [CrossRef]
35. Corbin, J.; Strauss, A. *Basics of Qualitative Research: Techniques and Procedures for Developing Grounded Theory*, 3rd ed.; SAGE Publications, Inc.: Thousand Oaks, CA, USA, 2008.
36. Billen, G. *Le PIREN-Seine: Un Programme De Recherche Né Du Dialogue Entre Scientifiques Et Gestionnaires*; CNRS Éditions; La Revue pour l'histoire du CNRS: Paris, France, 2001. [CrossRef]
37. Conseil Scientifique du Comité de Bassin Seine-Normandie. *Evaluation du Programme de Recherche PIREN-Seine*; Agence de l'Eau Seine-Normandie: Nanterre, France, 2008.
38. Wong, T.H.F.; Allen, R.; Brown, R.; Deletic, A.; Gangadharan, L.; Gernjak, W.; Jakob, C.; Johnstone, P.; Reeder, M.; Tapper, N.; Vietz, G.; Walsh, C. *Blueprint 2013—Stormwater Management in a Water Sensitive City*; Cooperative Research Centre for Water Sensitive Cities: Melbourne, Australia, 2013; ISBN 978-1-921912-01-6.
39. de Haan, F.J.; Ferguson, B.C.; Adamowicz, R.C.; Johnstone, P.; Brown, R.R.; Wong, T.H.F. The needs of society: A new understanding of transitions, sustainability and liveability. *Technol. Forecast. Soc. Chang.* **2014**, *85*, 121–132. [CrossRef]
40. Low, K.G.; Grant, S.B.; Hamilton, A.J.; Gan, K.; Saphores, J.-D.; Arora, M.; Feldman, D.L. Fighting drought with innovation: Melbourne's response to the Millennium Drought in Southeast Australia: Fighting drought with innovation. *Wiley Interdiscip. Rev. Water* **2015**, *2*, 315–328. [CrossRef]
41. Grant, S.B.; Fletcher, T.D.; Feldman, D.; Saphores, J.-D.; Cook, P.L.M.; Stewardson, M.; Low, K.; Burry, K.; Hamilton, A.J. Adapting Urban Water Systems to a Changing Climate: Lessons from the Millennium Drought in Southeast Australia. *Environ. Sci. Technol.* **2013**, *47*, 10727–10734. [CrossRef] [PubMed]
42. Heberger, M. Australia's millennium drought: Impacts and responses. In *The World's Water*; Island Press: Washington, DC, USA, 2012; pp. 97–125. ISBN 978-1-61091-048-4.
43. Brown, R.R.; Keath, N.; Wong, T.H.F. Urban water management in cities: Historical, current and future regimes. *Water Sci. Technol.* **2009**, *59*, 847–855. [CrossRef] [PubMed]
44. Ferguson, B.C.; Brown, R.R.; Frantzeskaki, N.; de Haan, F.J.; Deletic, A. The enabling institutional context for integrated water management: Lessons from Melbourne. *Water Res.* **2013**, *47*, 7300–7314. [CrossRef] [PubMed]
45. Qu'est-ce Que le PIREN Seine? Programme Interdisciplinaire de Recherche sur l'Environnement de la Seine. Available online: http://www.metis.upmc.fr/piren/?q=presentation_PIREN-Seine (accessed on 6 March 2017).
46. *Loi n° 92–3 du 3 Janvier 1992 sur l'eau*; Assemblée Nationale: Paris, France.
47. *Le SDAGE 2010–2015 du Bassin de la Seine et des Cours d'Eau Côtiers Normands*; Agence de l'Eau Seine-Normandie: Nanterre, France, 2010.
48. European Union. *Water Framework Directive*; European Commission: Brussels, Belgium, 2000.

49. Bach, P.M.; Rauch, W.; Mikkelsen, P.S.; McCarthy, D.T.; Deletic, A. A critical review of integrated urban water modelling—Urban drainage and beyond. *Environ. Model. Softw.* **2014**, *54*, 88–107. [CrossRef]

50. Carre, C.; Haghe, J.P.; De Coninck, A.; Becu, N.; Deroubaix, J.; Pivano, C.; Flipo, N.; Le Pichon, C.; Tallec, G. How to integrate scientific models in order to switch from, flood control river management to multifunctional river management? *Int. J. River Basin Manag.* **2014**, *12*, 231–249. [CrossRef]

51. Mouchel, J.-M. *Rapport de Synthèse 2007–2010—Introduction Générale*; PIREN-Seine: Paris, France, 2010.

52. Even, S.; Poulin, M.; Billen, G.; Garnier, J. *Modèles PROSE Et SENEQUE: Établissement De Versions De Référence Applicables Aux Études De Gestion*; PIREN-Seine: Paris, France, 1998.

53. Garnier, J.; Mouchel, J.-M. *Man and River Systems: The Functioning of River Systems at the Basin Scale*; Springer: Cham, The Netherlands, 1999; ISBN 978-94-017-2163-9.

54. Billen, G.; Garnier, J.; Mariotti, A. *Bilan Des Transferts D'azote Dans Le Bassin De La Seine: L'approche Du Modèle SENEQUE*; PIREN-Seine: Paris, France, 1998.

55. Ledoux, E. Modélisation Intégrée des Écoulements de Surface et des Écoulements Souterrains sur un Bassin Hydrologique. Doctoral Dissertation, Ecole des Mines, Paris, France, 1980.

56. Ledoux, E.; Girard, G.; de Marsily, G.; Villeneuve, J.P.; Deschenes, J. Spatially distributed modeling: Conceptual approach, coupling surface water and groundwater. In *Unsaturated Flow in Hydrologic Modeling*; Morel-Seytoux, H.J., Ed.; NATO ASI Series; Springer: Cham, The Netherlands, 1989; pp. 435–454. ISBN 978-94-010-7559-6.

57. Brisson, N.; Mary, B.; Ripoche, D.; Jeuffroy, M.H.; Ruget, F.; Nicoullaud, B.; Gate, P.; Devienne-Barret, F.; Antonioletti, R.; Durr, C.; et al. STICS: A generic model for the simulation of crops and their water and nitrogen balances. I. Theory and parameterization applied to wheat and corn. *Agron. Sustain. Dev.* **1998**, *18*, 36. [CrossRef]

58. Brisson, N.; Gary, C.; Justes, E.; Roche, R.; Mary, B.; Ripoche, D.; Zimmer, D.; Sierra, J.; Bertuzzi, P.; Burger, P.; et al. An overview of the crop model stics. *Eur. J. Agron.* **2003**, *18*, 309–332. [CrossRef]

59. Bilen, G.; Garnier, J.; Mouchel, J.-M.; Silvestre, M. The Seine system: Introduction to a multidisciplinary approach of the functioning of a regional river system. *Sci. Total Environ.* **2007**, *375*, 1–12. [CrossRef] [PubMed]

60. Even, S. *Description Du Logiciel ProSe, Version 4.1—Logiciel De Simulation De L'hydrodynamique, Du Transport Et Du Fonctionnement Biochimique D'un Écosystème Fluvial*; PIREN-Seine: Paris, France, 2007.

61. Even, S.; Poulin, M.; Garnier, J.; Billen, G.; Servais, P.; Chesterikoff, A.; Coste, M. River ecosystem modelling: Application of the PROSE model to the Seine river (France). *Hydrobiologia* **1998**, *373*, 27–45. [CrossRef]

62. Ruelland, D.; Billen, G. Riverstrahler, SENEQUE and SENECAM: Modelling tools for water resources management from regional to local scales. In Proceedings of the 6th International Conference of EWRA, France, Menton, 7–10 September 2005.

63. Thouvenot, M.; Billen, G.; Garnier, J. Denitrification in the Riverstrahler Model. In Proceedings of the Denitrification Modeling Workshop Agenda, New York, NY, USA, 28–30 November 2006.

64. Ruelland, D.; Billen, G.; Brunstein, D.; Garnier, J. SENEQUE: A multi-scaling GIS interface to the Riverstrahler model of the biogeochemical functioning of river systems. *Sci. Total Environ.* **2007**, *375*, 257–273. [CrossRef] [PubMed]

65. Ruelland, D.; Silvestre, M.; Thieu, V.; Billen, G. *Applicatif SENEQUE 3.4: Notice D'utilisation*; PIREN-Seine: Paris, France, 2007.

66. Thieu, V.; Billen, G.; Silvestre, M.; Garnier, J. *SENEQUE and Co: Développements Logiciels Et Améliorations Des Outils*; Rapport PIREN-Seine; PIREN-Seine: Paris, France, 2006.

67. Ledoux, E.; Gomez, E.; Monget, J.M.; Viavattene, C.; Viennot, P.; Ducharne, A.; Benoit, M.; Mignolet, C.; Schott, C.; Mary, B. Agriculture and groundwater nitrate contamination in the Seine basin. The STICS–MODCOU modelling chain. *Sci. Total Environ.* **2007**, *375*, 33–47. [CrossRef] [PubMed]

68. Viennot, P.; Ledoux, E. *Influence De L'augmentation Des Prélèvements Anthropiques En Formations Aquifères Sur Le Fonctionnement Hydrodynamique Du Bassin De La Seine*; PIREN-Seine: Paris, France, 2007.

69. Viennot, P.; Monget, J.-M.; Ledoux, E.; Schott, C. *Modélisation De La Pollution Nitrique Des Aquifères Du Bassin De La Seine: Intégration Des Bases De Données Actualisées Des Practiques Agricoles, Validation Des Simulations Sur La Période 1971–2004, Simulations Prospectives De Mesures Agro-Environnementales*; PIREN-Seine: Paris, France, 2006.

70. Habets, F.; Flipo, N.; Goblet, P.; Ledoux, E.; Monteil, C.; Philippe, E.; Querel, W.; Saleh, F.; Souhar, O.; Stouls, A.; et al. *Le Développement Du Modèle Intégré Des Hydrosystèmes Eau-Dyssée*; PIREN-Seine: Paris, France, 2009.

71. Saleh, F.; Ducharne, A.; Oudin, L.; Flipo, N.; Ledoux, E. *Hydraulic Modeling of Flow, Water Levels and Inundations: Serein River Case Study*; PIREN-Seine: Paris, France, 2009.

72. Coucheney, E.; Buis, S.; Launay, M.; Constantin, J.; Mary, B.; García de Cortázar-Atauri, I.; Ripoche, D.; Beaudoin, N.; Ruget, F.; Andrianarisoa, K.S.; et al. Accuracy, robustness and behavior of the STICS soil–crop model for plant, water and nitrogen outputs: Evaluation over a wide range of agro-environmental conditions in France. *Environ. Model. Softw.* **2015**, *64*, 177–190. [CrossRef]

73. Gomez, E.; Ledoux, E.; Mary, B. *La Démarche De Modélisation Régionale Des Écoulements D'eau, De La Production Et Du Transfert D'azote Sure Le Bassin De La Seine, Structure Du Modèle D'écoulement*; PIREN-Seine: Paris, France, 1998.

74. Gomez, E.; Ledoux, E.; Monget, J.-M.; De Marsily, G. Distributed surface-groundwater coupled model applied to climate or long term water management impacts at basin scale. *Eur. Water* **2003**, *1*, 3–8.

75. About the CRCWSC. CRC Water Sensitive Cities. Available online: https://watersensitivecities.org.au/about-the-crcwsc/ (accessed on 2 August 2017).

76. Howe, C.; Mitchell, C. *Water Sensitive Cities*; International Water Association (IWA) Publishing: London, UK, 2011; ISBN 978-1-84339-364-1.

77. Wong, T.H.F.; Brown, R.R. The water sensitive city: Principles for practice. *Water Sci. Technol.* **2009**, *60*, 673–682. [CrossRef] [PubMed]

78. Geiger, W.F. Sponge city and lid technology—Vision and tradition. *Landsc. Archit. Front.* **2015**, *3*, 10–22.

79. Li, X.; Li, J.; Fang, X.; Gong, Y.; Wang, W. Case studies of the sponge city program in China. In Proceedings of the World Environmental and Water Resources Congress 2016, West Palm Beach, FL, USA, 22–26 May 2016.

80. Xia, J.; Zhang, Y.; Xiong, L.; He, S.; Wang, L.; Yu, Z. Opportunities and challenges of the Sponge City construction related to urban water issues in China. *Sci. China Earth Sci.* **2017**, *60*, 652–658. [CrossRef]

81. Malekpour, S.; de Haan, F.J.; Brown, R.R. Marrying Exploratory Modelling to Strategic Planning: Towards Participatory Model Use. In Proceedings of the 20th International Congress on Modelling and Simulation (MODSIM2013), Adelaide, Australia, 1–6 December 2013.

82. Walker, W.E.; Haasnoot, M.; Kwakkel, J.H. Adapt or Perish: A Review of Planning Approaches for Adaptation under Deep Uncertainty. *Sustainability* **2013**, *5*, 955–979. [CrossRef]

83. Tewdwr-Jones, M.; Allmendinger, P. Deconstructing Communicative Rationality: A Critique of Habermasian Collaborative Planning. *Environ. Plan. A* **1998**, *30*, 1975–1989. [CrossRef]

84. Klosterman, R.E. Planning Support Systems: A New Perspective on Computer-Aided Planning. *J. Plan. Educ. Res.* **1997**, *17*, 45–54. [CrossRef]

85. Wong, T.H.; Fletcher, T.D.; Duncan, H.P.; Coleman, J.R.; Jenkins, G.A. A model for urban Stormwater improvement: Conceptualization. In *Global Solutions for Urban Drainage*; American Society of Civil Engineers: Reston, VA, USA, 2002; pp. 1–14.

86. Rauch, W.; Urich, C.; Bach, P.M.; Rogers, B.C.; de Haan, F.J.; Brown, R.R.; Mair, M.; McCarthy, D.T.; Kleidorfer, M.; Sitzenfrei, R.; et al. Modelling transitions in urban water systems. *Water Res.* **2017**, *126*, 501–514. [CrossRef] [PubMed]

87. Wong, T.H.F.; Fletcher, T.D.; Duncan, H.P.; Jenkins, G.A. Modelling Urban Stormwater Treatment—A Unified Approach. *Ecol. Eng.* **2006**, *27*, 58–70. [CrossRef]

88. eWater Water Quality Objectives—MUSIC Version 6 Documentation and Help—eWater Wiki. Available online: https://wiki.ewater.org.au/display/MD6/Water+Quality+Objectives (accessed on 16 May 2017).

89. *eWater Annual Report 2011–12*; eWater: Canberra, Australia, 2012.

90. Water sensitive cities modelling toolkit-CRC for Water sensitive cities. Available online: https://watersensitivecities.org.au/wp-content/uploads/2017/06/Fact-Sheet_Water-Sensitive-Cities-modelling-toolkit_Project-D1.5_V3.pdf (accessed on 2 August 2017).

91. Fletcher, T.D.; Walsh, C.J.; Bos, D.; Nemes, V.; RossRakesh, S.; Prosser, T.; Hatt, B.; Birch, R. Restoration of stormwater retention capacity at the allotment-scale through a novel economic instrument. *Water Sci. Technol.* **2011**, *64*, 494–502. [CrossRef] [PubMed]

92. Fletcher, T.D.; Mitchell, V.G.; Deletic, A.; Ladson, T.R.; Séven, A. Is stormwater harvesting beneficial to urban waterway environmental flows? *Water Sci. Technol.* **2007**, *55*, 265–272. [CrossRef] [PubMed]

93. Walsh, C.J.; Fletcher, T.D.; Bos, D.; RossRakesh, S.; Nemes, V.; Edwards, C.; O'Keefe, A. *Little Stringybark Creek: Environmental Benefit Calculator Technical Notes*; Melbourne Water; Department of Water and Environment, University of Melbourne: Melbourne, Australia, 2012.

94. Walsh, C.J.; Fletcher, T.D.; Ladson, A.R. Stream restoration in urban catchments through redesigning stormwater systems: Looking to the catchment to save the stream. *J. N. Am. Benthol. Soc.* **2005**, *24*, 690–705. [CrossRef]

95. Brookes, K.; Wong, T.H.F. The adequacy of stream erosion index as an alternate indicator of geomorphic stability in urban waterways. In Proceedings of the 6th Water Sensitive Urban Design Conference and Hydropolis, Perth, Australia, 5–8 May 2009.

96. Coutts, A.; Harris, R. *A Multi-Scale Assessment of Urban Heating in Melbourne during an Extreme Heat Event: Policy Approaches for Adaptation*; Monash University: Clayton, Australia, 2013.

97. Raut, B.; de la Fuente, L.; Seed, A.; Jakob, C.; Reeder, M. Application of a space-time stochastic model for downscaling future rainfall projections. In *Hydrology and Water Resources Symposium*; Engineers Australia: Barton, Australia, 2012; pp. 579–586.

98. Zhang, K.; Manuelpillai, D.; Raut, B.; Jakob, C.; Reeder, M.; Deletic, A.; Bach, P.M. Impact of future rainfall projections from ensemble GCMs on stormwater management. In Proceedings of the 14th International Conference on Urban Drainage (14ICUD), Prague, Czech Republic, 10–15 September 2017.

99. Rauch, W.; Bach, P.M.; Brown, R.; Rogers, B.; de Haan, F.J.; McCarthy, D.T.; Kleidorfer, M.; Mair, M.; Sitzenfrei, R.; Urich, C.; et al. Enabling change: Institutional adaptation. In *Climate Change, Water Supply and Sanitation*; IWA Publishing: London, UK, 2015.

100. Urich, C.; Bach, P.; Sitzenfrei, R.; Kleidorfer, M.; Mccarthy, D.; Deletic, A.; Rauch, W. Modelling cities and water infrastructure dynamics. *Proc. Inst. Civ. Eng. Eng. Sustain.* **2013**, *166*, 301–308. [CrossRef]

101. de Haan, F.J.; Rogers, B.C.; Brown, R.R.; Deletic, A. Many roads to Rome: The emergence of pathways from patterns of change through exploratory modelling of sustainability transitions. *Environ. Model. Softw.* **2016**, *85*, 279–292. [CrossRef]

102. Urich, C.; Rauch, W. Exploring critical pathways for urban water management to identify robust strategies under deep uncertainties. *Water Res.* **2014**, *66*, 374–389. [CrossRef] [PubMed]

103. Rossman, L.A. *EPANET 2 User Manual*; US Environmental Protection Agency: Cincinnati, OH, USA, 2000.

104. Rossman, L.A. *Storm Water Management Model—User's Manual*; Version 5.1; US Environmental Protection Agency: Cincinnati, OH, USA, 2004.

105. Löwe, R.; Urich, C.; Domingo, N.S.; Mark, O.; Deletic, A.; Arnbjerg-Nielsen, K. Assessment of urban pluvial flood risk and efficiency of adaptation options through simulations—A new generation of urban planning tools. *J. Hydrol.* **2017**, *550*, 355–367. [CrossRef]

106. Chesterfield, C.; Urich, C.; Beck, L.; Berge, K.; Charette-Castonguay, A.; Brown, R.; Dunn, G.; De Haan, F.; Lloyd, S.; Rogers, B. A Water Sensitive Cities Index—Benchmarking cities in developed and developing countries. In Proceedings of the International Low Impact Development Conference, Beijing, China, 26–29 June 2016.

107. Vogel, C.; Moser, S.C.; Kasperson, R.E.; Dabelko, G.D. Linking vulnerability, adaptation and resilience science to practice: Pathways, players and partnerships. *Glob. Environ. Chang.* **2007**, *17*, 349–364. [CrossRef]

108. Baker, R.; McKenzie, N. Troubled Waters: What is the Office of Living Victoria up to? *The Age*, 28 February 2014.

109. Department of Environment and Resource Management (DERM). *Urban Stormwater Quality Planning Guidlines 2010*; Queensland Government: Brisbane, Australia, 2010.

110. Department of Environment, Land, Water and Planning (DELWP). *Victoria Planning Provisions*; Department of Environment, Land, Water and Planning: Melbourne, Australia, 2017.

111. Evolving Water Management—eWater. Available online: http://ewater.org.au/ (accessed on 24 November 2017).

112. Argent, R.M. An overview of model integration for environmental applications—Components, frameworks and semantics. *Environ. Model. Softw.* **2004**, *19*, 219–234. [CrossRef]

113. Nilsson, M.; Jordan, A.; Turnpenny, J.; Hertin, J.; Nykvist, B.; Russel, D. The use and non-use of policy appraisal tools in public policy making: An analysis of three European countries and the European Union. *Policy Sci.* **2008**, *41*, 335–355. [CrossRef]

114. Fletcher, T.D.; Andrieu, H.; Hamel, P. Understanding, management and modelling of urban hydrology and its consequences for receiving waters: A state of the art. *Adv. Water Resour.* **2013**, *51*, 261–279. [CrossRef]

115. Brugnach, M.; Tagg, A.; Keil, F.; de Lange, W.J. Uncertainty Matters: Computer Models at the Science–Policy Interface. *Water Resour. Manag.* **2007**, *21*, 1075–1090. [CrossRef]

116. Hipel, K.W.; Ben-Haim, Y. Decision making in an uncertain world: Information-gap modeling in water resources management. *IEEE Trans. Syst. Man Cybern. Part C Appl. Rev.* **1999**, *29*, 506–517. [CrossRef]

117. Marlow, D.R.; Moglia, M.; Cook, S.; Beale, D.J. Towards sustainable urban water management: A critical reassessment. *Water Res.* **2013**, *47*, 7150–7161. [CrossRef] [PubMed]

118. Clark, W.C.; Tomich, T.P.; van Noordwijk, M.; Guston, D.; Catacutan, D.; Dickson, N.M.; McNie, E. Boundary work for sustainable development: Natural resource management at the Consultative Group on International Agricultural Research (CGIAR). *Proc. Natl. Acad. Sci.* **2011**, *113*, 4615–4622. [CrossRef] [PubMed]

119. O'Mahony, S.; Bechky, B.A. Boundary organizations: Enabling collaboration among unexpected allies. *Adm. Sci. Q.* **2008**, *53*, 422–459. [CrossRef]

120. Chong, N.; Bonhomme, C.; Deroubaix, J.-F.; Moilleron, R. *Production Et Usages Des Modèles Dans Le Cadre Du PIREN-Seine*; PIREN-Seine: Paris, France, 2016.

121. Johri, A. Boundary spanning knowledge broker: An emerging role in global engineering firms. In Proceedings of the 38th Annual Frontiers in Education Conference (FIE 2008), Saratoga Springs, NY, USA, 22–25 October 2008; p. S2E-7.

122. Turnhout, E.; Stuiver, M.; Klostermann, J.; Harms, B.; Leeuwis, C. New roles of science in society: Different repertoires of knowledge brokering. *Sci. Public Policy* **2013**, *40*, 354–365. [CrossRef]

123. Sverrisson, Á. Translation Networks, Knowledge Brokers and Novelty Construction: Pragmatic Environmentalism in Sweden. *Acta Sociol.* **2016**, *44*, 313–327. [CrossRef]

124. Meyer, M. The Rise of the Knowledge Broker. *Sci. Commun.* **2010**, *32*, 118–127. [CrossRef]

water

MDPI

Article

Water Experts' Perception of Risk for New and Unfamiliar Water Projects

Anna Kosovac [1,*] [ID], **Anna Hurlimann** [2] and **Brian Davidson** [1]

[1] Faculty of Veterinary and Agricultural Sciences, The University of Melbourne, Parkville, VIC 3010, Australia; b.davidson@unimelb.edu.au

[2] Melbourne School of Design, The University of Melbourne, Parkville, VIC 3010, Australia; anna.hurlimann@unimelb.edu.au

* Correspondence: kosovaca@student.unimelb.edu.au

Received: 31 October 2017; Accepted: 12 December 2017; Published: 15 December 2017

Abstract: In the context of a changing urban environment and increasing demand due to population growth, alternative water sources must be explored in order to create future water security. Risk assessments play a pivotal role in the take-up of new and unfamiliar water projects, acting as a decision-making tool for business cases. Perceptions of risk ultimately drive risk assessment processes, therefore providing insight into understanding projects that proceed and those that do not. Yet there is limited information on the risk perceptions water professionals have of new and unfamiliar water projects. In this study, 77 water professionals were surveyed from across the Melbourne metropolitan water industry to examine risk perceptions over a range of different, unfamiliar water projects. The qualitative data was thematically analysed, resulting in a number of risk perception factors for each hypothetical project. Risk factors that recurred most frequently are those that relate to community backlash and to the reputation of the organisation. These social risk perceptions occurred more frequently than other more technical risks, such as operational risks and process-related risks. These results were at odds with the existing literature assessing risk perceptions of business-as-usual projects, which presented cost as the key risk attribute. This study sheds light on the perceived nature of new and unfamiliar processes in the water sector, providing an understanding that public perceptions do matter to experts involved in water infrastructure decision-making.

Keywords: risk perception; water utilities; water experts; recycled water; community; fluoride

1. Introduction

The varied nature of risk, and how perceptions differ between stakeholders, has not been fully considered in the water sector, especially from a cultural and psychological perspective [1]. In particular, there is a lack of research regarding water utilities' perceived risks associated with new and unfamiliar projects. Perceptions of risk influence the decisions made about water infrastructure projects, and are, hence, an important consideration. Climate change, along with population growth and other socio-economic factors, is impacting the availability of freshwater resources [2]. This has implications for the need to consider water supply projects that are alternative to the traditional sources of supply. Hence, there is a need to understand utilities' risk perceptions of alternative water projects, so as to best design and facilitate their project implementation if, and when, necessary.

Placing this in the context of the special issue, public infrastructure, including water and sewage services, is essential to the efficient functioning of urban areas. Infrastructure provides the resources necessary to support urban economies, environments, and societies. Decisions surrounding the design and maintenance of public infrastructure (including water supply) involve the formal assessment of risk by experts (the word 'expert' is used in this paper to refer to a person with specialised knowledge in a particular area—in particular, those who have mastered certain skills they utilise in their careers. It is used interchangeably with 'professional' within this study). The importance of understanding public perception of risk associated with urban infrastructure projects has gained prominence over the past two decades. This is in part due to public and political opposition to project proposals, which ultimately led to the abandonment of plans in some cases. This includes, but is not limited to, proposed potable recycled water systems in Toowoomba, Australia [3] and San Diego, USA in the 1990s [4].

The aim of this paper is to explore the risks that water utility experts perceive as being associated with new and innovative projects, and how risk perception differs across projects. This aim is addressed through an empirical study of 77 experts working across four water utilities in Melbourne, Australia. An open-ended, qualitative approach is taken to identifying risk perceptions, leaving experts to highlight their own risk factors, rather than presenting predefined ones. This allows for a more exploratory approach to the question of risk perceptions. The research also differs from others through highlighting risk factors related to "new" or "unfamiliar" projects, (projects that have previously never been undertaken by the water authorities) and therefore explains risk attitudes to innovative approaches to water. The risks that were seen by water professionals as highest included community backlash and reputation to the organisation. These findings were unexpected considering the non-technical nature of the risks, highlighting a shift of the focus from the traditional public health aspect of water.

The paper begins by discussing key literature in the field of risk perception and water supply. The study's research method is then detailed, before the presentation and discussion of results.

2. Risk Assessment and Risk Perceptions

Risk plays an important role when considering the take-up of new methods and processes in the water industry. Risk assessment of a project often guides business cases and acts as a major decision-making tool for proposed schemes. As such, understanding perceptions of risk can guide us in understanding innovative or unfamiliar projects that proceed, compared to those that do not.

The epistemology of risk exhibits the tension between risk theories—in particular, the underlying element of whether risk is seen as a subjective or objective concept [5]. This refers to the notion of whether a "real" or "true" risk actually exists. Perceptions of risk, rather than whether a risk is seen as true or not, is what ultimately drives decision-making, and, therefore, we do not argue for or against the real, objective risk but instead highlight personal professional notions of perceived risk.

2.1. Risk Perceptions, and How These Differ between Laypeople and Experts

Extensive research has been undertaken in the space of risk perceptions and how they differ between laypeople and experts. Predominantly undertaken in the United States, this field of study expanded in the 1980s and continues to provide insight into understanding public perception, particularly in an environment in which participatory democracy is encouraged. Slovic and Fischoff are key proponents in this field, highlighting the importance of understanding individual psychological factors in the construction of perception [6,7]. They argue that the individual's psychological affiliation to a risk, based on personal experience, provides a key explanatory factor in their perception of the risk. Fundamental attributes, such as the "dread factor" (the sense of dread associated with the risk), or the "newness" of the risk (how familiar the risk is) are examples of explanatory elements of understanding risk perceptions [6].

When comparing public risk perceptions to those of experts, Slovic and Fischoff's study showed that of the laypeople surveyed, they generally ranked nuclear energy high in risk, because of the sense of dread associated with the effects of a negative radioactive fallout [8]. In comparison, they ranked the risk of driving a car as very low, despite figures showing the chance of being involved in a car accident is far higher than a nuclear one [6]. Experts instead had ranked driving a car as significantly higher risk than nuclear power, which the authors suggest is based on an objective understanding of the safety of nuclear energy, and, thus, the dread or familiarity factor is dissimilar to the layperson [6].

This evident difference in risk perceptions between lay people and experts becomes a challenge when the question of public policy, or deciding on public spending, enters the discussion. Prior to the 1960s, expert assessments were generally accepted by the public, with far less scrutiny than today [9]. As such, public consultations and public opinion polling did not carry as much relevance as they do now. Schon argues that the loss of faith in professional judgement began to occur between 1963 and 1981, much due to the unintended side effects of new technologies, and also conflicting recommendations within a professional field itself [10]. The consistent debunking of professional advice and theories also proved to be a large factor in the increasing distrust of professional experts [9,10]. Slovic also contends that this increase in distrust has also then flowed to distrust of risk assessments undertaken by these experts [7].

Similarly, a study by Schlosberg et al. [11] in Melbourne, Australia has shown a clear distinction between climate change discourses of government authorities and the public. In considering climate change adaptation, local government bodies used words like risk, water, control, event, and management [11]. In contrast, they found that community groups used language that focused more on the impacts of basic needs, such as "food", "community", "people", "energy", "water", and "local" [11]. The authors highlight the discordant nature of risk discourses between laypeople and experts, and posit that this may be due to lack of public engagement, while Guy and Kashima [12] theorise that it is a knowledge-based distinction.

2.2. Risk Perceptions Related to Water Projects

The supply of water to the public is an important service, given the dependence humans have on water for their survival, health, and economic prosperity. Additionally, the very nature of water's close link to the politically contentious issue of climate change creates an increase in public interest. The topic of water is not immune to the effects of a discourse on climate change, and, in particular, what adaptation measures should be undertaken—an area that is highly contested [13].

Climate change adaptation measures, including those necessary in the water sector, are highly politicised. This can be attributed to clear differences on the issue existing along political party lines in Australia [14]. Unsurprisingly, as an issue of public interest, media reports on water projects serve to sway opinion on climate change issues, as shown in a study on desalination delivery and public interest by Ettehad et al. [15].

Public sentiment—or backlash—can halt or stop projects altogether, and thus plays a key role in the project management process. Hurlimann and Dolnicar analysed the reasons why an indirect potable recycled water system proposed in 2006 for Toowoomba, Australia failed [3]. A plebiscite was held, and over 60% of the community voted against the scheme. The authors found that the failure of the plan was not just associated with community opposition, but also due to political involvement, the timing of the plebiscite, vested interests, and information manipulation. Public opposition also had a role to play in other potable reuse schemes including San Diego in the USA in the 1990s [4], and for desalination plant proposals, including Sydney in the 2000s [16]. Some authors [17] have attributed a decide, announce, defend (DAD) approach to the demise of these schemes [18].

Creating an alignment between public and expert risk perceptions is, thus, vital in a participatory democracy [7]. Yet there has been limited research conducted in this field, particularly applied to water infrastructure projects. Much literature focuses on public perceptions of risk in water [19–23]; however, there have been limited studies on expert risk perceptions for innovative, or new, water practices. The few studies that do investigate expert risk perceptions in water do not consider the Australian context [24,25]. Notable exceptions include studies by West et al. and Dobbie et al. on expert risk assessments in Australia [26,27]. In the study by West et al., the top three risk factors identified for the long-term viability of reuse schemes were unanticipated operational costs of the recycled water treatment plant, customer complaints, and regulatory requirements and approvals. Public health was seen as comparatively low risk (due to a low consequence score given by practitioners); hence, they call for a broadening of the traditionally considered risks in scheme assessments. The study by Dobbie et al. also considered expert risk perceptions of water projects, but focused on a varied number of projects, including stormwater harvesting, and also potable reuse [27]. The highest perceived risk for most of the alternative water projects was that of perceived cost, which Dobbie et al. put down to the traditional asset management that takes place within water authorities [27].

In another study by West et al. [28], 88 water experts across Australia were surveyed regarding the viability of residential recycled water systems. They found that a major risk to the long-term viability of recycled water systems was the inability to provide an incontestable business case. This, in turn, was found to be influenced by other risks including political, regulatory, organisational, financial, and community risk perception. Other studies in risk perceptions in the water industry in Australia mainly focus on the technical nature of risk assessments, determining their effectiveness. Very little research has been undertaken on understanding the psychological and cultural aspects of risk perception in the water industry [1].

3. Materials and Methods

3.1. Study Context

The study took place in Melbourne, Australia looking at three metropolitan water retailers (Yarra Valley Water, South East Water, and City West Water), and one bulk water provider (Melbourne Water, Melbourne, Australia). Melbourne's water is mainly supplied by dams in Melbourne's east. A drought in the 2000s brought water to the forefront of public discussion, as water restrictions were imposed and a controversial desalination plant proposed, as well as the installation of equally controversial infrastructure for inter-basin transfers. The drought has since broken, but there are still dire predictions based on increasing demand due to population growth, which are likely to place a strain on water resources in the future [29].

3.2. Overview of Risk Assessment Processes in the Water Industry in Melbourne

All four water authorities who participated in this research (Yarra Valley Water, City West Water, South East Water, and Melbourne Water) closely follow the standard approaches highlighted in AS/NZS ISO31000:2009 Risk Management: Principles and Guidelines [30]. Risk assessments are used within business cases, and as a tool to decide on a course of action, for example, within an options assessment. They can be used throughout a whole project; however, whether to fund a project is ordinarily decided upon at the concept phase.

Generally, the project delivery process runs from a concept phase, which could include an options assessment, through to preliminary design, functional design, detailed design, construction, and commissioning. The options assessment incorporates risk assessments as part of the early decision-making process, either in the format of a risk rating or statement, or in the form of a Cost–Benefit Analysis. Both of these can be argued to be heavily reliant on the risk assessor's perception and judgement. The projects that are approved based on these risk assessments are then allocated funding in future Water Plans. Fund allocations are determined by the water authority themselves, unless it is a large project of state political interest, which may have heavy political input. Other assessments, such as environmental impact statements, are ordinarily conducted at a later stage of the project, such as the Preliminary or Functional Design stage, and then may have less impact on whether the project is placed in the water plan for the future.

Yarra Valley Water, Melbourne Water, City West Water, and South East Water all have similar risk assessment approaches, with slight variations in risk appetites. Risks are categorised in the organisation's internal risk framework. These include elements such as safety risk, reputational risk, financial risk, customer risk, etc. In all three water authorities, they ensure that all of these risks and their examples of consequences are highlighted in a negative consequence table. Based on the ISO31000 standards, they all consider the consequence of a particular risk and, also, the likelihood of the risk occurring [30]. A score is given to each of these (the consequence and the likelihood separately) based on an assessment undertaken by the project manager.

Risk perceptions play a large role in driving forward new water projects at the project management level [31]. A risk assessor (the water professional) undertakes a risk assessment to determine the viability of selecting a certain course of action, such as in an options analysis. Their own risk assessments play a large role in understanding the projects that proceed, hence the importance of considering the implications of the risk perceptions observed in this study.

3.3. Data Collection

Surveys were undertaken at four metropolitan Melbourne water authorities (Melbourne Water, Yarra Valley Water, City West Water, and South East Water) between April and July 2017. Participants were recruited from within each of the water authorities through a single contact; in most cases, through the risk manager within the organisation. A total of 77 participants were recruited and completed the survey (34 from Yarra Valley Water, 16 from Melbourne Water, 14 from City West Water, and 13 from South East Water). The requirement for participation was to be a water professional who has previously undertaken a risk assessment on a water project. Participants from varied positions and employment levels were recruited, ranging from team members through to General Managers. Participants were of varied ages and gender. This ensured that a broad cross section of the workforce was surveyed.

Paper-based surveys were administered in sessions at each of the Water Authorities' main office. A pilot survey was undertaken prior to the main study commencing. The lead researcher was present to answer any questions and clarify any ambiguities.

As part of the surveys, participants were presented with three different projects: a recycled water treatment plant project (see Box 1), a new water treatment process (Box 2), and a project involving the removal of fluoride in the potable water supply (Box 3). Each project was entirely fictitious. Respondents were asked to derive a numerical risk assessment score for each project, and were then asked the following qualitative question: "What do you consider are the main risks arising from this project?" They were then asked a series of follow-up questions regarding risks associated with the proposed project, and risk perceptions in general. Demographic data was also collected.

The results were analysed to address this paper's aims, focusing on the answers to the open-ended risk perception question detailed above.

Box 1. Project 1—Using Recycled Water as Potable.

A Class A Recycled Water Treatment Plant in the historic area of Gumbark will be upgraded to allow for potable-quality water. This water will then be inserted directly into the potable supply, and mixed in with the existing potable water and delivered to residents within the area via the existing two-pipe system. Notifications and information sessions will be given to all residents prior to implementing the change. The total cost of the projects is expected to be $3.5million, delivering potable re-use water to approximately 1200 residential properties and businesses.

Box 2. Project 2—Using Radiation in Treatment of Drinking Water.

Research has been released, following extensive studies at various Water Collaborative Research Centres, highlighting a more effective way of treating drinking water that doesn't affect its taste and colour. The treatment process involves the use of high levels of radiation to exterminate any unsafe elements from drinking water. It results in a one-step process of treating water, and, ultimately, is cheaper than the existing methods of chlorination and fluoridation. Research has determined that the use of chlorine results in chemicals known as disinfection by-products. These by-products may be carcinogenic or have other toxicological effects associated with consumption.

The new radiation process ensures that the water doesn't carry these properties, that quality is always consistent regardless of the inflow water quality, and, furthermore, quality doesn't deteriorate the further the water travels (as is known to happen with chlorine). The research claims that the water post-radiation treatment is entirely safe for drinking.

Radiation stations will be fitted at the Gumbark Recycled Water Treatment Plant as the first testing scenario.

Box 3. Project 3—Removing Fluoride from Drinking Water Supply.

> Following a significant amount of lobbying from alternative health groups, the State government is under increasing pressure to remove fluoride from Melbourne's drinking water supply. In doing so, lobbyists argue that consumers could then make a choice about fluoride in water, rather than being forced to consume it.
>
> The Water Minister would like to phase out fluoride in the water supply over the course of 12 months through decommissioning each of Melbourne's fluoridation stations. The first location where fluoridation will be ceased is at the Gumbark Recycled Water Treatment Plant.

The scenarios (Boxes 1–3) were chosen as "unfamiliar" projects that hadn't previously been undertaken in the Melbourne context. All the projects have contentious elements to them, such as drinking treated wastewater; the use of radiation, which carries negative connotations [8]; and the removal of fluoride, which can be socially sensitive [32].

3.4. Data Analysis

This paper's aim is to explore the risks that water utility experts perceive to be associated with new and innovative projects, and how risk perception differs across projects. Hence, a qualitative content approach, via a thematic analysis, was utilised to explore the data, as highlighted by the Framework approach developed by the National Centre for Social Research (UK) [33]. The first step involved undertaking "open coding" on the data, through generating an index of terms. This assumes no preconceived codes, allowing them to be developed from the data directly in an exploratory way [34]. This is used to find connections between the codes, and developing hypotheses with regard to the themes. The responses were double coded by two of the researchers to verify the findings. There were a small number of minor differences between the coders. These were discussed, then revised as a result of the discussion. Given that the survey responses were in written form, many of the responses provided were succinct.

Thematic analysis was then undertaken to consider the group of codes, arranging them into core themes and even further into subthemes [33]. These themes were then compared to themes highlighted in the literature, to determine whether they verify or contradict existing research. When searching for themes, we considered topics that seemed to reoccur, thus quantitatively analysing repetition. A focus on repetition is the most common measure for establishing a thematic pattern in the data and, thus, was used in this research [34]. Each project was analysed in this way. The themes were compared across projects to determine whether there may be any overall trends.

4. Results and Discussion

4.1. Identified Perceived Risks

Through the use of a qualitative content analysis approach, fourteen key risk factors were observed across respondents for each hypothetical water project. A brief description of each risk factor is found below:

- Public health—direct: This category included a direct statement from respondents about the potential public health risk associated with the project. For example, with the fluoride project, *"possible risk of increased dental health issues, especially kids that don't practice [sic] proper dental hygiene"* (Female Team Leader, 36–45 years of age).
- Process-related risk with public health implication: This category was used when a respondent identified a process-related issue that would have a public health implication, but that the public health implication was not directly stated. For example, in relation to the potable recycled water system, *"off specification water reaching potable system"* (Female Project Manager, 26–35 years of age).
- Community Opposition: Responses relating to community opposition highlight a risk from public backlash due to the project.

- Reputation: This risk factor refers to the reputation of the water authority, as this is how it is framed in the risk management processes of each water authority: *"public concern over health impacts leading to reputational damage"* (Male Divisional Manager, 36–45 years of age).
- Safety: Safety refers to construction-based safety issues and worker safety, as well as any general mentions of safety.
- Questioning Research/Research Not Tested: This category was used for responses that critiqued studies presented. For example, in relation to the use of radiation in the water treatment process, *"what research has been done to support findings? Scientific evidence to back up? Risk of unknown and lack of evidence to support the project"* (Female, 26–35 years of age).
- Cost: Any responses that mentioned financial considerations of risk to the project.
- Governance and Regulation: This risk factor grouped any responses relating to regulatory approvals, for example, *"project could go over time, approvals may not be granted on time"* (Male Team Leader, 26–35 years of age).
- Environmental: This refers to any responses that highlighted any risks to the environment due to the project. *"Environmental [sic] will be impacted if system fails"* (Male Divisional Manager, 46–55 years of age).
- New Technology: Responses in this category highlighted the new or unfamiliar risk related to the project.
- Operational: This risk factor included responses that referred to the way the technology would be run or operated.

4.2. Risk Perceptions across the Hypothetical Water Projects

4.2.1. Risk Perceptions for Project 1—Potable Recycled Water Use

The perceived risks associated with the potable recycled water use scenario can be found in Table 1. As can be seen, community opposition was mentioned by the majority of respondents when considering the inclusion of recycled water in the potable supply. This is not an unfounded concern, as studies have shown that public perceptions relating to drinking recycled water are generally negative [21]. This was followed by risk to the reputation of the organisation, which could be closely linked to the impression that the public would not be in favour of the project. As discussed in the literature review section of this paper, this could also be as a result of an increasing public distrust in expert risk perceptions. This distrust can lead to an increased chance of public backlash, as project-based decisions are scrutinised. This points to whether water professionals seem to be cognisant of the effect of community action on projects, and its potential to become a politicised issue in the media.

Table 1. Expert risk perceptions associated with the potable recycled water project, and frequency of occurrence.

Risks Raised by Respondents (*n* = 77)	*n*
1. Community Opposition	50
2. Reputation	37
3. Public Health—Process-Related	37
4. Public Health—Illness	31
5. Cost	14
6. Governance and Regulation	12
7. Safety	10
8. Operational	9
9. Environment	2

Unsurprisingly, public health is also a highly cited risk category, fearing risk of illness as a result of the use of treated wastewater in the drinking water supply. References to the environment featured the least, with financial and regulatory risk having a moderate influence.

4.2.2. Risk Perceptions for Project 2—Radiation Treatment of Potable Water

Radiation was used in this project to draw out any historically negative connotations associated with the technology. Unsurprisingly, as shown in Table 2, safety features as the highest risk factor while regulation was mentioned the least. Community opposition and reputation appear high on the list, which may be a result of an expected negative public response to the technology, a theory confirmed in work by Fischoff et al. on risk perceptions [31].

Table 2. Expert risk perceptions associated with the radiation-based treatment system for water, and frequency of occurrence.

Risks Raised by Respondents ($n = 76$)	n
1. Safety	31
2. Community Opposition	29
3. Reputation	22
4. Questioning Research/Research Not Tested	21
5. Public Health—Illness	20
6. Public Health—Process-Related	17
7. Environment	11
8. New Technology	11
9. Operational	9
10. Governance and Regulation	2

The theme of questioning established research appeared as a risk factor for Project 2, which could be either interpreted as distrust of research, or healthy scrutiny of previous studies. In much the same way as the community expects to be engaged on projects, water professionals expect further information and reassurance from researchers. Interestingly, public health did not feature as high in responses, even though this was a new method altering the existing water treatment processes.

4.2.3. Risk Perceptions for Project 3—Removal of Fluoride from Potable Water Source

As shown in Table 3, project 3 had the highest mention of public health as a risk factor, which could be interpreted due to the nature of the removal of a dosing process, rather than the addition of a process (as occurs with the other projects). Most respondents were concerned about the effect on dental health if fluoride was removed, for example, "*possible risk of increased dental health issues, especially kids that don't practice proper dental hygiene*" (Female Team Leader, 36–45 years old).

Table 3. Expert risk perceptions associated with the removal of fluoride from the potable supply, and frequency of occurrence.

Risks Raised by Respondents ($n = 75$)	n
1. Public Health—Illness	46
2. Community Opposition	27
3. Reputation	25
4. Cost	12
5. Safety	8
6. Governance and Regulation	4
7. Environment	2
8. Public Health—Process-Related	2

Many respondents also mentioned the issue of pro-fluoride groups protesting, while others highlighted that they will be seen as "giving in", "giving luddites victory" (Male Project Manager, 36–45 years old), and heeding a movement not backed by evidence:

"It's not based on science, it's based on the opinion of people (generally) that have a poor understanding of science. These decisions need to be backed by science. It's a poor response to the problem and focuses on pleasing minorities at the expense of legitimate health benefits for everyone else. Why not allow minorities to use household based filters to remove the fluoride?" (Male, 18–25 years old)

These comments highlight the controversial nature of the project and the opinions of the vocal minority in the community.

4.3. Risk Perceptions—A Comparison between the Hypothetical Water Projects

As shown in Figure 1, "community opposition" and reputation featured consistently across all projects, while public health impacts also played a large role in risk perceptions.

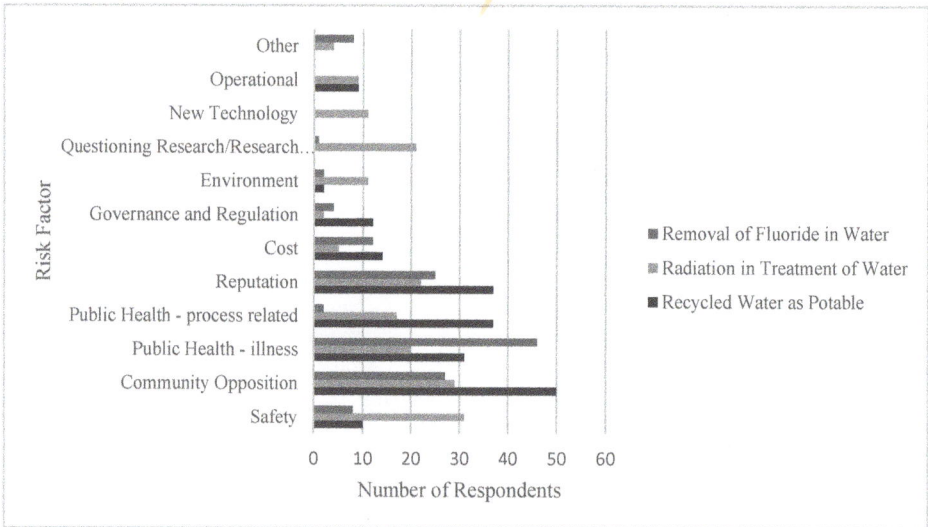

Figure 1. Water professionals' risk perceptions by project.

Community opposition was a risk factor that featured most prevalently across all projects. This was taken to refer to a negative public perception of the project that ultimately led to some form of protest or action. This is not surprising in a participatory democracy, especially as the water authorities have the state government as their key stakeholder. A focus towards active public decision-making has become a larger feature in the way water projects are managed in Australia [35], and, therefore, community opposition is shown to be front of mind in managing project risks.

While the fact that the majority of the respondents had technical backgrounds (either in engineering or in operations), two of the three main risks highlighted were not technical in nature, but social. The highest risk noted is the perception that community backlash could alter or stop new or unfamiliar projects from proceeding. This was raised for all of the three projects, but had more responses for the potable recycled water project scenario. To address this perceived risk, water practitioners could place a high emphasis on engaging the community early in the project. This also highlights the varied role of engineers and operational staff in considering public sentiment and social relations in their historically technical position. There is a potential need for greater in-house expertise on the public communications side of the project planning and implementation.

Risk to the reputation of the water authority also featured highly in the findings across each of the projects. Engaging government stakeholders early, while also managing media effectively, could help mitigate reputation-based risk. The risk to reputation is a factor that relates to trust in the water authority [22]. Conducting trust-building activities between water practitioners and the public provides a solution to counter the effects of distrust towards experts.

One area of difference between projects was for the issue of safety. For the radiation-treated water, this was raised by over 30 participants, but was raised by 10 or under for the other projects. The process of the radiation treatment of the water gave rise to a greater number of perceived risks associated with safety and

> *"operational risks, operators being exposed to high levels of radiation, long term effects vs. immediate effects of this on operators and in water supply"* (Female Project Manager 18–25 years of age)

> *"unknown long-term exposure to radiation catastrophic failure of radiation facility, risk of harm to the environment, fatality"* (Male, 26–35 years of age)

Cost does not feature highly in the risk perception findings, yet it plays a large role in business case development. This is in contrast to West's study [26,28], which highlighted cost as a major risk factor. An interpretation could be that the difference existed in the familiarity or unfamiliarity of projects presented.

4.4. Risk Perception—A Comparison with Existing Literature

The results differ to those of studies by both West et al. and Dobbie et al. on risk perceptions of water practitioners in Australia. Both studies showed cost to be the major risk factor identified in water projects, with customer complaints coming in at third. Boholm and Prutzer's Swedish study on the same topic found the two main risks identified by experts in water projects were delivery failure and quality failure [36]. Although "operational" issues did feature in our study, it was not a dominant risk factor (unlike in the Swedish context).

Interestingly, West's study did not include "safety" as a risk factor (despite 34 other factors being assessed). In our results, either construction/worker safety or general safety featured 31 times for project 2 (radiation), while in projects 1 and 3 the concept featured 10 and 9 times, respectively. The word "fatality", or similar, appeared three times in the data. The "availability heuristic" explored by Tversky and Kahneman might explain this focus on safety and fatalities [37]. Serious incidents in construction and operations within the water sector are communicated throughout the industry through either Worksafe (a state government safety authority) or through internal organisation processes. The availability heuristic highlights that if a particular issue comes to mind more readily, which occurs in the case of personal affiliation with a risk or experience with one, it is seen as riskier than those risks that do not carry this personal connection (even if both risks are seen as equally likely, with equal consequences) [37]. This may provide an explanation to the increased awareness of safety and the repeated mention of fatalities, if workers often either experience or speak about safety issues.

Public health also features in the risk attribute data; however, once again, it is not highlighted in West et al. as an area that water experts see as "high risk" [28]. Dobbie et al., instead, report public health to be a medium risk by experts in the case of potable reuse schemes [27]. Bolhom and Prutzer's study also found similar results [36]. West et al. argue that the high risk to public health is not revealed in comments because current risk assessment and management guidelines effectively address these issues [28]. As the issue of public health has arisen in all the projects assessed in this study, our interpretation is that this could be due to the risk assessor's unfamiliarity with the projects in question, as all are new and have never been undertaken at any of the water authorities that were part of the survey sample. A recycled water scheme, the issue addressed in West's study, is not an uncommon innovation across water authorities in Australia, and their sample came from across the country. Therefore, it could be argued that their familiarity with the project may reduce perceived risk [28]. Shrader-Frechette highlights the phenomenon of uncertainty acting as a large factor in how risks are perceived, with higher uncertainty leading to higher risk perceptions [38]. As all three projects presented have elements of uncertainty, it may explain the role of safety and public health as major factors in perceived risk by water experts. In particular, the greatest uncertainty across the three projects was for the radiation project, which is arguably the least familiar for the respondents.

Limitations to the research exist. The findings presented in this study are based on written surveys. If the research had been undertaken in a focus group, or verbally, respondents may have been provided with the opportunity to introduce more contextual elements to their responses and further probing would have been possible. This presents an opportunity for further investigation into understanding relative features of each risk perception attribute, to fine-tune these results. Broadening the study to a water context broader than Melbourne would also be of interest.

These findings add to existing research by providing further understanding of expert risk perceptions for new and unfamiliar projects within the water industry. Growing and urbanizing populations increase the demand on existing water infrastructure, while a changing climate alters water supply levels, leading to challenges to the delivery of sufficient water to residents. New and innovative water practices must be adopted to address these changes.

5. Conclusions

As water resources continue to become strained due to both demand and supply issues, new and alternative water sources must be sought and implemented to retain water security. However, new technologies and processes may be discriminated against, in favor of existing business-as-usual methods. Experts' risk perceptions of unfamiliar water projects have been the central element in this study. It was found that experts are sensitive to community sentiment. Experts have also indicated concern about the water authorities' reputation. Previous studies into risk perceptions highlighted the difference between the risks perceived by a layperson and an expert. This study indicates that experts within Melbourne water authorities have concerns regarding what this public risk perception is, especially as it is likely to differ from their own. Experts then, do consider laypeople's views on these new and unfamiliar water projects. This can be linked to the public's effect on the water authority's key stakeholder, the state government. Studies such as Sjoberg and Drottz-Sjoberg show that risk perceptions by politicians are very similar to the public, both of which generally differ from those of experts [39]. As the government—and, therefore, politicians—are a key stakeholder of the water authority, public sentiment and opinion plays a role in driving water policy. Due to the nature of a participatory democracy, moving with, rather than past, public opinion is preferred.

For new and unfamiliar projects, water experts were found to consider other elements such as public health as high risk, yet this factor does not appear in other studies of business-as-usual projects, such as building a recycled water scheme. Safety as a predominant risk factor appears in these projects (yet didn't appear in other studies), while cost was not a theme that featured prominently in the data. Highlighting these risk factors paves the way to understanding potential barriers or risk aversion towards the take-up and approval of new and unfamiliar water projects in the future. These can then be considered when determining an organisational risk appetite statement, as well as updated risk assessment processes, for new projects, especially if the authority would like to promote new projects to combat a changing climate.

Acknowledgments: Funding to undertake this research was received from Yarra Valley Water and a University of Melbourne Research Grant. The authors thank the participants for their time completing the survey, and to the water authorities for access to their workplaces and employees.

Author Contributions: Anna Kosovac and Brian Davidson conceived and designed the survey; Anna Kosovac conducted the surveys; Anna Kosovac and Anna Hurlimann analyzed the data; Anna Kosovac and Anna Hurlimann wrote the paper.

Conflicts of Interest: A.K. is a PhD candidate at the University of Melbourne and an employee of Yarra Valley Water. The funding sponsors had no role in the design of the study; in the collection, analyses, or interpretation of the data; in the writing of the manuscript, and in the decision to publish the results.

References

1. Kosovac, A.; Davidson, B.; Malano, H.; Cook, J. The varied nature of risk and considerations for the water industry: A review of the literature. *Environ. Nat. Resour. Res.* **2017**, *7*, 80–86. [CrossRef]
2. Kabat, P.; Ludwig, F.; van Schaik, H.; van der Valk, M. *Climate Change Adaptation in the Water Sector*; Taylor and Francis: Hoboken, NJ, USA, 2012.
3. Hurlimann, A.; Dolnicar, S. When public opposition defeats alternative water projects—The case of Toowoomba Australia. *Water Res.* **2010**, *44*, 287–297. [CrossRef] [PubMed]
4. Mills, R.A.; Karajeh, F.; Hultquist, R.H. California's Task Force evaluation of issues confronting water reuse. *Water Sci. Technol. J. Int. Assoc. Water Pollut. Res.* **2004**, *50*, 301–308.
5. Lupton, D. *Risk*, 2nd ed.; Routledge: New York, NY, USA; Oxon, UK, 2013.
6. Slovic, P.; Fischhoff, B.; Lichtenstein, S. Characterising perceived risk. In *Perilous Progress: Managing the Hazards of Technology*; Kates, R.W., Hohenemser, C., Kasperson, J.X., Eds.; Westview: Boulder, CO, USA, 1985; pp. 91–125.
7. Slovic, P. Perceived risk, trust, and democracy. *Risk Anal. Int. J.* **1993**, *13*, 675–682. [CrossRef]
8. Fischhoff, B.; Slovic, P.; Lichtenstein, S. "The Public" Vs. "The Experts": Perceived Vs. Actual Disagreements About Risks of Nuclear Power. In *The Analysis of Actual Versus Perceived Risks*; Covello, V.T., Flamm, W.G., Rodricks, J.V., Tardiff, R.G., Eds.; Plenum Press: New York, NY, USA; London, UK, 1983; pp. 235–249.
9. Nichols, T.M. *The Death of Expertise: The Campaign against Established Knowledge and Why It Matters*; Oxford University Press: New York, NY, USA, 2017.
10. Schon, D.A. *The Reflective Practitioner: How Professionals Think in Action*; Bookpoint Ltd.: Milton, UK, 1995.
11. Schlosberg, D.; Collins, L.B.; Niemeyer, S. Adaptation policy and community discourse: Risk, vulnerability, and just transformation. *Environ. Polit.* **2017**, *26*, 413–437. [CrossRef]
12. Guy, S.; Kashima, Y.; Walker, I.; O'Neill, S. Investigating the effects of knowledge and ideology on climate change beliefs. *Eur. J. Soc. Psychol.* **2014**, *44*, 421–429. [CrossRef]
13. Aerts, J.; Droogers, P. Adapting to climate change in the water sector. In *Climate Change Adaptation in the Water Sector*; Ludwig, F., Kabat, P., van Schaik, H., van der Valk, M., Eds.; Earthscan: London, UK, 2012; pp. 87–107.
14. Fielding, K.S.; Head, B.W.; Laffan, W.; Western, M.; Hoegh-Guldberg, O. Australian politicians' beliefs about climate change: Political partisanship and political ideology. *Environ. Polit.* **2012**, *21*, 712–733. [CrossRef]
15. Ettehad, E.; McKay, J.; Keremane, G. Public interest in desalination delivery in three Australian states: A newspaper content analysis. In Proceedings of the 2015 International Desalination Association World Congress on Desalination and Water Reuse, San Diego, CA, USA, 31 August 2015.
16. Davies, A. Desalination plant dumped: It was a stinker with voters, to be frank. Sydney Morning Herald. 8 February 2006. Available online: http://www.smh.com.au/news/national/desalination-plant-dumped-it-was-a-stinker-with-voters-to-befrank/2006/02/07/1139074234090.html (accessed on 10 November 2017).
17. Po, M.; Nancarrow, B.E. Literature review: Consumer perceptions of the use of reclaimed water for horticultural irrigation. In *CSIRO Land and Water*; CSIRO: Perth, Australia, 2004.
18. Forester, J. *The Deliberative Practitioner: Encouraging Participatory Planning Processes*; The MIT Press: Cambridge, MA, USA, 1999.
19. Hurlimann, A.C. Is recycled water use risky? An urban Australian community's perspective. *Environmentalist* **2007**, *27*, 83–94. [CrossRef]
20. Dolnicar, S.; Hurlimann, A.; Grün, B. What affects public acceptance of recycled and desalinated water? *Water Res.* **2011**, *45*, 933–943. [CrossRef] [PubMed]
21. Dolnicar, S.; Hurlimann, A. Drinking water from alternative water sources: Differences in beliefs, social norms and factors of perceived behavioural control across eight Australian locations. *Water Sci. Technol.* **2009**, *60*, 1433–1444. [CrossRef] [PubMed]
22. Ross, V.L.; Fielding, K.S.; Louis, W.R. Social trust, risk perceptions and public acceptance of recycled water: Testing a social-psychological model. *J. Environ. Manag.* **2014**, *137*, 61–68. [CrossRef] [PubMed]

23. Kandiah, V.; Binder, A.R.; Berglund, E.Z. An empirical agent-based model to simulate the adoption of water reuse using the social amplification of risk framework. *Risk Anal. Int. J.* **2017**, *37*, 2005–2022. [CrossRef] [PubMed]

24. Höllermann, B.; Evers, M. Perception and handling of uncertainties in water management—A study of practitioners' and scientists' perspectives on uncertainty in their daily decision-making. *Environ. Sci. Policy* **2017**, *71*, 9–18. [CrossRef]

25. McDaniels, T.L.; Axelrod, L.J.; Cavanagh, N.S.; Slovic, P. Perception of ecological risk to water environments. *Risk Anal. Int. J.* **1997**, *17*, 341–352. [CrossRef]

26. West, C.; Kenway, S.; Hassall, M.; Yuan, Z. Why do residential recycled water schemes fail? A comprehensive review of risk factors and impact on objectives. *Water Res.* **2016**, *102*, 271–281. [CrossRef] [PubMed]

27. Dobbie, M.F.; Brookes, K.L.; Brown, R.R. Transition to a water-cycle city: Risk perceptions and receptivity of Australian urban water practitioners. *Urban Water J.* **2014**, *11*, 427–443. [CrossRef]

28. West, C.; Kenway, S.; Hassall, M.; Yuan, Z. Expert opinion on risks to the long-term viability of residential recycled water schemes: An Australian study. *Water Res.* **2017**, *120*, 133–145. [CrossRef] [PubMed]

29. Infrastructure Australia. *Population Estimates and Projections: Australian Infrastructure Audit Background Paper*; Infrastructure Australia: Sydney, Australia, 2015.

30. Council of Standards Australia. *Risk Management—Principles and Guidelines*; Standards Australia: Sydney, Australia, 2009.

31. Fischhoff, B.; Slovic, P.; Lichtenstein, S.; Read, S.; Combs, B. How safe is safe enough—Psychometric study of Attitudes towards technological risks and benefits. *Policy Sci.* **1978**, *9*, 127–152. [CrossRef]

32. Rajagopal, R.; Tobin, G. Fluoride in drinking water: A survey of expert opinions. *Environ. Geochem. Health* **1991**, *13*, 3–13. [CrossRef] [PubMed]

33. Ritchie, J.; Lewis, J.; O'Connor, W. Carrying out qualitative analysis. In *Qualitative Research Practice: A Guide for Social Science Students and Researchers*; Ritchie, J., Lewis, J., Eds.; Sage Publications: London, UK; Thousand Oaks, CA, USA, 2003.

34. Bryman, A. *Social Research Methods*, 5th ed.; Oxford University Press: New York, NY, USA, 2016.

35. Yarra Valley Water. Citizens Jury to Help Determine Water Services and Pricing. 31 March 2017. Available online: https://www.yvw.com.au/about-us/news-room/citizens-jury-help-determine-water-services-and-pricing (accessed on 31 October 2017).

36. Boholm, Å.; Prutzer, M. Experts' understandings of drinking water risk management in a climate change scenario. *Clim. Risk Manag.* **2017**, *16*, 133–144. [CrossRef]

37. Tversky, A.; Kahneman, D. *Judgment under Uncertainty: Heuristics and Biases*; Oregon Research Institute: Eugene, OR, USA, 1973.

38. Shrader-Frechette, K.S. *Risk and Rationality: Philosophical Foundations for Populist Reforms*; Berkeley University of California Press: Oakland, CA, USA, 1991.

39. Sjoberg, L.; Sjoberg-Drottz, B. Risk perception by politicians and the public. *Energy Environ.* **2008**, *19*, 455–482. [CrossRef]

MDPI

St. Alban-Anlage 66

4052 Basel

Switzerland

Tel. +41 61 683 77 34

Fax +41 61 302 89 18

www.mdpi.com

Water Editorial Office

E-mail: water@mdpi.com

www.mdpi.com/journal/water

www.ingramcontent.com/pod-product-compliance
Lightning Source LLC
Chambersburg PA
CBHW051843210326
41597CB00033B/5761